T0184889

Communications in Computer and Information Science   1356

More information about this series at http://www.springer.com/series/7899

Huajun Chen · Kang Liu ·
Yizhou Sun · Suge Wang ·
Lei Hou (Eds.)

Knowledge Graph
and Semantic Computing

Knowledge Graph
and Cognitive Intelligence

5th China Conference, CCKS 2020
Nanchang, China, November 12–15, 2020
Revised Selected Papers

Springer

Editors
Huajun Chen
Zhejiang University
Hangzhou, China

Yizhou Sun
University of California
Los Angeles, CA, USA

Lei Hou
Tsinghua University
Beijing, China

Kang Liu
Chinese Academy of Sciences
Institute of Automation
Beijing, China

Suge Wang
Shanxi University
Taiyuan, China

ISSN 1865-0929   ISSN 1865-0937 (electronic)
Communications in Computer and Information Science
ISBN 978-981-16-1963-2   ISBN 978-981-16-1964-9 (eBook)
https://doi.org/10.1007/978-981-16-1964-9

This Springer imprint is published by the registered company Springer Nature Singapore Pte Ltd.
The registered company address is: 152 Beach Road, #21-01/04 Gateway East, Singapore 189721, Singapore

Preface

This volume contains the papers presented at CCKS 2020: the 14th China Conference on Knowledge Graph and Semantic Computing held during November 12–15, 2020, in Nanchang.

CCKS is organized by the Technical Committee on Language and Knowledge Computing of the Chinese Information Processing Society. CCKS is the merger of two relevant forums, i.e., the Chinese Knowledge Graph Symposium (KGS) and the Chinese Semantic Web and Web Science Conference (CSWS). KGS was previously held in Beijing (2013), Nanjing (2014), and Yichang (2015). CSWS was first held in Beijing in 2006 and has been the main forum for research on Semantic (Web) technologies in China for a decade. Since 2016, CCKS has brought together researchers from both forums and covered wider fields, including the knowledge graph, the Semantic Web, linked data, natural language processing, knowledge representation, graph databases, etc. It aims to become the top forum on knowledge graph and semantic technologies for Chinese researchers and practitioners from academia, industry, and government.

The theme of CCKS 2020 was *Knowledge Graph and Cognitive Intelligence*. Incorporating this theme, the conference scheduled various activities, including academic workshops, industrial forums, evaluation and competition, knowledge graph summit reviews, conference keynote reports, academic papers, etc. The conference invited Ruqian Lu (Academy of Mathematics and Systems Science of the Chinese Academy of Sciences), Hang Li (ByteDance AI Lab), and Barbara Tversky (Columbia University and Stanford University) to present the latest progress and development trends in the knowledge platform, pre-trained language models and cognitive intelligence. The forum also invited professional researchers in the industry to share their experience and promote industry-university-research cooperation.

As for the peer-reviewed papers, 173 submissions were received in the following five areas:

- Knowledge Graph Representation and Reasoning
- Knowledge Acquisition and Knowledge Graph Construction
- Linked Data, Knowledge Integration, and Knowledge Graph Storage Management
- Natural Language Understanding and Semantic Computing
- Knowledge Graph Applications (Semantic Search, Question Answering, Dialogue, Decision Support, and Recommendation)

Each submission was assigned to at least three Program Committee members. The committee decided to accept 63 full papers (32 papers in English and 31 in Chinese). The CCIS volume contains revised versions of 26 full papers in English.

The hard work and close collaboration of a number of people have contributed to the success of this conference. We would like to thank the Organizing Committee and Program Committee members for their support; and the authors and participants who are the primary reason for the success of this conference. We also thank Springer for their trust and for publishing the proceedings of CCKS 2020.

Finally, we appreciate the sponsorships from Huawei as chief sponsor, Meituan, Alibaba Ant Group, and Tencent as diamond sponsors, JD Cloud, PingAn, Global Tone Communication Technology as platinum sponsors, Xiaomi, Baidu, and Yidu Cloud as gold sponsors, and Zhipu.ai as the silver sponsor.

March 2021   Huajun Chen
                                                                            Kang Liu
                                                                          Yizhou Sun
                                                                          Suge Wang
                                                                            Lei Hou

Organization

CCKS 2020 was organized by the Technical Committee on Language and Knowledge Computing of the Chinese Information Processing Society.

General Chair

Huajun Chen   Zhejiang University, China

Program Committee Chairs

Kang Liu   Institute of Automation, Chinese Academy of Sciences, China

Yizhou Sun   University of California, Los Angeles, USA .

Local Chair

Mingwen Wang   Jiangxi Normal University, China

Publicity Chairs

Yanghua Xiao   Fudan University, China

Junyu Lin   Institute of Information Engineering, Chinese Academy of Sciences, China

Publication Chairs

Suge Wang   Shanxi University, China

Lei Hou   Tsinghua University, China

Tutorial Chairs

Haofen Wang   Tongji University, China

Jing Yuan   Huawei, China

Evaluation Chairs

Xianpei Han   Institute of Software, Chinese Academy of Sciences, China

Jun Yan   Yidu Cloud, China

Top Conference Reviewing Chair

Gong Cheng   Nanjing University, China

Young Scholar Forum Chairs

Jun Zhao   Institute of Automation, Chinese Academy of Sciences,
                              China
Bing Qin   Harbin Institute of Technology, China

Poster/Demo Chairs

Wenliang Chen   Soochow University, China
Feiliang Ren   Northeastern University, China

Sponsorship Chairs

Guilin Qi   Southeast University, China
Yubo Chen   Institute of Automation, Chinese Academy of Sciences,
                              China

Industry Track Chairs

Yi Cai   South China University of Technology, China
Changliang Li   Kingsoft, China

Website Chair

Wenting Li   Institute of Automation, Chinese Academy of Sciences,
                              China

Area Chairs

Knowledge Graph Representation and Reasoning

Jianfeng Du   Guangdong University of Foreign Studies, China
Quan Wang   Baidu Inc., China

Knowledge Acquisition and Knowledge Graph Construction

Wenliang Chen   Soochow University, China
Ming Liu   Harbin Institute of Technology, China

Linked Data, Knowledge Integration, and Knowledge Graph Storage Management

Xiaowang Zhang   Tianjin University, China
Weiguo Zheng   Fudan University, China

Natural Language Understanding and Semantic Computing

Xianling Mao	Beijing Institute of Technology, China
Daojian Zeng	Changsha University of Science and Technology, China

Knowledge Graph Applications (Semantic Search, Question Answering, Dialogue, Decision Support, and Recommendation)

Gong Cheng	Nanjing University, China
Shizhu He	Institute of Automation, Chinese Academy of Sciences, China

Program Committee

Yi Cai	South China University of Technology, China
Pengfei Cao	Institute of Automation, Chinese Academy of Sciences, China
Yixin Cao	National University of Singapore, China
Bo Chen	Institute of Software, Chinese Academy of Sciences, China
Wenliang Chen	Soochow University, China
Yubo Chen	Institute of Automation, Chinese Academy of Sciences, China
Gong Cheng	Nanjing University, China
Hong Cheng	The Chinese University of Hong Kong, China
Cunli Mao	Kunming University of Science and Technology, China
Ruoyao Ding	Guangdong University of Foreign Studies, China
Jianfeng Du	Guangdong University of Foreign Studies, China
Liangda Fang	Jinan University, China
Shi Feng	Northeastern University, China
Binbin Gu	University of California, Santa Cruz, USA
Shu Guo	National Internet Emergency Center, China
Xianpei Han	Institute of Software, Chinese Academy of Sciences, China
Yanchao Hao	Institute of Automation, Chinese Academy of Sciences, China
Shizhu He	Institute of Automation, Chinese Academy of Sciences, China
Liang Hong	Wuhan University, China
Lei Hou	Tsinghua University, China
Baotian Hu	Harbin Institute of Technology, Shenzhen, China
Linmei Hu	Beijing University of Posts and Telecommunications, China
Renfen Hu	Beijing Normal University, China
Wei Hu	Nanjing University, China
Pingping Huang	Baidu Inc., China

Jianmin Ji	University of Science and Technology of China, China
Wenbin Jiang	Baidu Inc., China
Zhuoxuan Jiang	Peking University, China
Daifeng Li	Sun Yat-sen University, China
Huiying Li	Southeast University, China
Jianxin Li	Deakin University, Australia
Lishuang Li	Dalian University of Technology, China
Moses Li	Jiangxi Normal University, China
Qiang Li	Alibaba Inc., China
Yuan-Fang Li	Monash University, Australia
Hongyu Lin	Institute of Software, Chinese Academy of Sciences, China
Cao Liu	Institute of Automation, Chinese Academy of Sciences, China
Jian Liu	Institute of Automation, Chinese Academy of Sciences, China
Kang Liu	Institute of Automation, Chinese Academy of Sciences, China
Ming Liu	Harbin Institute of Technology, China
Qingxia Liu	Nanjing University, China
Yongbin Liu	University of South China, China
Yaojie Lu	Institute of Software, Chinese Academy of Sciences, China
Zhunchen Luo	PLA Academy of Military Science, China
Yinglong Ma	North China Electric Power University, China
Yue Ma	Université Paris Sud, France
Xianling Mao	Beijing Institute of Technology, China
Peng Peng	Hunan University, China
Feiliang Ren	Northeastern University, China
Hao Shao	Gowild Inc., China
Wei Shen	Nankai University, China
Yatian Shen	Henan University, China
Yuming Shen	Guangdong University of Foreign Studies, China
Yuping Shen	Sun Yat-sen University, China
Wei Shi	Institute for Infocomm Research, Singapore
Xiao Sun	Hefei University of Technology, China
Yawei Sun	Nanjing University, China
Yizhou Sun	University of California, Los Angeles, USA
Jintao Tang	National University of Defense Technology, China
Jin Ting	Hainan University, China
Hai Wan	Sun Yat-sen University, China
Huaiyu Wan	Beijing Jiaotong University, China
Jing Wan	Beijing University of Chemical Technology, China
Hongzhi Wang	Harbin Institute of Technology, China
Junhu Wang	Griffith University, Australia
Meng Wang	Southeast University, China

Quan Wang	Institute of Information Engineering, Chinese Academy of Sciences, China
Senzhang Wang	Beihang University, China
Xiao Wang	Beijing University of Posts and Telecommunications, China
Xiaoyang Wang	Zhejiang Gongshang University, China
Xin Wang	Tianjin University, China
Zhe Wang	Griffith University, Australia
Zhichun Wang	Beijing Normal University, China
Zhigang Wang	Tsinghua University, China
Zhuoyu Wei	Microsoft, China
Gang Wu	Northeastern University, China
Guohui Xiao	Free University of Bozen-Bolzano, Italy
Tong Xiao	Northeastern University, China
Zhipeng Xie	Fudan University, China
Xin Xin	Beijing Institute of Technology, China
Benfeng Xu	University of Science and Technology of China, China
Bo Xu	Donghua University, China
Fan Xu	Jiangxi Normal University, China
Kang Xu	Nanjing University of Posts and Telecommunications, China
Yulan Yan	IBM, Japan
Liang Yang	Dalian University of Technology, China
Qiang Yang	King Abdullah University of Science and Technology, Saudi Arabia
Dong Yu	Beijing Language and Culture University, China
Ye Yuan	Northeastern University, China
Daojian Zeng	Changsha University of Science and Technology, China
Fu Zhang	Northeastern University, China
Meishan Zhang	Heilongjiang University, China
Xiang Zhang	Southeast University, China
Xiaowang Zhang	Tianjin University, China
Yuanzhe Zhang	Institute of Automation, Chinese Academy of Sciences, China
Zhizheng Zhang	Southeast University, China
Xin Zhao	Renmin University of China, China
Weiguo Zheng	Fudan University, China

Sponsors

Chief Sponsor

Diamond Sponsors

Platinum Sponsors

Gold Sponsors

Silver Sponsor

Contents

Knowledge Extraction: Event

Knowledge Applications: Question Answering, Dialogue, Decision Support, and Recommendation

Knowledge Extraction: Lexical and Entity

Chinese Punctuation Prediction with Adaptive Attention and Dependency Tree

Zelong Yan, Jianzong Wang$^{(\boxtimes)}$, Ning Cheng, Tianbo Wu, and Jing Xiao

Ping An Technology (Shenzhen) Co., Ltd., Shenzhen, China
{yanzelong632,wangjianzong347,chengning211,
wutianbo484,xiaojing661}@pingan.com.cn

Abstract. Punctuation prediction, as a key step of connecting speech recognition and natural language processing, has a profound impact on subsequent tasks. Although methods of bidirectional long short term memory and conditional random fields (BiLSTM+CRF) are proposed and remain advanced for a long time, it still suffers from the lack of ability of capturing long-distance interactions among words and extracting useful semantic information, especially in Chinese punctuation prediction. In this paper, considering the characteristic of Chinese punctuation symbols, we propose a novel method of punctuation standardization. In our BiLSTM+CRF based network, adaptive attention and dependency parsing tree are utilized to capture the long distance interactions and extract useful semantic information, and thus enhancing the word representation. As for the performance, the first proposal of Chinese punctuation prediction outperforms BiLSTM+CRF with a gain of 0.292% and 0.127% on accuracy in two datasets respectively. The second proposal outperforms existing methods with a gap of above 4.5% of accuracy and reaches state-of-the-art performance in two datasets.

Keywords: Chinese punctuation prediction · Conditional random fields · Adaptive attention · Dependency parsing tree.

1 Introduction

In many applications, there are numerous texts without punctuation and sentence boundary from speech recognition, social network and so on. The missing punctuation makes the free texts difficult to read and creates barriers to various downstream natural language processing tasks. Punctuation prediction is designed to insert useful information like punctuation or sentence boundaries and improve the readability of the text greatly for the subsequent tasks. Current methods are devoted to learning based algorithms and have been achieved the good performance [1]. These end-to-end models work better than traditional methods which need manual feature engineering and achieve a satisfactory result close to human in various short and simple sentences. Now BiLSTM+CRF is one of the most popular approaches to sequence labelling [2,3]. However, if a sentence

© Springer Nature Singapore Pte Ltd. 2021
H. Chen et al. (Eds.): CCKS 2020, CCIS 1356, pp. 3–14, 2021.
https://doi.org/10.1007/978-981-16-1964-9_1

is profound with complicated grammatical structure, such methods are not in a position to meet requirements for the lack of ability of capturing long-distance interactions among words and extracting useful semantic information.

Recently, more and more approaches have been proposed to address such problem. Some research concatenates word embedding and more information from texts like char embedding, lexicon features and so on and feeds them to neural network together [2,4,5]. These methods aim to collect more related information from texts thus subsequent network can make full use of them for inferring. Some research tries different combinations of neural networks to learn which combination of block is better in sequence labelling [6–8]. For example, some researchers replace block of LSTM with varieties of convolution neural network(CNN) like dilated CNN [9] or adopts more block in parallel. These adjustments are intended to extract kinds of features in different levels and perspectives which may be of great help in neural networks. Unfortunately, all such adjustments only bring limited improvement and are not enough to tackle the problem. Especially in Chinese punctuation prediction, the lack of its corresponding research and complicated usage make it still a difficult problem. Surprisingly, dependency parsing tree is able to capture functional semantic contextual information within texts. Now corresponding research efforts have derived useful discrete features from dependency parsing tree like corresponding related word, sematic relation and so on [10–12]. But the question of how to make good use of the rich semantic relational information as well as complex long-distance interactions among words generated by the dependency parsing tree for sequence labelling remains to be answered.

A great amount of work has been done in sequence labelling and a related task named sentence boundary detection [13,14], which only predicts sentence ends. Some previous research explores both lexical and acoustic features [15,16]. However, such acoustic signal is not be readily available in many applications and only lexical features are utilized in much research.

In this paper, we call our approaches CPPAA for "Chinese Punctuation Prediction with Adaptive Attention" and CPPAADT for "Chinese Punctuation Prediction with Adaptive Attention and Dependency Tree". Our contributions can be listed as follows:

(a) We propose a novel data preprocessing method of symbol standardization according to the characteristics of Simplified Chinese text.
(b) We introduce adaptive multi-head self-attention in front of LSTM to enhance the ability of capturing long distance interaction among words. Specifically, different positions of adaptive attention have been tried to figure out the difference of its position.
(c) We utilize dependency parsing tree in a surprising way to generate multi-scale embedding thus enrich word representation largely. Moreover, we evaluate CPPAADT against existing methods based on BiLSTM+CRF. CPPAADT achieves consistently best performance in two public datasets. In addition, the significance and influence of several CPPAADT's components are verified separately.

2   Related Work

Huang and Zweig [17] proposed a method based on the maximum entropy model and utilized it to insert punctuation marks into English speech. Liu et al. [13] proposed a method based on conditional random fields. Combining the advantages of the hidden Markov model and the maximum entropy model, it is utilized to find the sentence boundary and improve readability. Lu et al. [18] utilized dynamic conditional random fields in joint tasks of sentence boundary and punctuation prediction. The special case, the factor condition random fields, was found work well on the English and Chinese datasets. Also, the addition of the corresponding sentence sign gives a very obviously different result.

Collobert et al. [6] brought in a deep learning model based on convolutional neural networks + conditional random fields, and achieved potential results in many sequence labelling tasks. Zhang et al. [19] succeeded in utilizing syntax parsing to cope with punctuation prediction. Graves et al. [20] proposed a deep learning network method in sequence labeling and found that the increase of the depth of the added network is more essential than that of the units in the hidden layer. Wang et al. [21] utilized conditional random fields in joint tasks for punctuation and disfluency prediction and found that these two tasks interact with each other. That is, the information of one task is useful for the other task. Tilk and Alumac [22] proposed a two-stage recurrent neural network based model using long short-term memory units. It only utilized textual features in the first stage and the pause durations are added in the second stage and corresponding results showed its improvements. Huang et al. [7] compared various methods on the part-of-speech tagging of texts, including long short term memory networks, conditional random fields and so on. Xu et al. [1] tried a large variety of long short term memory networks in punctuation prediction and found that bidirectional long short term memory networks that utilize both past information and future information are significantly better than single original long short term memory networks and conditional random fields of 1 order language model. Ma et al. [3] built an end-to-end model by combining bidirectional LSTM, CNN and CRF. By using both word-level and character-level representation, it is more suitable for a wide range of tasks of sequence labeling on different languages and achieved a good performance in the task of named entity recognition. Dependency parsing tree was employed by Jie and Lu [23] to capture long distance interactions and semantic relationships among words in a sentence in the task of named entity recognition and find its significance.

3   Method

In this section, we make a brief description of the task of Chinese punctuation prediction and several architectures we proposed. The task of Chinese punctuation prediction can be illustrated briefly as follows. Given an input sentence of words, we label each word based on the punctuation after this word. In detail, we label each word with comma, period and blank(non-punctuation). In addition,

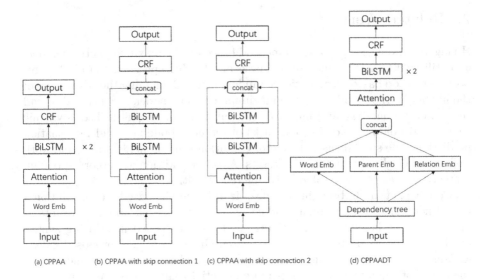

Fig. 1. The general architecture of proposed models.

Fig. 1 shows an overview of several architectures we would discuss in the following part. The left is the structure of CPPAA, the middle one and the middle two are variants of CPPAA with different types of skip connection, named CPPAA with skip connection 1 and CPPAA with skip connection 2, respectively, and the right is CPPAADT.

3.1 CPPAA

Attention mechanism, a resource allocation model that enables neural networks to have the ability to focus on certain subsets of features. By weighting the source data sequence in a comprehensive natural way, the whole system's performance is effectively improved. In most NLP tasks, attention mechanism is able to model dependencies without regard to their distance in the input or output sequences [24]. Self-attention, which relies on dot-products between elements of the input sequence to derive a weighted sum, has also been a critical ingredient in modern NLP architectures. CPPAA utilizes multi-head self-attention as illustrated in Fig. 1(a) and is able to focus on tagging consistency across multiple instances of the same token [25]. With the help of adaptive attention, every input in different time is connected directly thus encoder in every time is able to fully make use of complete contextual information. Though adaptive attention focuses on word-level representation, our proposed CPPAA is still a sentence-level method.

The core component of adaptive multi-head self-attention can be illustrated as follows. A set of queries simultaneously can be packed together into a matrix Q [26]. The keys and values are also packed together into matrices K and V. $\frac{1}{\sqrt{d}}$ and $head_i$ are the scaling factor and i_{th} attention head. The number of heads is adjusted and decided by its real significance in real time. W^O, W_i^Q, W_i^K, W_i^V are projection matrices respectively. The corresponding output can be written as follows.

$$Multihead(Q, K, V) = Concat(head_1, ...head_n)W^O \qquad (1)$$

$$head_i = softmax(\frac{QW_i^Q KW_i^K}{\sqrt{d}} VW_i^V) \qquad (2)$$

In CPPAA, a very important step is how to choose an optimal label sequence from all possible label sequences. The way of choosing an optimal label sequence is to score for every possible label sequence and finally return an optimal one with the highest score. For a given input $X = (X_1, X_2, ..., X_T)$ with length T, if its corresponding output label sequence can be represented as $Y = (Y_1, Y_2, ..., Y_T)$ with length T, its corresponding score can be calculated by the following formula (3), (4), (5). C is the transition matrix produced by CRF. $C_{i,j}$ denotes the score of transition from label i to label j.P is the position matrix. $P_{i,j}$ denotes the score of i_{th} word is labeled as j_{th} label. The transition matrix C is independent of position but the position matrix P is dependent of position. The elements in C and P are determined by training. Also ,y_0 and y_{T+1} denote the start and the end of the label sequence. For any given input $X = (X_1, X_2, ..., X_T)$ with length T, the optimal output label sequence Y^* must be the most possible sequence with the highest score.

$$score(X, Y) = \sum_{i=0}^{T} C_{y_i, y_{i+1}} + \sum_{i=1}^{T} P_{i, y_i} \qquad (3)$$

$$P(Y|X) = \frac{exp(score(X, Y))}{\sum_{Y_\theta} exp(score(X, Y_\theta))} \qquad (4)$$

$$Y^* = \arg\max_{\theta} P(Y_\theta | X) \qquad (5)$$

In the neural network end with CRF, the CRF layer models the relation between neighbouring labels which leads to better results than simply predicting each label separately based on the outputs from BiLSTM [27]. CRF only takes bigram information into consideration, thus its corresponding dynamic algorithm, the Viterbi algorithm can be calculated effectively and output the optimal label sequence quickly.

3.2 CPPAA with Skip Connection

In most cases, the increase of depth of the neural network comes with the improvement of performance. But the increase of the depth of the network

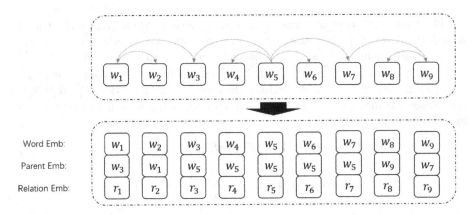

Fig. 2. How dependency parsing tree generates multi-scale embedding.

without limit brings many problems, such as vanishing gradient, exploding gradient and so on. Due to this reason, skip connection mechanism comes into existence [28]. It combines various outputs of layers instead of that only comes from the former layer and can be expressed as a linear superposition of the nonlinear transformation of the input and input. When concatenating different feature maps from different layers, the model is supposed not to underperform corresponding network without concatenating such different feature maps.

Based on CPPAA mentioned above, skip connection is added to figure out its effect. CPPAA network is combined with skip connection in two different ways as shown in Fig. 1(b) and (c). One way is only concatenating the output of attention layer and the last BiLSTM layer, and the other way is concatenating all the output from attention layer to the last BiLSTM layer. The reason why making attempts with two variants of skip connection is to figure out which connection with the CRF layer is more important and would give a useful suggestion for further research. The skip connection works well in relatively deeper networks with information loss. When the network is shallow, the skip connection may not make some difference but cause parameter redundancy.

3.3 CPPAADT

It is known that neural networks can be more robust and smart by enriching useful features. For each word in a complete sentence, dependency parsing tree can find a corresponding parent word, and also a structural relationship between it [29,30]. As the current word is driven by another word, this another word is called its parent word. By introducing dependency parsing tree, CPPAADT is proposed as illustrated in Fig. 1(d) and such dependency parsing information is fully utilized to generate novel multi-scale features thus enrich the word representation [31]. CPPAADT is robust for it is not sensitive for adjustment of words [32]. In short, if modifying a complete sentence by dropping, deleting, adding or swapping, the dependency parsing tree is still likely to generate a highly similar sematic and syntactic tree compared to the that of unmodified one thus it is commonly accepted with a strong ability of generalization.

$$u_t = [w_t^1, w_t^2, r_t] \tag{6}$$

$$r_t = Relation(w_t^1, w_t^2) \tag{7}$$

In CPPAADT, a multi-scale embedding is generated as illustrated in formula (6), (7). Input of BiLSTM u_t in moment t can be denoted by concatenating w_t^1, w_t^2 and r_t. Here w_t^1, w_t^2 are word embedding of current word and its parent word, r_t is the corresponding relation between w_t^1 and w_t^2. In CPPAA, the layer of LSTM after embedding accept w_t^1 as input in time t, but in CPPAADT, it accepts multi-scale embedding u_t as input. Compared to the original word representation, multi-scale embedding in CPPAADT is generated by concatenating the embedding of the current word, the embedding of its corresponding parent word and the embedding of its inner dependency relationship between it and its parent word, as illustrated in Fig. 2. The red line means the related relationship from a parent word to a current word and it is highly likely that several words share the same parent word. Also, every word has a unique parent word which can be related with several words. The embedding of inner relationship is randomly initialized and updated properly during training. By taking additional semantic contextual information into consideration, CPPAADT enhances word representation greatly and becomes able to captures long distance inner relation among context.

4   Experiments

4.1   Data

Our experiments use two datasets in Simplified Chinese from the second International Chinese Word Segmentation Conference Backoff as illustrated in Table 1. One of the datasets comes from Microsoft Research Institute, which is recorded as dataset A. The other dataset comes from Peking University, which is recorded as dataset B. For the train data, every sentence has punctuation marks, which have been processed by word segmentation and separated by spaces. For the test data, every sentence has its corresponding punctuation, but no word segmentation.

Table 1. Details of datasets.

Name	Origin	Word's type	# of train data	# of test data
A	Microsoft Research	88,119	86,925	3,986
B	Peking University	55,303	19,057	1,946

Compared to English text, a rule of symbol standardization for Simplified Chinese is proposed by taking all usage of Chinese symbols into consideration seriously, as illustrated in Table 2. Depending on this rule, all punctuation marks are sorted into three conventional symbols by removing or replacing. Also, to eliminate the influence of word segmentation in train data, the same tokenizer is employed to split text into words in both train data and test data. Because English texts are just split by space, it doesn't need such word segmentation. But Chinese word carries more sematic information than character, so Chinese texts need word segmentation.

Table 2. Rule of symbol standardization.

Before	After
—" "" (()) 、 《》	blank( )
," ； ；： ：	comma(,)
.o ． …! !? ?	period(.)

4.2 Main Results

In our experiments, models' performance is judged by the accuracy and F1 score. For simplicity, acc, $F(c)$, $F(p)$ are used to denote total accuracy, F1 score of comma, F1 score of the period respectively. As label blank is easy to distinguish for most methods, we leave out the metric of label blank.

In this paper, we call LSTM for unidirectional LSTM and BiLSTM for bidirectional LSTM. The subscript after LSTM or BiLSTM denotes the depth of the corresponding type of LSTM. We compare our proposed CPPAA and CPPAADT with some widely used methods, as shown in Table 3. Obviously, CPPAA and CPPAADT are competitive than methods of BiLSTM+CRF and CPPAADT shows its superiority and advance. Moreover, methods of BiLSTM are much better than that of LSTM. $BiLSTM_2$+CRF outperforms $LSTM_2$+CRF with a gap of 1.9% and 2.6% of accuracy in two datasets respectively and CPPAA outperforms Attention+$LSTM_2$+CRF with a gap of 0.7% and 1.4% of accuracy respectively. Because BiLSTM can utilize both historical information and future information among text which LSTM can't.

4.3 The Effect of Adaptive Attention

It's obvious that the model with adaptive attention has uniformly best performances on all public datasets compared to ones without attention. CPPAA outperforms $BiLSTM_2$+CRF with a gap of 0.23% and 0.29% of accuracy in two datasets. And Attention+$LSTM_2$+CRF outperforms $LSTM_2$+CRF with a gap of 1.5% of accuracy in both datasets. By allocating different words with different weights, in every time t, adaptive attention would generate a more reasonable output and pass it to LSTM. It indicates that by considering every word in the whole sentence with different weights, LSTM is able to take indeed in current

Table 3. Experimental results on 2 datasets.

Model	Dataset A			Dataset B		
	acc	F(c)	F(p)	acc	F(c)	F(p)
LSTM$_2$+CRF	91.622	47.939	99.431	88.006	34.487	55.170
Attention+LSTM$_2$+CRF	93.171	63.901	99.446	89.577	47.173	65.022
Attention+BiLSTM$_1$+CRF	93.769	66.899	99.302	90.371	54.302	69.973
BiLSTM$_2$+CRF	93.581	67.401	98.886	90.643	54.673	70.291
BiLSTM$_2$+Attention+CRF	93.947	70.667	99.230	90.663	57.078	69.517
CPPAA	93.816	66.957	99.446	90.935	54.796	71.342
CPPAADT	**98.422**	**91.489**	**99.579**	**95.470**	**77.973**	**71.373**

time as its current input, which is better than just pay attention to the current word. The improvements of employing adaptive multi-head self-attention are highly consistent no matter based on BiLSTM$_2$+CRF or LSTM$_2$+CRF. That is, adaptive multi-head self-attention is much beneficial for Chinese punctuation prediction. In addition, adaptive attention can obtain more improvements based on LSTM than on BiLSTM. It is not such sensitive for employing adaptive multi-head self-attention based on BiLSTM compared to LSTM because BiLSTM is able to pay attention to both historical words and future words together.

4.4   The Effect of Position of Adaptive Attention

In order to explore an optimal position of adaptive attention, we compare networks with different positions of attention. One way CPPAA utilizes is setting attention block between the embedding layer and BiLSTM layer, the other way BiLSTM$_2$+Attention +CRF utilizes is setting attention block between BiLSTM layer and CRF layer. In both approaches, networks with adaptive attention outperform BiLSTM+CRF methods without attention. But when looking for the optimal position of setting adaptive attention, it is hard to judge which way is better for the performance. For the performance, CPPAA works better in predicting label period and BiLSTM$_2$+Attetnion+CRF works better in label common. That is, CPPAA network is good for predicting long sentences and BiLSTM$_2$+Attention+CRF is good for predicting short sentences. Before encoding the text, the position information and the word information is likely to be very bright and apparent. So the attention can view words in a position further away and deal with long sentences better before the encoder LSTM. On the contrary, after encoding the text, the position information becomes weak and the word information is mixed in BiLSTM$_2$+Attention+CRF.

4.5   The Effect of Dependency Parsing Tree

From the experimental results, it is clear that dependency parsing tree makes a great deal of improvement in sequence labelling and enables CPPAADT to be a

Table 4. Experimental results with skip connection.

Model	Dataset A			Dataset B		
	acc	F(c)	F(p)	acc	F(c)	F(p)
CPPAA	93.816	66.957	99.446	90.935	54.796	71.342
CPPAA with skip connection 1	93.991	66.648	99.375	90.938	55.086	71.039
CPPAA with skip connection 2	94.101	67.684	99.116	90.985	57.250	**71.592**
CPPAADT	**98.422**	**91.489**	**99.579**	**95.470**	**77.973**	71.373

very competitive method. CPPAADT outperforms other existing methods with a gap of above 4.5% of accuracy in two datasets mentioned above. The result between CPPAA and CPPAADT shows that additional contextual dependency information generated by dependency parsing tree is critical for improving the model's ability. Compared to the original word embedding, CPPAADT accompanies with stronger word representation by generating the multi-scale embedding. When dealing with the present word, if the model can find its parent word and its corresponding relation, it can pay more attention to these important contextual semantic and syntactic information. So its word representation can be enhanced greatly by related contextual information which original network can't extract. It inspires us to utilize dependency parsing tree subtly when countering with NLP tasks which need contextual information.

4.6   The Effect of Skip Connection

As indicated in Table 4, skip connection doesn't make a slight difference based on CPPAA network. Because skip connection just brings the outputs of lower layers to the final layer directly but the vast majority of such information can be passed to higher layers in CPPAA. In dataset B, the difference of accuracy between methods with and without skip connection is no more that 0.05%. The way of skip connection just concatenates the hidden information in different layers, such information maybe similar to each other and results in much useless redundancy so its improvement is limited compared to the way of concatenating information in multi scales like using dependency parsing tree subtly.

5   Conclusions

In this paper, we propose networks with adaptive attention and dependency parsing tree in Chinese punctuation prediction. Our proposed CPPAADT outperforms existing methods with a gap of above 4.5% of accuracy and reaches state-of-the-art performance in two datasets. Also we find out the effect of several CPPAADT's component clearly so our attempts have certain guiding significance for more in-depth research in Chinese punctuation prediction. As experiments show that adaptive attention and dependency parsing tree works well in capturing long-distance interactions among words and extracting useful semantic information in sequence labeling, future work includes how to make good use of such methods in a smart way based on other advanced frame like graph neural network and so on.

Acknowledgement. This paper is supported by National Key Research and Development Program of China under grant No. 2018YFB1003500, No. 2018YFB0204400 and No. 2017YFB1401202.

References

1. Xu, K., Xie, L., Yao, K.: Investigating LSTM for punctuation prediction. International Symposium on Chinese Spoken Language Processing (2017)
2. Lample, G., Ballesteros, M., Subramanian, S., Kawakami, K., Dyer, C.: Neural architectures for named entity recognition. arXiv preprint arXiv:1603.01360 (2016)
3. Ma, X., Hovy, E.: End-to-end sequence labeling via bi-directional LSTM-CNNS-CRF. In: Proceedings of ACL (2016)
4. Santos, C.N.D., Zadrozny, B.: Learning character-level representations for part-of-speech tagging. In: International Conference on Machine Learning (2014)
5. Passos, A., Kumar, V., Mccallum, A.: Lexicon infused phrase embeddings for named entity resolution. In: Computer Science (2014)
6. Collobert, R., Weston, J., Bottou, L., Karlen, M., Kavukcuoglu, K., Kuksa, P.: Natural language processing (almost) from scratch. J. Mach. Learn. Res. **12**, 2493–2537 (2011)
7. Huang, Z., Xu, W., Yu, K.: Bidirectional LSTM-CRF models for sequence tagging. In: Computer Science (2015)
8. Wang, L., Li, S., Wong, D.F., Chao, L.S.: A joint chinese named entity recognition and disambiguation system. In: Proceedings of Empirical Methods in Natural Language Processing (2015)
9. Strubell, E., Verga, P., Belanger, D., Mccallum, A.: Fast and accurate sequence labeling with iterated dilated convolutions. arXiv preprint arXiv:1702.02098 (2017)
10. Cucchiarelli, A., Velardi, P.: Unsupervised named entity recognition using syntactic and semantic contextual evidence. Comput. Linguist. **27**, 123–131 (2001)
11. Sasano, R., Kurohashi, S.: Japanese named entity recognition using structural natural language processing. International Joint Conference on Natural Language Processing (2008)
12. Ling, X., Weld, D.S.: Fine-grained entity recognition. In: AAAI (2012)
13. Liu, Y., Stolcke, A., Shriberg, E., Harper, M.: Using conditional random fields for sentence boundary detection in speech. In: Proceedings of the 43rd Annual Meeting on Association for Computational Linguistics, pp. 451–458 (2005)
14. Xu, C., Xie, L., Huang, G., Xiao, X., Chng, E.S., Li, H.: A deep neural network approach for sentence boundary detection in broadcast news. International Speech Communication Association, pp. 2887–2891 (2014)
15. Kim, J.-H., Woodland, P.C.: The use of prosody in a combined system for punctuation generation and speech recognition. In: International Speech Communication Association, pp. 2757–2760 (2001)
16. Christensen, H., Gotoh, Y., Renals, S.: Punctuation annotation using statistical prosody models. In: International Symposium on Computer Architecture (2001)
17. Huang, J., Zweig, G.: Maximum entropy model for punctuation annotation from speech. In: Seventh International Conference on Spoken Language Processing (2002)

18. Lu, W., Ng, H.T.: Better punctuation prediction with dynamic conditional random fields. In: Proceedings of the 2010 Conference on Empirical Methods in Natural Language Processing, pp. 177–186 (2010)
19. Zhang, D., Wu, S., Yang, N., Li, M.: Punctuation prediction with transition-based parsing. In: Proceedings of the 51st Annual Meeting of the Association for Computational Linguistics, vol. 1, pp. 752–760 (2013)
20. Graves, A., Mohamed, A.-.R, Hinton, G.: Speech recognition with deep recurrent neural networks. In: 2013 IEEE International Conference on Acoustics, Speech and Signal Processing (2013)
21. Wang, X., Sim, K.C., Ng, H.T.: Combining punctuation and disfluency prediction: an empirical study. In: Proceedings of the 2014 Conference on Empirical Methods in Natural Language Processing, pp. 121–130 (2014)
22. Tilk, O., Alumäe, T.: LSTM for punctuation restoration in speech transcripts. In: Sixteenth Annual Conference of the International Speech Communication Association (2015)
23. Jie, Z., Lu, W.: Dependency-guided LSTM-CRF for named entity recognition. In: Proceedings of Empirical Methods in Natural Language Processing (2019)
24. Bahdanau, D., Cho, K., Bengio, Y.: Neural machine translation by jointly learning to align and translate. Computer Science (2014)
25. Luo, L., et al.: An attention-based BiLSTM-CRF approach to document-level chemical named entity recognition. Bioinformatics 34, 1381–1388 (2017)
26. Vaswani, A., et al.: Attention is all you need (2017)
27. Wang, X., Jiang, Y., Bach, N., et al.: Structure-level knowledge distillation for multilingual sequence labeling. In: Processings of ACL (2020)
28. He, K., Zhang, X., Ren, S., Jian, S.: Deep residual learning for image recognition. In: IEEE Conference on Computer Vision and Pattern Recognition (2016)
29. Zhang, Y., Yang, J.: Chinese NER using lattice LSTM. In: Proceedings of ACL (2018)
30. Zhang, Y., Li, Z., Zhang, M.: Efficient second-order TreeCRF for neural dependency parsing. In: Proceedings of ACL (2020)
31. Sun, K., Zhang, R., Mensah, S., Mao, Y., Liu, X.: Aspect-level sentiment analysis via convolution over dependency tree. In: Proceedings of Empirical Methods in Natural Language Processing (2019)
32. Vania, C., Kementchedjhieva, Y., Søgaard, A., Lopez, A.: A systematic comparison of methods for low-resource dependency parsing on genuinely low-resource languages. In: Proceedings of Empirical Methods in Natural Language Processing (2019)

Sememe Tree Prediction for English-Chinese Word Pairs

Baoju Liu[1,2], Xuejun Shang[3], Liqing Liu[3], Yuanpeng Tan[4], Lei Hou[1,2(✉)], and Juanzi Li[1,2]

[1] Department of Computer Science and Technology, BNRist, Beijing, China
[2] KIRC, Institute for Artificial Intelligence, Tsinghua University, Beijing 100084, China
liu-bj17@mails.tsinghua.edu.cn, {houlei,lijuanzi}@tsinghua.edu.cn
[3] Tianjin Electric Power Science and Research Institute, Tianjin 300130, China
xuejun.shang@tj.sgcc.com.cn
[4] Artificial Intelligence Application Department, China Electric Power Research Institute, Beijing 100192, China
tanyuanpeng@epri.sgcc.com.cn

Abstract. Sememe is the minimum unambiguous semantic unit in human language. The semantics of word senses are encoded and expressed by sememe trees in sememe knowledge base. Sememe knowledge benefits many NLP tasks. But it is time-consuming to construct the sememe knowledge base manually. There is one existing work that slightly involves sememe tree prediction, but there are two limitations. The first is they use the word as the unit instead of the word sense. The second is that their method only deals with words with dictionary definitions, not all words. In this article, we use English and Chinese bilingual information to help disambiguate word sense. We propose the Chinese and English bilingual sememe tree prediction task which can automatically extend the famous knowledge base HowNet. And we propose two methods. For a given word pair with categorial sememe, starting from the root node, the first method uses neural networks to gradually generate edges and nodes in a depth-first order. The second is a recommended method. For a given word pair with categorial sememe, we find some word pairs with the same categorial sememe and semantically similar to it, and construct a propagation function to transfer sememe tree information of these word pairs to the word pair to be predicted. Experiments show that our method has a significant effect of F1 84.0%. Further, we use the Oxford English-Chinese Bilingual Dictionary as data and add about 90,000 word pairs to HowNet, nearly expanding HowNet by half.

Keywords: Sememe tree · Bilingual word pair · Tree generator · Label propagation

1 Introduction

Sememes are defined as the minimum unambiguous indivisible semantic units of human languages in the field of linguistics [1]. Sememe knowledge has improved

© Springer Nature Singapore Pte Ltd. 2021
H. Chen et al. (Eds.): CCKS 2020, CCIS 1356, pp. 15–27, 2021.
https://doi.org/10.1007/978-981-16-1964-9_2

many natural language processing tasks, e.g., word sense disambiguation with sememes [2], pre-train word embedding enhancement [4], semantic composition modeling [5], event detection [16], relation extraction augement [6], sentiment analysis [3] and textual adversarial attack [8].

HowNet is a Chinese and English bilingual sememe knowledge base constructed by human language experts. In HowNet, there are 2214 sememes defined by human language experts and 116 semantic relations used to express the semantics of word senses. Each word sense in HowNet is described by a sememe tree. The nodes of each sememe tree are sememes, and the edges are semantic relations. The root node of each tree is a special sememe, called the categorial sememem of the word sense, which expresses the basic category the word sense belongs to. For example, the sememe tree for word sense "outlet|零售店" in Fig. 1 corresponds to the natrual language that" An outlet is an institute place in the commerce domain where people buy things and it sells things in the manner of small quantity".

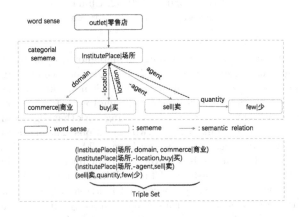

Fig. 1. Sememe tree for word "outlet"

New words appear constantly, and existing words may also have new word senses. But artificially constructing the sememe tree is a very time-consuming task. For example, human language experts spend about 30 years to complete the current famous sememe knowledge base HowNet. At present, [17] has studied to automatically construct the sememe tree. They use BiLSTM to predict sememes for words with dictionary definitions. But there are two limitations in this work. The first is that the work predict the sememe tree for each word, but sememe knowledge base should construct a sememe tree for each sense. The second limitation is that they process the words with dictionary definitions. But for newly emerged words, they have often not been compiled into the official dictionary.

Inspired by [9,10], we use cross-lingual information to help disambiguate word senses. Statistics on HowNet's 208,276 English-Chinese word pairs show that 98.04% of the English-Chinese word pairs are unambiguous. Therefore, we assume that a English-Chinese word pair represents an unambiguous word sense. A few words that do not follow this assumption are reserved for future work. Furthermore, the categorial sememe of the word sense implies important information. As the root node of the sememe tree, once it is wrong, it may cause serious error propagation. We have studied the categorial sememe prediction and achieved very good results close to manual annotation. Therefore, we propose sememe tree prediction for bilingual word pair with categorial sememe.

We propose two methods for automatically constructing the sememe tree. The first method is a generative method. For a given word pair with categorial sememe, starting from the root node, we use Recurrent Neural Network (RNN) and Multi-Layer Perceptron (MLP) to gradually generate edges and nodes in depth-first order. The second is a recommended method. For a given word pair with categorial sememe, we find some word pairs with the same categorial sememe and semantically similar to it, and construct a propagation function to transfer sememe tree information of these word pairs to the word pair to be predicted.

The contribution of this article can be summarized in the following items:

- We first propose the English-Chinese word pair sememe tree prediction task, and use bilingual information for word sense disambiguation.
- We propose a generative method. For a given word pair with categorial sememe, starting from the root node, we use RNN and MLP to gradually generate edges and nodes in depth-first order.
- We propose a recommendation-based method. For a given word pair with categorial sememe, we find some word pairs with the same categorial sememe and semantically similar to it, and construct a propagation function to transfer sememe tree information of these word pairs to the word pair to be predicted.
- We used the Oxford English-Chinese Bilingual Dictionary to add 90,000 new word pairs to HowNet, expanding HowNet by almost half.

2   Task Formalization

In this section, we will formally define sememe, categorial sememe, semantic relation, and sememe tree, and then introduce the **STP** task.

Definition 1 (Sememe). *A sememe is an unambiguous indivisible semantic unit in human languages, e.g., "buy |买", "few |少". Let $S = \{s_1, s_2, \ldots, s_n\}$ denote the sememe set in HowNet, where $n = 2,214$.*

Definition 2 (Categorial Sememe). *Each word sense sen has one categorial sememe $\tilde{s}_{sen}$ which indicates the basic category the sense belongs to, e.g., the categorial sememe of word sense "outlet|零售店" is "InstitutePlace|场所" in Fig. 1.*

Definition 3 (Semantic Relation). *Semantic relations are used to describe the semantics of the word senses, e.g., domain. We use $R = \{r_1, r_2, \ldots, r_m\}$ to denote the semantic relation set, where $m = 233$. We use (InstitutePlace |场所, -agent, sell |卖) to replace (sell |卖, agent, 场所| InstitutePlace) in Fig. 1, In order to make the directions of all edges in the sememe tree from top to bottom. So we introduce the inverse relationship. e.g., -agent. In addition, the "null" relationship is introduced as the label for the classifier to judge the end of generation.*

Definition 4 (Sememe Tree). *A bilingual parallel word pair denotes a word sense sen = $(w^e|w^z)$ and w^e and w^z denote the English word and Chinese word respectively. Each word sense sen in HowNet is defined by a sememe tree tree whose root node is the categorial sememe $\tilde{s}_{sen} \in S$. For each edge r in the tree, e.g. in Fig. 1, the head entity h is above the edge, and the tail entity t is below the edge. The tree of word sense sen can be parsed into a triple set $T_{sen} = \{(h, r, t)|h, t \in S; r \in R\}$. There are a total of 7825 different triplets in HowNet. Figure 1 illustrates the sememe tree and the triple set of word sense "outlet|零售店".*

Definition 5 (Sememe Tree Prediction). *We use $H^s = \{sen_i\}_{i=1}^{N_1}$ and $D^s = \{sen_j\}_{j=1}^{N_2}$ to represent the HowNet **word pair** set and target **word pair** set (without sememe tree information). $H^s \cap D^s = \emptyset$. Given S, R and HowNet, for each word sense sen $\in D^s$ and its categorial sememe $\tilde{s}_{sen}$, sememe tree prediction(**STP**) is to predict the sememe tree for sen.*

3   Methodology

We propose effective methods for **STP** tasks. In this section, we will introduce two methods, path generation and label propagation, respectively.

3.1   Path Generation

A sememe tree is composed of two parts: nodes and edges, and the root node is the categorial sememe. So we constructed the edge generator and the node generator respectively. Starting from the root node, we use the node generator and the edge generator to gradually generate the sememe tree in a depth-first manner. First, we introduce the edge generator and the node generator, and then show how to use these two generators to generate the whole sememe tree.

Edge Generator: In depth-first manner, for each node, the edge generator needs to generate all edges with this node as the head entity. For each node in a sememe tree, let "head-edges" denote all edges with this node as the head entity in the tree. When generating the head-edges for each node, in addition to the information of the node itself, the structure and position of the node in the sememe tree also need to be considered. Therefore, our node generator uses the path from the root node to the current node to enhance input information for the edge generator. For node s_i, let $P_{s_i} = [s_1, r_1, s_2, r_2, \ldots, s_j, r_j, \ldots, s_{i-1}, r_{i-1}, s_i]$[1,2] denote the path from the root node to s_i. For different nodes, the path length may be different, but s_j, r_j can be used as a cyclic unit. We use RNN to model the path information. We model sememe and relation as one-hot vectors, i.e., $s_j \in R^n$, $r_j \in R^m$. For sen we concatenate the English pre-training word vector w^e and Chinese pre-training word vector w^z to represent the sen, denoted as $sen, sen \in R^o$.

$$state_{i-1}^e = RNN_e([s_1, r_1, s_2, r_2, \ldots, s_j, r_j, \ldots, s_{i-1}, r_{i-1}])$$

---

[1] $P_{s_i} + r_i = [s_1, r_1, \ldots, s_{i-1}, r_{i-1}, s_i, r_i]$.

[2] $P_{s_i} + r_i + s_{i+1} = [s_1, r_1, \ldots, s_{i-1}, r_{i-1}, s_i, r_i, s_{i+1}]$.

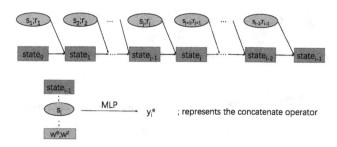

Fig. 2. Edge generator

$state^e_{i-1}$ is the final state obtained from RNN_e. Then we use a classifier, input $state^e_{i-1}, s_i, sen$, and output the scores of all relations. The score of each relation represents the probability that the relation becomes one of the head-edges of node s_i.

$$y^e_i = softmax(W^e * [state^e_{i-1}; s_i; sen] + b^e)$$

The symbol ; represents the concatenate operator.

$$state^e_{i-1} \in R^m, y^e_i \in R^m, b^e \in R^m, sen \in R^o, W^e \in R^{(m+n+o)*m}$$

After obtaining the vector y^e_i of the relations, they are normalized according to L_1 norm. Since one relation corresponds to one dimension, all relations can be sorted according to the score from high to low. When the scores of the top k+1 relations accumulate beyond the threshold m_e, we truncate and select the first k relations as the output result.

The model needs to make a prediction for each node in the sememe tree. Each prediction can be seen as a multi-classification problem, and the answer is the set of all correct edges. So the answer is a multi-hot vector $\tilde{y}^e_i$ for node s_i, $\tilde{y}^e_i \in R^m$. Since each relationship corresponds to a dimension one by one, the value in the dimension of the correct edge is 1, otherwise 0 in $\tilde{y}^e_i$. The loss function is

$$L^e = ||\tilde{y}^e_i - y^e_i||_2$$

Node Generator: Given a node and one head-edge of the node, there may be multiple tail nodes, e.g. (huamn|人, HostOf, wisdom|智慧), (huamn|人, HostOf, name|姓名). When generating all the tail nodes of an edge, in addition to the information of the edge itself, the structure and position of the edge in the sememe tree also need to be considered. Therefore, our node generator uses the path from the root node to the current edge to augment the input information of node generator. For edge r_i, let $P_{r_i} = [s_1, r_1, s_2, r_2, \ldots, s_j, r_j, \ldots, s_i, r_i]$ denote he path from root node to r_i. For different edges, the path length may be different. But s_j, r_j can be used as a cyclic unit. We use RNN to model the path information.

$$state^n_i = RNN_n([s_1, r_1, s_2, r_2, \ldots, s_j, r_j, \ldots, s_i, r_i])$$

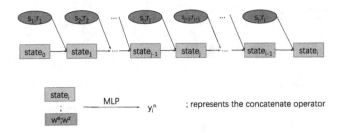

Fig. 3. Node generator

$state_i^n$ is the final state obtained from RNN_n. Then we use a classifier, input $state_i^n$, sen, and output the scores of all sememes. The score of each sememe represents the probability that the sememe becomes one of the tail nodes of edge r_i.

$$y_i^n = softmax(W^n * [state_i^n; sen] + b^n)$$

The symbol : represents the concatenate operator.

$$state_i^n \in R^n, y_i^n \in R^n, b^n \in R^n, sen \in R^o, W^n \in R^{(m+n+o)*n}$$

After obtaining the vector y_i^n of the sememes, they are normalized according to L_1 norm. Since one sememe corresponds to one dimension, all sememes can be sorted according to the score from high to low. When the scores of the top k+1 sememes accumulate beyond the threshold m_n, we truncate and select the first k sememes as the output result.

The model needs to make a prediction for each edge in the sememe tree. Each prediction can be seen as a multi-classification problem, and the answer is the set of all correct sememes. So the answer is a multi-hot vector $\tilde{y_i^n}$ for edge r_i, $\tilde{y_i^n} \in R^n$. Since each sememe corresponds to a dimension one by one, the value in the dimension of the correct sememe is 1, otherwise 0 in $\tilde{y_i^n}$. The loss function is

$$L^n = ||\tilde{y_i^n} - y_i^n||_2$$

Tree Generator: Starting from the root node, we use a depth-first method to generate each sememe tree. We have designed a recursive Algorithm 1. The input is word sense sen, the current node s_i, and the path P_{s_i}. The algorithm generates the part of sememe tree below the node s_i. So if we take sen, $\tilde{s}_{sen}$ and $[\tilde{s}_{sen}]$ as input, the tree generator can generate the whole tree.

3.2 Lable Propagation

Pre-trained word vectors play an important role in natural language processing tasks. Words with similar word vectors are more likely to have similar semantics, and thus are more likely to have similar sememe trees. Therefore, we adopt the idea of collaborative filtering. For a word sense sen_i without sememe tree, we select

---

Algorithm 1: Tree Generator

Input: word sense sen, a sememe tree $tree$, a node s_i, the path from root to the node P_{s_i}

Output: a sememe tree $tree$

1   E=$Edge\ Generator(sen, s_i, P_{s_i}))$ generate a list of all correct head-edges
2   Sort the output correct results according to the score from high to low
3   **if** $E_0 == "null"$ **then**
4 | return $tree$
5   **else**
6 | **for** e_j in E **do**
7 | $N_i = Node\ Generator(w_s, e_j, P_{s_i} + e_j)$ generate a list of all correct tail nodes
8 | **for** n_k in N_i **do**
9 | | $Tree\ Generator(sen, tree, n_k, P_{s_i} + e_j + n_k)$
10 | **end**
11 | **end**
12 **end**

---

some word senses whose semantics are similiar with sen_i in our existing sememe knowledge base, and pass the sememe tree information of these word senses to sen_i.

First, In order to express the similar relationship between words and pave the way for subsequent label propagation, we construct a word sense graph. In this graph, semantically similar word meanings are connected to each other, and each edge in the graph is weighted, representing the similarity degree of the two word senses. Then based on the word sense graph, we designed a propagation method to generate the sememe tree for the word sense without sememe tree.

Word Sense Graph: For a word sense sen_i in $H^s \cup D^s$, We use SEN_i to represent the word senses in H whose categorial sememe is $s_{\tilde{sen}_i}$. $SEN_i = \{sen_j | s_{\tilde{sen}_j} = s_{\tilde{sen}_i}, sen_j \in H\}$. $sen_i = [w_i^e; w_i^z]$, w_i^e and w_i^z are the pretrained word vectors of w_i^e and w_i^z. We use $cos(w_i^e, w_j^e) + cos(w_i^z, w_j^z)$ to represent the semantic similarity of sen_i and sen_j. We choose to select the top N_k word senses in SEN_i with the highest similarity to sen_i and connect them to sen_i. The weight value of the edge is the similarity between the two word senses.

Label Propagation Method: Each sememe tree of sen_j can be parsed into a set of triples T_{sen_j}. We associate each triplet with a dimension, and a triplet set is expressed as a multi-hot vector l_j. For a word sense sen_i to be predicted, we use N_i to denote its neighbor node set and construct a model to pass sememe information from N_i to sen_i. sen_i obtains information from neighbors and we need to determine the weight of each neighbor. $f(sen_i, sen_j)$ is used to represent the weight of sen_j to sen_i.

$$\tilde{l}_i = g\left(\sum_{sen_j \in N_i} f(sen_i, sen_j) * l_j \right)$$

Algorithm 2: Triple Set to a Tree

    Input: a triple set T, a categorial sememe $\tilde{s}_{sen}$
    Output: a sememe tree *tree*
1 Build a *tree* with root node $\tilde{s}_{sen}$
2 Create a set $N_{tree} = \{\tilde{s}_{sen}\}$ represents the set of all nodes of the tree
3 $l_{past} = 0$
4 **while** $len(N_{tree}) > l_{past}$ **do**
5 $l_{past} = len(N_{tree})$
6   **for** *(h,r,t) in T* **do**
7   **if** *(h $\in N_{tree}$)* **then**
8 N_{tree}.add(t)
9   add(h,r,t) to *tree* at node h
10   **end**
11   **end**
12   **return** *tree*
13 **end**

$$f(sen_i, sen_j) = \frac{e^{a*(cos(w_i^e, w_j^e) + cos(w_i^z, w_j^z))}}{\sum_{sen_j \in N_i} e^{a*(cos(w_i^e, w_j^e) + cos(w_i^z, w_j^z))}}$$

g is an activation function, the **tanh** function is used in the experiment. The loss function is:

$$L = ||\tilde{l}_i - l_i||_2$$

$\tilde{l}_i$ is the predicted triple vector, l_i is the actual triple vector. After obtaining the vectors of the triplets, they are normalized according to L_1 norm. Since one triplet corresponds to one dimension, all triplets can be sorted according to the score from high to low. When the scores of the top k+1 triples accumulate beyond the threshold m_t, we truncate and select the first k triplets as the output result.

Convert a Triplet Set to a Tree: We designed an algorithm with a triple set T and categorial sememe $\tilde{s}_{sen}$ as inputs and a tree *tree* as output.

4 Experiment

In the experiment, we evaluate our mothods on the **STP** task. We first introduce the data set used in the experiment and the experiment settings in the model. Next, we show the experimental results, and finally analyze the results.

4.1 Dataset

STP is a new task that we propose, so we create an evaluation data set based on HowNet. Our model requires pre-trained word vectors for English words and

Chinese words, so we look for word vectors that are publicly available. We choose Glove [14][3] vectors which contains 1.9 million words and phrases for English words and Tencent AI [15] Lab[4] which consists of about 8 million words and phrases for Chinese words. Due to the limited training corpus, some words in HowNet have no pre-trained vectors. And because it is very time-consuming to construct the sememe knowledge base manually, HowNet is not complete. Some word pairs in HowNet have no sememe tree. We make the following statistics in Table 1 for HowNet data.

Table 1. Datasets statistics

Words		Word Pairs		
Chinese	English	Total	With vectors	With vectors and sememe trees
104,027	118,347	208,276	93,081	44393

4.2 Experiment Settings

We denote the model using RNN to model path information as **P-RNN**, and the model using label propagation as **LP**. The dimension of the Chinese pre-trained word vector is 200, and the dimension of the English word vector is 300. We divide the data into training set, validation set set, and test set according to the ratio of 8:1:1.

We use stochastic gradient descent to optimize the target loss L^e, L^n, L separately. When the loss on the validation set does not drop, we stop training. The thresholds of m_e, m_n and m_t are 0.5, 0.4 and 0.8 respectively. N_k is 3. The parameters of RNN_n and RNN_e are randomly initialized between $[-1, 1]$. W^e, b^e, W^n, b^n, a are initialized to 0. The learning rate of RNN_e is 1.0 and the batch size is 100. The learning rate of RNN_n is 0.5 and the batch size is 200. The learning rate of LP is 0.01 and the batch size is 100.

4.3 Evaluation Protocol

Since each tree can be decomposed into a set of triples, we can view the task as a multi-classification task, using the triple set F1 value as the verification metric. For example, $tree$ and $\tilde{tree}$ are the correct tree and the predicted tree, and we calculate the F1 value between set T_{tree} and set $T_{\tilde{tree}}$.

---

[3] https://nlp.stanford.edu/projects/glove/.
[4] https://ai.tencent.com/ailab/nlp/embedding.html.

4.4   Overall Results

Table 2. Overal results

Method	Edge generator	Node generator	P-RNN	LP
P	0.824	0.765	0.576	0.818
R	0.853	0.739	0.542	0.863
F	0.838	0.752	0.558	0.840

Table 2 shows the experimental overall results of **P-RNN** and **LP**. From the results, we can see:

- The generative method **P-RNN** is not as effective as **LP**. We guess one reason is that due to the complex structure of the sememe tree and the large number of sememes and semantic relations. It is difficult for the machine to understand the 2214 sememes and distinguish them accurately. Another reason may be that the tree generator is composed of two parts: an edge generator and a node generator, where error propagation exists. Table 2 shows the experiment results of the two generators. The generative method has lower results than the recommended method, but it can produce triples that did not exist in the previous knowledge base.
- The **LP** method works very well, which shows the rationality of our previous assumptions. Words with similar word vectors are more likely to have similar word senses, and similar word senses are more likely to have similar sememe trees.

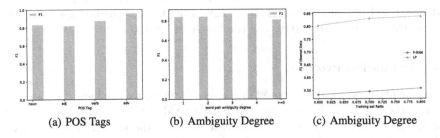

(a) POS Tags   (b) Ambiguity Degree   (c) Ambiguity Degree

Fig. 4. Results on different POS Tags, Ambiguity Degree and Training set Ratios.

4.5   Results on Different POS Tags

Table 3. POS tags statistics of HowNet test data

POS tag	Noun	adj	Verb	Adv
# word pair	3,605	350	422	37
# c_{sen}	1,273	1,107	790	448

Figure 4(a) shows the experimental results of **LP** under different POS tags. Each part-of-speech has many word senses belonging to this part-of-speech, and each word sense has a categorial sememe. We counted the number of different categorial sememes that each part-of-speech has as # c_{sen} in Table 3.

According to the knowledge of linguistics, we guess that the more categorial sememes a part of speech contains, the smaller its classification granularity. And it is more difficult to distinguish and predict for the machine. The experimental results of the four parts of speech almost all conform to this rule, but the nouns are slightly abnormal and the prediction results are slightly higher. We guess that because the semantics of nouns are not particularly abstract compared to other parts of speech, it is easier for the machine to predict.

4.6   Results on Different Ambiguity Degrees

A word may have multiple word senses, so a word may have multiple categorial sememes. We use the number of different categorial sememes a word has as the ambiguity degree of the word. For a word pair, there is an English word and a Chinese word. We take the ambiguity degree of the word with the larger ambiguity degree as the ambiguity degree of the word pair.

Figure 4(b) shows the experimental results of word pairs with different ambiguity degrees by **LP**. The experimental results predicted by the model did not decrease with the increase of ambiguity degree. We speculate that the reason may be that the bilingual information and the categorial sememe information are sufficient enough for the machine to disambiguate. With such input information, the factors restricting the prediction effect of the sememe tree may no longer be the problem of ambiguity.

4.7   Resuults on Different Training Set Ratios

Figure 4(c) shows the experimental results of **P-RNN** and **LP** under different training data ratios. From the experimental results, we can see that the experimental effects of the two methods have not declined rapidly with the decrease of training data. The methods of **P-RNN** and **LP** are relatively stable, and their generalization effects are relatively good.

The **LP** method drops slightly faster than the **P-RNN** method. For each word pair to be predicted, the **LP** method looks for the first 3 words with the semantics closest to it in the training set. Reducing the size of the training set may change first 3 words. So we guess that this may be the reason why **LP** degrades slightly faster than **P-RNN** performance.

5   Related Work

At present, only [17] has studied sememe tree prediction. Their task is to predict sememe trees for words in the dictionary that have dictionary definitions. They use BiLSTM to predict the sememe set for each word. After each word has a sememe set, a multi-layer neural network is used to predict an edge between every two sememes. Then they gradually delete edges with lower confidence until there are no rings. There are two limitations in their task. The first is that the sememe source knowledge base should construct a semantic tree for each word sense, not each word. The Chinese-English sememe knowledge base HowNet has 221,540 word senses expressed by 104,027 Chinese words and 118,347 English words. They construct a sememe tree for each word. The second is that their method requires that each word to be predicted has a dictionary definition, but the newly emerging word senses are often not included in past dictionaries. It is time-consuming to manually annotate dictionary definitions for a large number of words.

We use bilingual information to help disambiguate and proposes the task of bilingual word pair sememe tree prediction. The word to be predicted do not need to have a dictionary definition. And HowNet is bilingual, our task helps to expand HowNet.

6   Conclusion and Future Work

In this paper, we first proposed the Chinese and English bilingual sememe tree prediction task, and proposed a depth-first generative method and a recommendation-based method to predict the sememe tree. We use HowNet as experimental verification data, and the results show that our method has a significant effect of F1 84.0%. Further, we used Oxford Bilingual Dictionary to add 90,642 word pairs to HowNet, expanding HowNet by almost half.

We will explore the following directions in the future: 1. About 2% of the Chinese and English word pairs are polysemy, and we will explore to obtain more information to help disambiguate word sense or explore better methods of disambiguation. 2. English is the bridge among all languages in the world. If there is a rich sememe knowledge base of Chinese and English, we can explore the construction of the sememe knowledge base of other languages.

Acknowledgements. This work is supported by the Key Technology Develop and Research Project (SGTJDK00DWJS1900242) in STATE GRID Corporation of China.

References

1. Bloomfield, L.: A set of postulates for the science of language. Language **2**(3), 153–164 (1926)
2. Zhang, Y., Gong, L., Wang, Y.: Chinese word sense disambiguation using HowNet. In: International Conference on Advances in Natural Computation (2005)
3. Dang, L., Zhang, L.: Method of discriminant for Chinese sentence sentiment orientation based on HowNet. In: Application Research of Computers (2010)
4. Gu, Y., et al.: Language modeling with sparse product of sememe experts. In: EMNLP 2018: 2018 Conference on Empirical Methods in Natural Language Processing, pp. 4642–4651 (2018)
5. Qi, F., et al.: Modeling semantic compositionality with sememe knowledge. In: ACL 2019 : The 57th Annual Meeting of the Association for Computational Linguistics, pp. 5706–5715 (2019)
6. Li, Z., Ding, N., Liu, Z., Zheng, H., Shen, Y.: Chinese relation extraction with multi-grained information and external linguistic knowledge. In: Proceedings of the 57th Annual Meeting of the Association for Computational Linguistics, pp. 4377–4386 (2019)
7. Sun, J.G., Cai, D.F., LV, D., Dong, Y.: HowNet based Chinese question automatic classification. J. Chin. Inf. Process. **21**(1), 90–95 (2007)
8. Zang, Y., et al.: Textual adversarial attack as combinatorial optimization. arXiv: Computation and Language (2019)
9. Adriani, M.: Using statistical term similarity for sense disambiguation in cross-language information retrieval. Inf. Retrieval **2**(1), 71–82 (2000)
10. Balkova, V., Sukhonogov, A., Yablonsky, S.: Russian wordnet. In: Proceedings of the Second Global Wordnet Conference (2004)
11. Dong, Z., Dong, Q.: HowNet - a hybrid language and knowledge resource. In: 2003 Proceedings of International Conference on Natural Language Processing and Knowledge Engineering, pp. 820–824 (2003)
12. Du, J., Qi, F., Sun, M., Liu, Z.: Lexical sememe prediction using dictionary definitions by capturing local semantic correspondence. arXiv preprint arXiv:2001.05954 (2020)
13. Xie, R., Yuan, X., Liu, Z., Sun, M.: Lexical sememe prediction via word embeddings and matrix factorization. In: International Joint Conference on Artificial Intelligence (2017)
14. Pennington, J., Socher, R., Manning, C.D.: Glove: global vectors for word representation. In: Empirical Methods in Natural Language Processing (EMNLP), pp. 1532–1543 (2014)
15. Song, Y., Shi, S., Li, J., Zhang, H.: Directional skip-gram: Explicitly distinguishing left and right context for word embed-dings. In: Proceedings of the 2018 Conference of the North American Chapter of the Association for Computational Linguistics: Human Language Technologies, Volume 2 (Short Papers), pp. 175–180. Association for Computational Linguistics, New Or-leans, Louisiana, June 2018
16. Ding, N., Li, Z., Liu, Z., Zheng, H., Lin, Z.: Event detection with trigger-aware lattice neural network. In: Proceedings of the2019 Conference on Empirical Methods in Natural Language Processing and the 9th International Joint Conference on Natural Language Processing (EMNLP-IJCNLP), pp. 347–356, January 2019
17. Du, J., Qi, F., Sun, M., Liu, Z.: Lexical sememe prediction by dictionary definitions and local semantic correspondence. J. Chin. Inf. Process. **34**(5), 1–9 (2020)

FGN: Fusion Glyph Network for Chinese Named Entity Recognition

Zhenyu Xuan, Rui Bao, and Shengyi Jiang[✉]

School of Information Science and Technology,
Guangdong University of Foreign Studies, Guangdong, China

Abstract. As pictographs, Chinese characters contain latent glyph information, which is often overlooked. In this paper, we propose the FGN (https://githup. com/AidenHuen/FGN-NER), Fusion Glyph Network for Chinese NER. Except for encoding glyph information with a novel CNN, this method may extract interactive information between character distributed representation and glyph representation by a fusion mechanism. The major innovations of FGN include: (1) a novel CNN structure called CGS-CNN is proposed to capture glyph information and interactive information between the neighboring graphs. (2) we provide a method with sliding window and attention mechanism to fuse the BERT representation and glyph representation for each character. This method may capture potential interactive knowledge between context and glyph. Experiments are conducted on four NER datasets, showing that FGN with LSTM-CRF as tagger achieves new state-of-the-art performance for Chinese NER. Further, more experiments are conducted to investigate the influences of various components and settings in FGN.

Keywords: Glyph · Name entity recognition · Interactive knowledge

1 Introduction

Named entity recognition (NER) is generally treated as sequence tagging problem and solved by statistical methods or neural networks. In the field of Chinese NER, researches generally adopt character-based tagging strategy to label named entities [1, 2]. Some researches [3, 4] explicitly compared character-based methods and word-based methods for NER, confirming that character-based methods avoid the error from word segmentation stage and perform better. When using character-based methods for NER, the effect of character-level knowledge representation may greatly affect the performance of Chinese NER model.

Currently, distributed representation learning has become the mainstream method to represent Chinese characters, especially after the raise of BERT [5], which raised the baselines for almost all fields of NLP. However, these methods overlooked the information inside words or characters like Chinese glyph. There have been studies, focusing on internal components of words or characters. In English field, researchers [6] used Convolutional Neural Network (CNN) to encode the spelling of words for sequence tagging task. This method is not suitable for Chinese NER, as Chinese is not alphabetical language but hieroglyphic language. Chinese characters can be further

H. Chen et al. (Eds.): CCKS 2020, CCIS 1356, pp. 28–40, 2021.
https://doi.org/10.1007/978-981-16-1964-9_3

segmented into radicals. For example, character "抓" (grasp) is constitutive of "扌" (hand) and "爪" (claw). Study on radical-based character embedding [7] confirmed the effectiveness of these components in Chinese.

Further, researchers turned attention to regard Chinese characters as graphs for glyph encoding. Some researchers [8, 9, 25] tried running CNNs to capture glyph information in character graphs. However, these works just obtained neglectable improvement on trial. Avoiding the shortcomings of previous works, Meng et al. [2] proposed a glyph-based BERT model called Glyce, which achieved SOTA performances in various NLP tasks including NER. They adopted Tianzige-CNN to encode seven historical and contemporary scripts of each Chinese character. Tianzige is a traditional form of Chinese calligraphy, which conforms the radical distribution inside a Chinese character. Then Transformer [10] was used as sequence encoder in Glyce. Further, Sehanobish and Song [11] proposed a glyph-based NER model called GlyNN, which encoded only Hei Ti font of each character to offer glyph information and used BiLSTM-CRF as sequence tagger. Moreover, representations of non-Chinese characters were taken into consideration carefully in GlyNN. Compared with Glyce, GlyNN with BERT achieved comparable performance in multiple NER datasets, using less glyph resource and smaller CNN. It proved that historical scripts are meaningless for NER to some extent. We suspect this is because the types and numbers of entities in modern Chinese are far more abundant and complex than the ones in ancient times.

The above works just encoded the glyph and distributed representation independently. They ignored the interactive knowledges between glyphs and contexts, which have been studied in the field of multimodal deep learning [12–14]. Moreover, as the meaning of Chinese character is not complete, we suspect that encoding each character glyph individually is not an appropriate approach. In fact, interactive knowledge between the glyphs of neighboring characters maybe benefit the NER task. For example, characters in tree names like "杨树" (aspen), "柏树" (cypress) and "松树" (pine tree) have the same radical "木" (wood), but characters of an algorithm name "决策树" (decision tree) have no such pattern. There are more similar patterns in Chinese language, which can be differentiated by interactive knowledge between neighboring glyphs.

Therefore, we propose the FGN, Fusion Glyph Network for Chinese NER. The major innovations in FGN include: (1) a novel CNN structure called CGS-CNN, Character Graph Sequence CNN is offered for glyph encoding. CGS-CNN may capture potential information between the glyphs of neighboring characters. (2) We provide a fusion method with out-of-sync sliding window and Slice-Attention to capture interactive knowledge between glyph representation and character representation.

FGN is found to improve the performance of NER, which outperforms other SOTA models on four NER datasets (Sect. 4.2). In addition, we verify and discuss the influence of various proposed settings in FGN (Sect. 4.3).

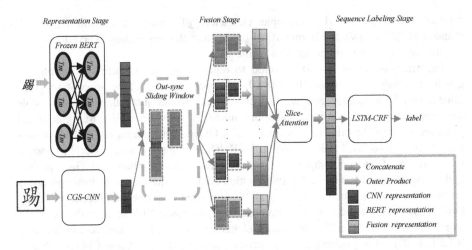

Fig. 1. Architecture of the FGN for named entity recognition.

2   Related Work

Our work is related to neural network for NER. Ronan et al. [15] proposed the CNN-CRF model, which obtained competitive performance to various best statistical NER models. LSTM-CRF [16] has been the mainstream component in subsequent NER models at present. To enhance word-level representation, Ma and Hovy [6] proposed the LSTM-CNN-CRF structure for sequence labeling, which adopted CNNs to encode the spelling of each English word for semantic enhancement. Further, a coreference aware representation learning method [17] was proposed, which was combined with LSTM-CNN-CRF for English NER. In Chinese field, Dong et al. [18] organized radicals in each character as sequence and used LSTM network to capture the radical information for Chinese NER. Zhang et al. [19] proposed a novel NER method called lattice-LSTM, which skillfully encoded Chinese characters as well as all potential words that match a lexicon. Drawing on Lattice-LSTM, Word-Character LSTM (WC-LSTM) [20] was proposed, which added word information into the start and the end characters of a word to alleviate the influence of word segmentation errors.

Our work is also related to some multimodal works. Currently, knowledge from vision has been widely-used in NLP. We simply divide these relative researches into two categories according to the source of vision knowledge: glyph representation learning and multimodal deep learning. The Former is scarce as mentioned earlier. We transform the input sentences to graph sequences for 3D encoding. To our knowledge, we are the first to encode character glyph in sentence-level by 3D convolution [21], which was mostly proposed to encode video information. The latter is current hotspot in various NLP fields. Zhang et al. [12] proposed an adaptive co-attention network for tweets NER, which adaptively balanced the fusion proportions of image representation and text representation from a tweet. With reference of BERT, a multimodal BERT [13] was proposed for target-oriented sentiment classification. Multiple self-attention layers [9] were used in this model to capture interactive information after concatenating BERT and visual representation.

Further, Mai et al. [14] proposed a fusion network with local and global perspective for multimodal affective computing. They provided a sliding window to slice multimodal vectors and fused each slice pair by outer product function. And attentive Bi-directional Skip-connected LSTM was used to combine slice pairs. Our method borrows the ideas of above-mentioned methods for multimodal fusion. Different from their work that fused the sentence-level representation, we focus on character-level fusion for Chinese NER.

3 Model

In this section, we introduce the FGN in detail. As shown in Fig. 1, FGN can be divided into three stages: representation stage, fusion stage and tagging stage. We follow the strategy of character-based sequence tagging for Chinese NER.

3.1 Representation Stage

Here we discuss the representation learning for Chinese character including character representation from BERT and glyph representation form CGS-CNN. Detail of these representations are as followed.

BERT. BERT is a multi-layer Transformer encoder, which offers distributed representations for words or characters. We use the Chinese pre-trained BERT to encode each character in sentences. Different from the normal fine-tuning strategy, we first fine-tune BERT on training set with a CRF layer as tagger. Then freeze the BERT parameters and transfer them to FGN. experiment in Sect. 4.3 shows the effectiveness of this strategy.

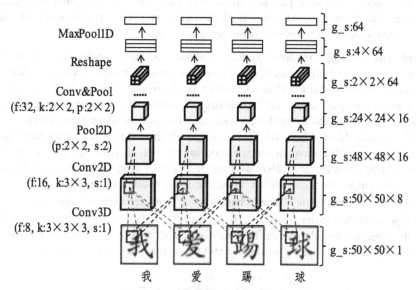

Fig. 2. Architecture of CGS-CNN with a input sample "我爱踢球" (I love playing football). "f", "k", "s", "p" stand for kernel number, kernel size, stride, and pooling window size. "g_s" represents the tensor size of output from each layer.

CGS-CNN. Figure 2 depicts the architecture of CGS-CNN. We only choose the simple Chinese script to generate glyph vectors, as the past work [11] showed that using only one Chinese script achieved comparative performance as well as seven scripts. The input format for CGS-CNN is character graph sequence. We first convert sentences to graph sequences, in which characters are replaced with 50×50 gray-scale graphs. Characters which are not Chinese may be given corresponding initialize matrices with parameters between 0 and 1. Then we provide two $3 \times 3 \times 3$ 3D convolution layers to encode graph sequence and output each 50×50 graph with 8 channels. 3D convolution can extract feature from both spatial and temporal dimensions, which means each glyph vector may obtain additional glyph information from the neighboring graphs. Using padding on the dimension of graph sequence, we may keep the length of graph sequence constant after passing through 3D convolution, which is necessary for character-based tagging. Then the output of 3D convolution may pass through several groups of 2D convolution and 2D max pooling to compress each graph to 2×2 Tianzige-structure with 64 channels. In order to filter noises and blank pixels, we flatten the 2×2 structures and adopt a 1D max pooling to extract glyph vector for each character. The size of glyph vectors is set to 64, which is much smaller than the size of Tianzige-CNN output (1024 dimension).

Different from Glyce that sets image classification task to learn glyph representation, we learn the parameters of CGS-CNN while training whole NER model in domain datasets.

3.2 Fusion Stage

We provide a sliding window to slide through both BERT and glyph representations. In sliding window, each slice pair is computed by outer product to capture local interactive features. Then Slice-Attention is adopted to balance the importance of each slice pair and combine them to output fusion representation.

Out-of-sync Sliding Window. Sliding window has been applied in multimodal affective computing [14] as mentioned above. The reason for using sliding windows is that directly fusing vectors with outer product would exponentially expand vector size, which increases space and time complexity for subsequent network. However, this method requires the multimodal representations to have the same size, which is not suitable to slide through both BERT vector and glyph vector. Because character representations of BERT have richer semantic information than glyph representations, requiring a bigger vector size. Here we provide an out-of-sync sliding window that can satisfy different vector sizes while keeping the same number of slices.

Assume that we have one Chinese character with character vector defined as $c\_v \in \mathbb{R}^{d^c}$ and glyph vector defined as $g\_v \in \mathbb{R}^{d^g}$. Here d^c and d^g stand for the sizes of two vectors. To keep the same number of the slices of these two vectors after passing through the sliding window, the setting of sliding window needs to meet the following limitation:

$$n = \frac{d^c - k^c}{s^c} + 1 = \frac{d^g - k^g}{s^g} + 1, \ n \in \mathbb{N}^* \tag{1}$$

Where n is a positive integer, standing for slice number of two vectors; k^c and s^c respectively stand for window size and stride of character vector. k^g and s^g respectively represent window size and stride for glyph vector. The strategy we use to satisfy this condition is to limit the hyper-parameters of sliding window such that d^c, k^c and s^c are respectively an integral multiple of d^g, k^g and s^g.

To get slice pairs, we first calculate the left border index of sliding window at each stride:

$$i \in \{1,2,3\cdots,n\} \tag{2}$$

$$p^c_{(i)} = s^c(i-1) \tag{3}$$

$$p^g_{(i)} = s^g(i-1) \tag{4}$$

Where $p^c_{(i)}$ and $p^g_{(i)}$ represent the boundary index of sliding window respectively for character and glyph vector at the i th stride. Then we can obtain each slice during the following formula:

$$c\_s_{(i)} = \left\{ c\_v_{(p^c_{(i)}+1)}, c\_v_{(p^c_{(i)}+2)} \cdots, c\_v_{(p^g_{(i)}+k^c)} \right\} \tag{5}$$

$$g\_s_{(i)} = \left\{ g\_v_{(p^g_{(i)}+1)}, g\_v_{(p^g_{(i)}+2)} \cdots, g\_v_{(p^g_{(i)}+k^g)} \right\} \tag{6}$$

Where $c\_s_{(i)}$ and $g\_s_{(i)}$ represent the i th slices respectively from two vectors; $c\_v_{(p^c_{(i)}+1)}$ stands for the value at $(p^c_{(i)}+1)$ th dimension of c_v.

In order to fuse two slices in a local perspective, outer product is adopted to generate an interactive tensor, as shown in the formula:

$$
\begin{aligned}
m_i &= Outer\left(c\_s_{(i)}, g\_s_{(i)}\right) \\
&= \begin{bmatrix}
c\_v_{p^c_{(i)}+1} g\_v_{p^g_{(i)}+1}, & \cdots & c\_v_{p^c_{(i)}+1} g\_v_{p^g_{(i)}+k^g} \\
\vdots & \ddots & \vdots \\
c\_v_{p^g_{(i)}+k^c} g\_v_{p^g_{(i)}+1}, & \cdots & c\_v_{p^g_{(i)}+k^c} g\_v_{p^g_{(i)}+k^g}
\end{bmatrix}
\end{aligned}
\tag{7}
$$

Where $m_i \in \mathbb{R}^{d^c \times d^g}$ stands for fusion tensor of the i th slice pair; $c\_v_{p^c_{(i)}+1} g\_v_{p^g_{(i)}+1}$ represent product result between the $p^c_{(i)}+1$ th value in c_v and the $p^g_{(i)}+1$ th value in g_v. During outer product, we may obtain all product result among elements from two vectors.

Then we flatten each tensor m_i to vector $m_i' \in \mathbb{R}^{d^c d^g}$. Representation of slices for one character can be represented as:

$$m' = \left\{ m_1', m_2', \cdots m_{n-1}', m_n' \right\}, \quad m' \in \mathbb{R}^{n \times (k^c k^g)} \tag{8}$$

Where m' contains n fusion vectors of slice pairs. The size of each vector is $k^c k^g$.

Slice-Attention. Outer product offers interactive information for character-level representation at the same time generates more noises, as many features are irrelevant. With reference to attention mechanism, we propose the Slice-Attention, which may adaptively quantify the importance of each slice pair and combined them to represent a character. Importance of slice pair can be quantified as:

$$a_i = \frac{exp\left(\sigma(v)\sigma\left(W^{slice}m_i' + b^{slice}\right)\right)}{\sum_{i=1}^{n} exp(\sigma(v)\sigma(W^{slice}m_i' + b^{slice}))} \tag{9}$$

Where a_i stands for importance value of the i th slice pair; σ is Sigmoid function. Sigmoid function here may limit the value range in vectors between 0 and 1, which ensures subsequent dot product computing meaningful. $W^{slice} \in \mathbb{R}^{(k^c k^g) \times (k^c k^g)}$ and $b^{slice} \in \mathbb{R}^{k^c k^g}$ stand for initialized weight and bias. $v \in \mathbb{R}^{(k^c k^g)}$ imitates the query in self-attention [9], which is another initialized weight.

Finally, we fuse the vectors of slice pairs by weighted average computation and obtain fusion vector f_v for a character:

$$f\_v = \sum_{i=1}^{n} a_i m_i' \tag{10}$$

3.3   Tagging Stage

We concatenate each vector in character-level before tagging. The final representation of a sentence can be defined as $x = \{x_1, x_2 \ldots, x_\tau\}$, where τ stands for the length of sentence. Then BiLSTM is adopted as sequence encoder and CRF is adopted as decoder for named entity tagging.

BiLSTM. LSTM (Long Short Terms Memory) units contain three specially designed gates to control information transmission along a sequence. To encode sequence information of , we use a forward LSTM network to obtain forward hidden state and a backward LSTM network to obtain backward hidden state. Then the two hidden states are combined as:

$$h = \overrightarrow{LSTM}(x) + \overleftarrow{LSTM}(x) \tag{11}$$

Here $h = \{h_1, h_2 \ldots, h_\tau\}$ is the hidden representation for characters. We sum the corresponding values between two hidden states to create the h.

$$P(y|s) = \frac{exp(\sum_{i=1}^{\tau}\left(W_{l_i}^{crf}h_i + b_{(l_{i-1},l_i)}^{crf}\right))}{\sum_{y'}exp\left(\sum_{i=1}^{\tau}\left(W_{l_i'}^{crf}h_i + b_{(l_{i-1}',l_i')}^{crf}\right)\right)} \tag{12}$$

Where y' represents a possible label sequence; $W_{l_i}^{crf}$ represents the weight for l_i; and $b_{(l_{i-1},l_i)}^{crf}$ is the bias from l_{i-1} to l_i.

After CRF decoding, we use first-order Viterbi algorithm to find the most probable label sequence for a sentence. Assume that there is a labeled set $\{(s_i, y_i)\}|_{i=1}^{N}$, we minimize the below negative log-likelihood function to train the whole model:

$$L = -\sum_{i=1}^{N} log(P(y_i|s_i)) \tag{13}$$

4   Experiments

In Sect. 4.1 and Sect. 4.2, we introduce the situation of datasets we use and some setting of the follow-up experiments. The main experiment result can be found in Sect. 4.2, where we set a comparison of our model and various SOTA models. FGN we proposed are tested for 10 times in each dataset to compute the average Precision (P), Recall (R), F1-socre (F1). In Sect. 4.3, we test some main components in FGN and each component is also test for 10 times to compute the average metrics.

4.1   Experimental Settings

Dataset. Four widely-used NER datasets are chosen for experiments, including Onto-Notes 4 [22], MSRA [23], Weibo [24] and Resume [19]. All of these Dataset is annotated with a BMES tagging scheme. Among them, OntoNotes 4 and MSRA are in news domain; Weibo is annotated from Sina Weibo, a social media in China. These three datasets only contain traditional name entities, such as location, personal name and organization. Resume was annotated from personal resumes with 8 types of named entities.

Hyper-parameter Setting. We use dropout mechanism for both character representation and glyph representation. Dropout rate of CGS-CNN is set to 0.2 and the one of radical self-attention is set to 0.5. The hidden size of LSTM is set to 764 and the dropout rate of LSTM is set to 0.5. We used the Chinese BERT which was pre-trained by Google2. Following the default configuration, output vector size of each character is set to 764. Character graphs we used are collected from *Xinhua Dictionaries*3 with the number of 8630. We covert these graphs to 50×50 gray-scale graph. As mentioned in Sect. 3.2, window size and stride in sliding window of character vector are respectively an integer multiple of the ones for glyph vectors. Thus, we set size and stride of the former to 96 and 8, and the later to 12 and 1 according to empirical study. Adam is adopted as optimizer for both BERT fine-tuning and NER model training. Learning rates for fine-tuning condition and training condition are different. The former one is set to 0.00001, and the latter one is set to 0.002.

4.2 Main Result

Table 1 and Table 2 show some detailed statistics of FGN, which is compared with other SOTA models on four NER datasets. Here FGN represents the proposed glyph model with LSTM-CRF as tagger; Lattice LSTM [19] and WC-LSTM [20] are the SOTA model without BERT, combining both word embedding and character embedding. BERT-LMCRF represent the BERT model with BiLSTM-CRF as NER tagger. Glyce [2] is the SOTA BERT-based glyph network as mentioned earlier. GlyNN [11] is another SOTA BERT-based glyph network. Especially, we select the average F1 of GlyNN for comparison as we also adopt the average F1 as metric. For other baselines, we select their result shown in trial, as they have not illustrated whether they used the average F1 or not.

As can be seen, FGN outperforms other SOTA models in all four datasets. Compared with BERT-LMCRF, F1 of FGN obtains obvious boosts of 3.13%, 2.88%, 1.01% and 0.84% respectively on Weibo, OntoNote 4, MSRA and Resume. Further, FGN outperformed some SOTA glyph-based NER model like Glyce and GlyNN. However, FGN did not achieve significant improvement on Resume and MSRA dataset as BERT-LMCRF can already recognize most of the entities on these two datasets. In fact, the datasets Weibo and OntoNote4 are more difficult for NER, as the entity types and entity mentions are more diverse. For example, some interesting and extraordinary entity words in Weibo and OntoNote4 like "铼德" (company name) and "啊滋猫" (milk tea shop), which were successfully identified only by FGN. We guess the reason is because the character "铼" contain the radical "钅" which means "metal" and the character "滋" contains the radical "氵" which means "water". These radicals are related to the products of their companies. In fact, this phenomenon is common in various Chinese entities including company, personal name and location, which are deeply influenced by the naming culture of Chinese people. Combined the contextual information with the above glyph information, FGN may capture extra feature to recognize some extraordinary named entities in some cases.

Table 1. Detailed statistics of FGN on Weibo and OntoNote 4

Model	Weibo			OntoNote 4		
	P	R	F1	P	R	F1
Lattice-LSTM	53.04	62.25	58.79	76.35	71.56	73.88
WC-LSTM	52.55	67.41	59.84	76.09	72.85	74.43
BERT-LMCRF	66.88	67.33	67.12	78.01	80.35	79.16
Glyce	67.68	67.71	67.70	80.87	80.40	80.62
GlyNN	N/A	N/A	69.20	N/A	N/A	N/A
FGN	**69.02**	**73.65**	**71.25**	**82.61**	**81.48**	**82.04**

Table 2. Detailed statistics of FGN on Resume and MSRA.

Model	Resume			MSRA		
	P	R	F1	P	P	F1
Lattice-LSTM	93.57	92.79	93.18	93.57	92.79	93.18
WC-LSTM	95.27	95.15	95.21	94.58	92.91	93.74
BERT-LMCRF	96.12	95.45	95.78	94.97	94.62	94.80
Glyce	**96.62**	96.48	96.54	**95.57**	95.51	95.07
GlyNN	N/A	N/A	95.66	N/A	N/A	95.21
FGN	96.49	**97.08**	**96.79**	95.45	**95.81**	**95.64**

4.3   Ablation Study

Here we discuss the influences of various settings and components in FGN. The components we investigate contain: CNN structure, named entity tagger and fusion method. Weibo dataset is used for these illustrations.

Effect of CNN Structure. As shown in Table 3, we investigate the performances of various CNN structures while keeping other settings of FGN unchanged. In this table, "2d" represents the CGS-CNN with no 3D convolution layer. "avg" represents that 1D max pooling in CGS-CNN is replaced by 1D average pooling. 2D CNN represents the CNN structure with only 2D convolution and 2D pooling layers. Tianzige-CNN is proposed from Glyce.

As can be seen, the common 2D-CNN structure obtains the worse result, as it completely overlooks the information of Tianzige structure and neighbor character glyph. Comparing with Tianzige-CNN, using CGS-CNN introduces a boost of 0.66% in F1, as CGS-CNN may capture interactive information between the character glyph. Compared with 2D convolution, Using FGN with 3D convolution introduces a boost of 1.14% in F1, which confirmed the benefit from adjacent glyph information of phrases or words. Otherwise, max pooling works better than average pooling when capture feature in Tianzige structure. As mentioned earlier, max pooling here may filter some blank pixels and noises in character graphs.

Table 3. Performances of various CNN structures on Weibo dataset.

CNN-type	P	R	F1
CGS-CNN^{2d}	68.56	71.45	70.01
CGS-CNN^{avg}	69.13	71.35	70.22
2D-CNN	67.75	72.45	69.93
Tianzige-CNN	**70.94**	70.24	70.59
CGS-CNN	69.02	**73.65**	**71.25**

Effect of Named Entity Tagger. Some widely-used sequence taggers are chosen to replace BiLSTM-CRF in FGN for discussion. Table 4 shows the performances of various chosen taggers. As can be seen, methods that based on LSTM and CRF outperform Transformer [9] encoder in NER task. In fact, Most of the SOTA NER methods [11, 19, 20] prefer to use BiLSTM rather than Transformer as their sequence encoder. Compared with only CRF, LSTM-CRF introduces a boost of 0.43% in F1. In addition, bidirectional LSTM introduces a further boost of 0.56% in F1. In this experiment, LSTM-CRF performed better than Transformer in NER task.

Table 4. Performances of various taggers on Weibo dataset.

tagger-type	P	R	F1
CRF	7044	70.10	70.26
LSTM-CRF	70.77	70.60	70.69
BiLSTM-CRF	69.02	**73.65**	**71.25**
Transformer	**72.14**	66.08	68.98

Effect of Fusion Method. We investigate the performances of different setting in fusion stage as shown in Table 5. In this table, "concat" represents concatenating glyph and BERT representation without any fusion; "no freeze" represents FGN with trainable BERT; "avg pool" and "max pool" represent that Slice-Attention in FGN is respectively replaced by pooling or max pooling. In addition, we reset the window size to (196, 16), (48, 4) and the stride to (24, 2) in sliding window respectively for character and glyph representations to test the FGN.

Compared to directly concatenating vectors from glyph and BERT, FGN introduces a boost of 0.82% in F1, which confirms the effectiveness of our fusion strategy. FGN with the strategy of fine-tuning and freezing BERT in different stages outperforms the FGN with a trainable BERT. We consider is because that fine-tuning BERT only requires minimal gradient values when updating the BERT parameters, but LSTM-CRF need to set a larger learning rate to adjusting the initialized parameter with suitable gradient values. Using Slice-Attention outperforms using average pooling or max pooling in FGN, as Slice-Attention adaptively balances information of each slices and pooling layer only filter information statically. Otherwise, sliding window with the setting in Sect. 4.1 slightly outperforms the ones with other hyper-parameter settings.

Table 5. Performances of different fusion settings on Weibo dataset.

fusion-type	P	R	F1
concat	69.13	71.35	70.43
no freeze	66.92	**74.87**	70.67
avg pool	69.00	73.61	70.11
max pool	69.60	71.40	70.64
w(196, 16)	**70.58**	71.10	70.84
w(48, 4)	70.25	71.22	70.73
s(24, 2)	69.07	73.00	70.98
FGN	69.02	**73.65**	**71.25**

5   Conclusion

In this paper, we proposed the FGN for Chinese NER. In FGN, a novel CNN structure called CGS-CNN was applied to capture both glyph information and interactive information between the neighboring graphs. Then a fusion method with out-of-sync sliding window and Slice-Attention were adopted to fuse the output representations from BERT and CGS-CNN, which may offer extra interactive information for NER tasks. Experiments are conducted on four NER datasets, showing that FGN with LSTM-CRF as tagger obtained SOTA performance on four datasets. Further, influences of various settings and components in FGN are discussed during ablation study.

Acknowledgments. This work was supported by the National Natural Science Foundation of China (No. 61572145) and the Major Projects of Guangdong Education Department for Foundation Research and Applied Research (No. 2017KZDXM031). The authors would like to thank the anonymous reviewers for their valuable comments and suggestions.

References

1. Yanan, L., Yue, Z., Dong-Hong, J.: Multi-prototype Chinese character embedding. In: Conference on Language Resources and Evaluation, pp. 855–859 (2016)
2. Yuxian, M., Wei, W., Fei, W., et al.: Glyce:Glyph-vectors for Chinese character representations. In: Advances in Neural Information Processing Systems, pp. 2742–2753 (2019)
3. Haibo, L., Masato, H., Qi, L., Heng, J.: Comparison of the impact of word segmentation on name tagging for Chinese and Japanese. In: International Conference on Language Resources and Evaluation, pp. 2532–2536 (2014)
4. Liu, Z., Zhu, C., Zhao, T.: Chinese named entity recognition with a sequence labeling approach: based on characters, or based on words? In: Huang, D.S., Zhang, X., Reyes García, C.A., Zhang, L. (eds.) ICIC 2010. LNCS (LNAI), vol. 6216, pp. 634–640. Springer, Heidelberg (2010). https://doi.org/10.1007/978-3-642-14932-0_78
5. Jacob, D., Ming-Wei, C., Kenton, L., Kristina, T.: BERT: pre-training of deep bidirectional transformers for language understanding. In: Conference of the North American Chapter of the Association for Computational Linguistics, pp. 4171–4186 (2019)
6. Xuezhe, M., Eduard, H.: End-to-end Sequence Labeling via Bi-directional LSTM-CNNs-CRF. In: Proceedings of the Annual Meeting of the Association for Computational Linguistics, pp. 1064–1074 (2016)
7. Sun, Y., Lin, L., Yang, N., Ji, Z., Wang, X.: Radical-enhanced Chinese character embedding. In: Loo, C.K., Yap, K.S., Wong, K.W., Teoh, A., Huang, K. (eds.) ICONIP 2014. LNCS, vol. 8835, pp. 279–286. Springer, Cham (2014). https://doi.org/10.1007/978-3-319-12640-1_34
8. Falcon, D., Cai, Z.: Glyph-aware embedding of chinese characters. In: Proceedings of the First Workshop on Subword and Character Level Models in NLP, pp. 64–69 (2017)
9. Yan, S., Christian, H., Jorg, T., Joakim, N.: Character-based joint segmentation and pos tagging for chinese using bidirectional rnn-crf. arXiv preprint arXiv:1704.01314, Computer Science (2017)
10. Ashish, V., Noam, S., Niki, P., et al.: Attention is all you need. In: Advances in Neural Information Processing Systems, pp. 5998–6008 (2017)

11. Arijit, S., Chan, S.: Using Chinese glyphs for named entity recognition. arXiv preprint arXiv:1909.09922, Computer Science (2019)
12. Qi, Z,, Jinlan, F., Xiaoyu, L., Xuanjing, H.: Adaptive co-attention network for named entity recognition in tweets. In: Proceedings of the AAAI Conference on Artificial Intelligence, pp. 5674–5681 (2018)
13. Jianfei, Y., Jing, J.: Adapting BERT for target-oriented multimodal sentiment classification. In: Proceedings of the International Joint Conference on Artificial Intelligence, pp. 5408–5414 (2019)
14. Sijie, M., Haifeng, H., Songlong, X.: Divide, Conquer And Combine: Hierarchical Feature Fusion Network With Local And Global Perspectives For Multimodal Affective Computing. In: Annual Meeting of the Association for Computational Linguistics, pp. 481–492 (2019)
15. Ronan, C., Jason, W., Leon, B., et al.: Natural language processing (almost) from scratch. J. Mach. Learn. Res. **12**(Aug), 2493–2537 (2011)
16. Huang, Z., Xu, W., Yu, K.: Bidirectional LSTM-CRF models for sequence tagging. arXiv preprint arXiv: 1508.01991, Computer Science (2015)
17. Dai, Z., Fei, H., Li, P.: Coreference aware representation learning for neural named entity recognition. In: Proceedings of the International Joint Conference on Artificial Intelligence, pp. 4946–4953 (2019)
18. Dong, C., Zhang, J., Zong, C., Hattori, M., Di, H.: Character-based LSTM-CRF with radical-level features for chinese named entity recognition. In: Lin, C.-Y., Xue, N., Zhao, D.-Y., Huang, X., Feng, Y. (eds.) ICCPOL/NLPCC -2016. LNCS (LNAI), vol. 10102, pp. 239–250. Springer, Cham (2016). https://doi.org/10.1007/978-3-319-50496-4_20
19. Yue, Z., Jie, Y.: Chinese NER using lattice LSTM. In: Proceedings of the Annual Meeting of the Association for Computational Linguistics, pp.1554–1564 (2018)
20. Wei, L., Tongge, X., Qinghua, X.: An encoding strategy based word-character LSTM for Chinese NER. In: Proceedings of Conference of the North American Chapter of the Association for Computational Linguistics, pp. 2379–2389 (2019)
21. Shuiwang, J., Wei, X., Ming, Y., et al.: 3D convolutional neural networks for human action recognition. IEEE Trans. Pattern Anal. Mach. Intell. **35**(1), 221–231 (2012)
22. Ralph, W., Sameer, P., Lance, R., et al.: Ontonotes release 4.0. LDC2011T03, Philadelphia, Penn.: Linguistic Data Consortium (2011)
23. Gina-Anne, L.: The third international Chinese language processing bakeoff: Word segmentation and named entity recognition. In: Proceedings of the Fifth SIGHAN Workshop on Chinese Language Processing, pp. 108–117 (2006)
24. Nanyun, P., Mark, D.: Named entity recognition for chinese social media with jointly trained embeddings. In: Conference on Empirical Methods in Natural Language Processing, pp. 548–554 (2015)
25. Su, T., Hungyi, L.: Learning Chinese word representations from glyphs of characters. preprint arXiv:1708.04755, Computer Science (2017)

Semantic Label Enhanced Named Entity Recognition with Incompletely Annotated Data

Yunke Zhang[1,2(✉)], Long Yu[2], and Zhanfeng Liu[2]

[1] Queen's University, Kingston, Canada
[2] Qihoo360, Beijing, China
{yulong,liuzhanfeng}@360.cn

Abstract. Named Entity Recognition (NER) is a fundamental task in natural language processing, and the output of which is usually utilized as the input for numerous downstream tasks, especially for dialogue system, question answering system, etc. Generally speaking, entities are assumed to have labels in most previous works. Nevertheless, it is unrealistic to obtain adequate labeled data in real circumstances. In this paper, a novel semantic label enhanced named entity recognition model is proposed to tackle with NER problems with incompletely annotated data. Significant improvements are achieved through extensive experiments. The proposed algorithm improves the F1 score on the Youku dataset by around 8% than the baseline model. Besides, our experiments prove the crucial role of semantic label in NER with incomplete annotations.

Keywords: Information extraction · Named entity recognition

1 Introduction

Information Extraction (IE) is a task of natural language processing that involves extracting structure information [1] from texts. IE technology is of considerable significance since most information on the Internet is stored in the unstructured format. Information extraction itself is a crucial task consisting of several sub-tasks such as named entity recognition, relation extraction, event extraction, etc. For Named Entity Recognition (NER) tasks, a primary concern is to recognize specific named entities from plain texts. Commonly, the data with complete annotations are provided for handling NER tasks. Most existing approaches for named entity recognition deal with the problems in a supervised manner, where fully annotated entity information is assumed to be available during the training phase. Nevertheless, obtaining high-quality annotations is usually laborious and time-consuming in industrial scenarios. Different from conventional tasks, named entity recognition with incompletely annotated data focus on dealing with the scenario that most annotated entities are correct, but not all entities are identified and no trustworthy examples of the negative class exist. More concretely,

© Springer Nature Singapore Pte Ltd. 2021
H. Chen et al. (Eds.): CCKS 2020, CCIS 1356, pp. 41–53, 2021.
https://doi.org/10.1007/978-981-16-1964-9_4

Fig. 1 indicates four example annotations in the dataset that our named entity recognition model is trained on, where A1 represents the golden annotation for the sentence, and A2 to A4 represent different types of incomplete annotations. Due to sufficient generality and robustness to data, named entity recognition with incomplete annotations is an essential technique of entity recognition under low resource for both academic and industry scenario.

	Ellen	Show		Seanson 14				American	talk show		
	艾	伦	秀	第	十	四	季	美	国	脱	口	秀
A1:	B-TV	I-TV	E-TV	B-NUM	I-NUM	I-NUM	E-NUM	O	O	O	O	O
A2:	B-PER	I-PER	O	B-NUM	I-NUM	I-NUM	E-NUM	O	O	O	O	O
A3:	B-TV	I-TV	E-TV	O	O	O	O	O	O	O	O	O
A4:	O	O	O	O	O	O	O	O	O	O	O	O

Fig. 1. Four types of annotations that appears in our incompletely annotated NER dataset.

Generally, supervised-style named entity recognition models often adopt convolutional or recurrent neural architectures, incorporated with a CRF layer [2–4]. Recently, deep contextualized representations relying on BiLSTM [5], transformer [6,7], or contextual string embedding [8] have also reached state-of-the-art performance on NER tasks. When the data annotations are incomplete, those algorithms assume the incomplete annotations can be obtained by merely removing either word-level labels or span-level labels. The above assumption has the obvious pitfalls with modeling incomplete annotations and makes sub-optimal assumptions on the available labels. Hence, a proper way to estimate rational label distribution is required. To alleviate the predicament, a cross-validation based approach is proposed [9], which is considered as a common solution for this task and is treated as the baseline model in this paper. Within this algorithm, specific labels are assigned to words without labels to form complete label sequences. Then a k-fold cross-validation is performed on the training set for approximating golden label distribution. On the basis of learned label distribution, the baseline model is able to recognize named entities through incomplete annotations.

Despite there are some applications of NER with incomplete annotations, they still failed to gain satisfactory results due to insufficient new entity discovery and inaccurate entity border prediction, which are the toughest parts for this task. Based on the intuition that additional prior knowledge is beneficial to boost model performance, an innovative semantic label enhanced named entity recognition solution is proposed in this study. Firstly, a matching based entity existence detection model is trained to generate semantic labels for each sample. Subsequently, a semantic label enhanced cross-validation approach is applied for obtaining an elementary NER model. Through introducing entity existence detection result in the form of semantic label as a weak constraint, easier new entity discovery is achieved, which makes the NER model absorbed in accurate entity border prediction. Moreover, in order to alleviate the impact of context on entity recognition, inspired by the AdaSampling algorithm in Positive-Unlabeled (PU) learning [10], an adaptive fine-tuning algorithm is implemented to acquire further amelioration. The experiment results indicate that the semantic label enhanced model achieves crucial improvements than the baseline method in F1 score. Our study raises the idea that through mining latent semantic information and introducing adaptive optimization, satisfactory results on NER with incomplete annotations task can be obtained. Generally, our key contributions are as follows:

1) A novel semantic label enhanced NER algorithm is proposed to perform named entity recognition task with incompletely annotated data.
2) The introduction of semantic labels in NER with incomplete annotations task is proved to lead to a more capable model.
3) An adaptive fine-tuning algorithm is proposed to mitigate the context's negative impact on named entity recognition, making the model performance forward.
4) Compared to the baseline model, the proposed method improves the F1 score by about 8%, which is a significant improvement.

2   Related Work

We distribute the previous research efforts into two categories, the method based on conditional random field(CRF) [11] and the method based on semi-supervised learning. The CRF-based methods mainly include the linear chain CRF method on incompletely annotated data, which proposed by Bellare and McCallum [12], also known as Partial CRF [13] or EM Marginal CRF [14]. Although partial CRF proposed a solution to the task, it merely received restricted improvement due to confined sequence representation and its merely focus on English. In subsequent work, Mayhew [15] and Jie [9] proposed models based on the marginal CRF framework. Jie [9] proposed a cross-validation training method to estimate the golden distribution, and Mayhew [15] combined the concept of PU learning based on the marginal CRF framework. There are several similarities between the two methods. For example, during each iteration, they both train a model to assign a soft gold-labeling to each token. Yet, differences still exist between the two works. The work of [9] uses cross-validation training in each iteration, which leverages the discrepancy of data distribution for training a model that fitting golden label distribution. On the other hand [15] follows the work of [16]. Each round of training they conduct is comprehensive training. The main shortcoming of the foregoing methods is insufficient sentence level modelling because their training objective is all based on token level.

The semi-supervised learning method is distributed into a Cross-view training method and a PU learning [17] method. The cross-view training method on NER was proposed by Clark [18]. They are also based on the BILSTM+CRF framework, which mainly lies in optimizing the representation of BiLSTM-Encoder. Cao [19] proposed an NER model that combines a classification model and a sequence labeling model. The main role of the classification model is to optimize the representation of the BiLSTM layer, which is used as the encoder of entity recognition model. Nevertheless, the classification model here merely provides a pre-trained encoder and the latent relation between sentence and entity tags is not fully explored. PU learning was first proposed to tackle the problem on document classification [17, 20–22]. Peng [23] expressed the NER task on incompletely annotated data as a PU learning problem, and a novel PU learning algorithm is proposed to perform the task. In their work, they explored ways to perform NER tasks using only unlabeled data and named entity dictionaries. The advantages of PU learning for NER with incompletely annotated data task is the utilization of unlabeled sequences. But the deficiency of the works above is missing the combination of PU learning with other effective methods, which is implemented in our proposed approach.

3   Methodology

In conventional NER task, the goal is to label a given word sequence x as a tagging sequence y. Generally, y follows the tagging scheme like BIOE or BIOES [24]. Given a labeled training dataset $\mathcal{D}$ that consists of multiple instances (x^i, y^i), a bidirectional long short-term memory network [25] with CRF [11] is a classical approach to solve the problem [26]. In this method, loss functions can be written as the following formula:

$$\mathcal{L}(w) = -\sum_i logp_w(y^i|x^i)) \tag{1}$$

where $\mathcal{L}$ represents the loss value and w represents the parameters in the algorithm. Instead of a complete label sequence y^i, we get an incompletely annotated label sequence y_p^i in NER with incompletely annotated data task. Additionally, wrong labels may appear in y_p^i. By utilizing instance (x^i, y_p^i), all possible fully annotated sequences that fit this partial annotated sequence can be obtained. On this occasion, loss function should be modified in a new way [9]:

$$\mathcal{L}(w) = -\sum_i log \sum_{y \in C(y_p^i)} q\mathcal{D}(y|x^i))p_w(y|x^i)) \tag{2}$$

where $C(y_p^i)$ represents the set of all complete label sequences that compatible with y_p^i. The formula (2) suggests that $q\mathcal{D}(y^i|x^i))$ can be regarded as the token level weights of $p_w(y^i|x^i)$, and the distribution of $q\mathcal{D}$ also defines a soft golden label for $p_w(y^i|x^i)$. Therefore, the estimation of $q\mathcal{D}(y^i|x^i))$ is a principal part of this task. Motivated by the previous work [9], k-fold cross-validation is used as a foundational method to acquire q distribution in this study. In detail, we separate training data into k folds and build a BERT [6] based model with a standard

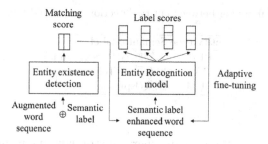

Fig. 2. The framework of the purposed method. The augmented word sequence is used to train a matching based entity existence detection model. According to entity detection result, semantic label enhanced data is constructed as the input of entity recognition model. The prediction of entity recognition model is leveraged in adaptive fine-tuning stage for alleviating the negative impact of context.

linear chain CRF stacking over its output for every k folds. After training the model for $k - 1$ fold, a constrained Viterbi procedure is implemented for predicting label sequences in the remaining fold to estimate q distribution. Subsequently, a BERT-CRF model is performed for estimating the final pw distribution.

Improvement Through Semantic Labels. In our foundational mechanism, k-fold cross-validation is a general method that leverages the discrepancy of label distribution between k folds, which does not have any specific optimization for NER with incomplete annotations. The cross-validation mechanism merely learns from different k-fold data and we believe the information inside is limited. This constraint leads to a laborious training process and an inferior performance on both new entity discovery and entity border prediction, which are the two hardest task in NER with incompletely annotated data. Intuitively, an acceptable result can not be received through a cross-validation method without extra information(and we will prove that in our experiments). Hence, a semantic label enhanced model is built for better performance when fitting both $q\mathcal{D}$ and p_w. The framework of our approach is shown in Fig. 2. In the first place, a new dataset $\mathbb{D}_c$ is constructed for training an entity existence detection model. The goal of entity detection model is to detect whether a certain kind of entity exists in word sequence x^i. From this perspective, $\mathbb{D}_c$ can be regarded as a matching task or multi-label classification task dataset. We implement a matching based model for detecting different entity existence and will illustrate the model in Sect. 3.1. Moreover, a table that maps all entity types to a set of semantic labels is created and denoted as $\mathbb{C}_s$, which is shown in Table 1. The objective of this step is to merge entity detection results with x^i for the next entity recognition task. For example, if entity detection model shows there is a person entity in x^i, this result will be mapped to a semantic label "人物姓名 (Person Name) ". After semantic label generation, we merge initial word sequence x^i with a generated semantic label:

$$x_l^i = f(x^i, l_s^j) \tag{3}$$

Table 1. The mapping between entity detection result and semantic labels.

Entity type	Semantic label
PER	人物姓名(Person name)
TV	电视节目(Telecast)
NUM	剧集(Episode)

where $l_s^j \in \mathbb{C}_s$ and f is the recursive merging strategy. In our implementations, the merging strategy is concatenating semantic label and word sequence with a BERT separator. By applying semantic label generation in all word sequence x_l^i in $\mathcal{D}$, a new semantic label enhanced dataset is acquired, which is marked as $\mathcal{D}_l$. Under this circumstance, entity existence knowledge is fused into original data, which can be seen as prior knowledge for downstream k-fold cross-validation $q\mathcal{D}$ and pw estimation. Subsequently, x_l^i is used as the input of the cross-validation based entity recognition model:

$$h_l^i = BERT(x_l^i)_{hid} \tag{4}$$

$$F(h_l^i) = w^T h_l^i + b \tag{5}$$

After getting $F(h_l^i)$ as the hidden representation for word sequence x^i, a CRF algorithm is utilized for obtaining final score of a sequence along with a path of tags:

$$s(x_l^i, p^i, \theta) = \sum_{t=1}^{T} (A_{[p]_t, [p]_{t-1}} + F(h_l^i)_{[p]_t, t}) \tag{6}$$

where $A_{[p]_t, [p]_{t-1}}$ denotes the transaction score that model the transition from t-th state to $t-1$-th for a pair of consecutive time steps. With our semantic label based approach, the loss function can be rewritten as follow:

$$\mathcal{L}(\theta) = -\sum_i log \sum_{y \in C(y_{pl}^i)} q\mathcal{D}_l(y|x_l^i))p_\theta(y|x_l^i)) \tag{7}$$

Based on the observation of $q\mathcal{D}_l$ and pw, we find that the recognized entity depends heavily on its context. Even the same entity cannot be accurately predicted in different context, which leads to an inferior recall score. This observation indicates that a proper way to fine-tuning the predicted distribution is needed. Inspired by adasampling algorithm [10], an adaptive fine-tuning strategy is proposed. The core idea of this approach is to adaptive enriching entity dictionary and fine-tuning the estimated distribution until getting a stable result. The detail of adaptive fine-tuning will be introduced in Sect. 3.2.

3.1   Entity Existence Detection

In this work, motivated by BERT behaviour in ranking [27], a matching based entity existence detection model is proposed. Firstly, text data augmentation is applied on the original dataset. The approach that we implement contains synonyms replace, randomly insert, randomly swap, and randomly delete [28]. The scale of the data is effectively increased through our augmentation. We denote each word sequence in the augmented dataset $\mathcal{D}^+$ as x^i, and $\mathbb{C}_s$ as the set of all semantic labels. The input of the entity existence detection model x_{ed}^i is constructed by concatenating x^i and every element l^n in $\mathbb{C}_s$.

$$x_{ed}^i = x^i \oplus l^n \tag{8}$$

The interaction between word sequence and semantic labels is obtained through the approach above. Up to now, entity existence detection dataset $\mathbb{D}_c$ is constructed. A deep neural model based on BERT is built for entity existence detection:

$$h_{ed}^i = BERT(x_{ed}^i)_{cls}^{last} \tag{9}$$

$$F(h_{ed}^i) = w^T h_{ed}^i + b \tag{10}$$

$$score(x_{ed}^i) = softmax(F(h_{ed}^i)) \tag{11}$$

The output of BERT's last layer on "[CLS]" position is utilized as the entity existence features and combined linearly with weight w^T. The existence score (i.e., matching score in text matching task) from model includes all term pair interactions between the word sequence and semantic label via its transformer's cross-match attentions [29]. The loss function of matching based entity existence detection model is as follows:

$$\mathcal{L}(w) = -\sum_i y_{ed}^i log(score(x_{ed}^i)) + \lambda L_2(w) \tag{12}$$

where all parameters in the model are represented by w, and L_2 indicates the L2 regularization term.

3.2 Adaptive Fine-Tuning

To alleviate the problem that the prediction of entity depends heavily on context, an adaptive fine-tuning algorithm is implemented. More concretely, we first train a semantic label enhanced NER model and extract entities from original dataset through the model. Most entities in original dataset could be recognized in this stage. Furthermore, a pre-trained word segmentation model [30] that trained on general corpus is used to segment every word sequence in dataset. If an entity recognized by the semantic label enhanced model also occurs in the segmentation result from the same word sequence, it is considered as a valid entity and will be added into the entity dictionary.

After enriching the entity dictionary, training data is labeled through the enriched entity dictionary and utilized for iteratively training the next semantic label enhanced model. The iteration stops while the number of entity dictionary is stable. We believe the final model could adapt to more contexts and is able to reach a high recall score now. In the final step, the initial semantic label enhanced model and the last model are combined as an ensemble one as our final estimation.

4   Experiments

In this section, we empirically study 4 aspects:

- the overall performance of our proposed approach;
- the effectiveness of semantic label compared with non-semantic label;
- the capability of matching based entity existence detection model;
- the influence of adaptive fine-tuning.

4.1   Datasets

All experiments are conducted on the Youku dataset[1]. The training dataset contains 10000 samples, the validation dataset contains 1000 samples, and 2000 samples are involved in the test dataset. There are three types of labels in all samples, which are TV, PER and NUM. The annotation of all data follows BIE scheme. Our statistic indicates that there are 4331 annotated data in training dataset, which merely takes up 43.3%. The detail statistics of Youku dataset are shown in Table 2.

Table 2. Statistics of Youku dataset utilized in our experiments.

Dataset	Sentences	Characters	PER entities	TV entities	NUM entities
Training	10000	164880	1266	3245	239
Validation	1000	15867	423	851	95
Testing	2000	33123	863	1722	185

4.2   Comparative Methods

Four baseline methods are compared with our semantic label enhanced model. First, a BERT-CRF baseline that follows cross-validation approach [9] is built[2]. We train this model on the original dataset $\mathcal{D}$ and use the term **Baseline** to represent it. The second model employs the same architecture with our proposed method but using a symbolic label alternatively. More concretely, the initial word sequence x^i is merged with a single word, non-semantic label rather than a semantic label. Instead of $\mathcal{D}_l$, we construct a new dataset $\mathcal{D}_{syl}$ as the training data for NER with incomplete annotations. The goal of the model is to verify the influence of semantic information within labels. The abbreviation **SymEM**(i.e., symbolic label enhanced model) is utilized to refer to the second model. For the third model, a multi-label classification algorithm is implemented for entity

---

[1] Dataset available at https://github.com/ZhuiyiTechnology/AutoIE/tree/master/data.

[2] Model available at https://github.com/ZhuiyiTechnology/AutoIE/tree/master/baseline.

existence detection, while the NER method remains the same. We mark the third model as **CSemEM**. In CSemEM, entity existence detection is viewed as a multi-label classification task. The multi-label classification model that we implement is also based on BERT:

$$h^i = BERT(x^i)_{cls}^{last} \tag{13}$$

$$F(h^i) = w^T h^i + b \tag{14}$$

$$score(x^i) = \delta(F(h^i)) \tag{15}$$

$$\mathcal{L}(w_m) = -\sum_i y^i log(score(x^i)) + (1 - y^i)log(1 - score(x^i)) + \lambda L_2(w_m) \tag{16}$$

where δ here represents the activation function, w_m indicates all parameters in the classification model and $\mathcal{L}$ represents the loss function. The object of this model is to compare the performance of matching based and multi-label classification based entity existence detection. The fourth model called **SemEM** follows our semantic label enhanced approaches without applying adaptive fine-tuning. For the model that applies every mechanism we mentioned in Sect. 3, we denote it as **SemEMA** (i.e., Semantic Enhanced Model with Adaptive fine-tuning).

4.3   Implementation Detail

Specifically, Chinese RoBERTa [31] with 12 layers, 768 hidden size, 12 heads, and 110M parameters is chosen as our pre-trained BERT model. We employ Adam algorithm for both entity existence detection model and entity recognition model. For the entity existence detection model, we use 0.00005 as the initial learning rate and 128 as the batch size. Best model is selected through loss value on validation dataset. An early stop tactics is applied if the loss value does not decrease for 500 batches. In NER model training, we select 0.0001 as the learning rate and the batch size is 32. The number of iterations for model is 10. We set k equals 2 in cross-validation process follows [9]. For entity recognition model, the best model is selected by F1 score on validation set. All the parameters are fixed according to previous work in the literature without grid-search adjustments for our Youku dataset. To avoid randomness in our results, all mentioned models are trained three times and final results are presented by averaging them.

4.4   Main Results

General Performance. Table 3 shows model performance of each entity type and overall performance on 2000 test samples. What stands out in the table is every model that we implement receives more than 1.5% improvement compared with the Baseline model. Especially, SemEMA model reaches F1 score 71.96 with a 7.98 absolute increment on baseline model. Furthermore, a case study is also completed on our approach and baseline model. As shown in Table 4, it is apparent that our SemEMA model predicts all entities correctly and performs better

Table 3. Main results of different models on test dataset.

Model name	PER			TV			NUM			ALL		
	P.	R.	F.	P.	R.	F.	P.	R.	F.	P.	R.	F.
Baseline	73.93	64.07	68.65	69.04	56.21	61.97	71.42	54.05	61.53	70.07	58.51	63.98
SymEM	83.92	74.39	78.86	76.81	47.90	59.01	72.72	47.46	57.51	79.33	56.13	65.75
CSemEM	79.28	79.83	79.56	70.52	55.13	63.45	78.46	55.13	64.76	74.11	64.44	68.92
SemEM	78.22	82.85	**80.47**	67.04	63.48	64.68	75.17	58.29	**66.03**	71.31	69.48	70.87
SemEMA	74.07	85.39	79.33	64.03	74.85	**69.02**	75.00	55.13	63.55	67.68	76.82	**71.96**

on both new entity discovery and entity border prediction. More concretely, "火影忍者 (Naruto)" is a TV entity that does not exist in initial entity annotations, which is discovered successfully by our method. Besides, the rest of cases in Table 4 indicates that a more accurately entity border prediction is achieved compared with baseline model.

We believe the improvement of our approach can be explained from three perspectives: **1)** a well-trained matching based model is implemented for entity existence detection task, which leads to an efficient new entity discovery. **2)** The result of entity existence detection can be seen as prior knowledge for entity recognition task. Prior knowledge alleviates the problem that the regular model should learn both entity existence and entity position, which makes entity recognition model focus on precise entity border prediction. In addition, the semantic label information can be seen as a restriction of NER model, which forces the model to converge quickly and accurately. **3)** A better context independent new entity discovery is achieved through adaptive fine-tuning.

Table 4. Case study of Baseline and SemEMA.

Word sequence	Baseline	SemEMA
火影忍者357预告	None	火影忍者(TV)
Naruto 357 advance notice	None	Naruto(TV)
旋风少女第17集电视剧全集	旋风(TV)	旋风少女(TV)
Cyclone Girl episode 17 complete TV series.	Cyclone(TV)	Cyclone Girl(TV)
练习刘德华	练(PER)	刘德华(PER)
Practice Dehua Liu	Pra(PER)	Dehua Liu(PER)

Table 5. Entity existence detection result of matching based and multi classification based model on test dataset.

Model	F1 score
Classification based	73.78
Matching based	**74.97**

Effectiveness of Semantic Label. Table 3 also illustrates that SymEM has about 2% absolute increment than baseline, which is restricted compared with SemEM or SemEMA model. Two essential facts are suggested by this phenomenon: **1)** the fusion of entity existence detection result and word sequence through concatenate labels with initial word sequence is a general way to improve NER with incompletely annotated data result. **2)** Compared with the symbolic label, transferring entity existence detection result into semantic label shows more compatibility and could reach a better result.

Performance of Matching Based Detection. We then study the effectiveness of matching based entity existence detection. As we mentioned in Sect. 4.1, a multi-classification based model is built for the same task. Table 5 shows the result of two approaches on the test dataset, which indicates that matching based method performs 1.19% absolute improvement than the multi-classification based model. Moreover, what stands out in Table 3 is the F1 score of multi-classification based NER model (i.e., CSemEM) drops around 2% compared with SemEM on all entity types. The overall result reveals that matching based approach is better than conventional multi-classification method on entity existence detection task since matching based approach focus on the consistency between word sequence and semantic label.

Influence of Adaptive Fine-Tuning. From Table 3, it is obvious that SemEMA has the best overall result among all models. Nevertheless, the F1 score of PER and NUM entity decrease from SemEM (i.e., no fine-tuning model), and the improvements mainly come from the better recall of TV entity. The appearance suggests that although adaptive fine-tuning leads to overall improvements, false positive predictions are also brought. While the SemEMA model tends to recognize more entities and brings the improvement of recall score, it fails to ensure the correctness of every prediction. According to our observation, the capability of SemEMA on entity border prediction is poorer than SemEM. We believe there are two reasons for the phenomenon. Firstly, during the iteration, the model learns some entities that has unique context, which would lead incorrect predictions in most circumstances. Second, when error occurs in both entity recognition stage and word segmentation stage, false training data will be brought for the next iteration and cause wrong predictions.

5   Conclusions and Future Work

The principal objective of the current study is to tackle the NER problem with incompletely annotated data. In this work, a novel and easy-to-implement semantic label enhanced model is proposed, which includes a matching based method to detect entity existence and generate semantic labels, and a semantic label enhanced approach to recognize entities. Besides, by adding adaptive fine-tuning modules, the model gains further improvements. Extensive experiments and analysis have shown that our approach acquires remarkable improvements.

Notably, the importance of semantic label is proved through controlled trials. Although our method focuses on the named entity recognition task, we expect that the proposed framework will play an essential role in other scenarios where imperfect annotated data is provided.

References

1. Kumar, S.: A survey of deep learning methods for relation extraction. arXiv preprint arXiv:1705.03645 (2017)
2. Chiu, J.P.C., Nichols, E.: Named entity recognition with bidirectional LSTM-CNNs. Trans. Assoc. Comput. Linguist. **4**, 357–370 (2016)
3. Lample, G., Ballesteros, M., Subramanian, S., Kawakami, K., Dyer, C.: Neural architectures for named entity recognition. arXiv preprint arXiv:1603.01360 (2016)
4. Yadav, V., Bethard, S.: A survey on recent advances in named entity recognition from deep learning models. arXiv preprint arXiv:1910.11470 (2019)
5. Peters, M.E., et al.: Deep contextualized word representations. arXiv preprint arXiv:1802.05365 (2018)
6. Devlin, J., Chang, M.-W., Lee, K., Toutanova, K.: BERT: pre-training of deep bidirectional transformers for language understanding. arXiv preprint arXiv:1810.04805 (2018)
7. Yan, H., Deng, B., Li, X., Qiu, X.: TENER: adapting transformer encoder for name entity recognition. arXiv preprint arXiv:1911.04474 (2019)
8. Akbik, A., Bergmann, T., Vollgraf, R.: Pooled contextualized embeddings for named entity recognition. In: Proceedings of the 2019 Conference of the North American Chapter of the Association for Computational Linguistics: Human Language Technologies, (Long and Short Papers), vol. 1, pp. 724–728 (2019)
9. Jie, Z., Xie, P., Lu, W., Ding, R., Li, L.: Better modeling of incomplete annotations for named entity recognition. In Proceedings of the 2019 Conference of the North American Chapter of the Association for Computational Linguistics: Human Language Technologies, (Long and Short Papers), vol. 1, pp. 729–734 (2019)
10. Yang, P., Liu, W., Yang, J.: Positive unlabeled learning via wrapper-based adaptive sampling. In: IJCAI, pp. 3273–3279 (2017)
11. Lafferty, J., McCallum, A., Pereira, F.C.N.: Conditional random fields: probabilistic models for segmenting and labeling sequence data (2001)
12. Bellare, K., McCallum, A.: Learning extractors from unlabeled text using relevant databases. In Sixth International Workshop on Information Integration on the Web (2007)
13. Carlson, A., Gaffney, S., Vasile, F.: Learning a named entity tagger from gazetteers with the partial perceptron. In: AAAI Spring Symposium: Learning by Reading and Learning to Read, pp. 7–13 (2009)
14. Greenberg, N., Bansal, T., Verga, P., McCallum, A.: Marginal likelihood training of BILSTM-CRF for biomedical named entity recognition from disjoint label sets. In: Proceedings of the 2018 Conference on Empirical Methods in Natural Language Processing, pp. 2824–2829 (2018)
15. Mayhew, S., Chaturvedi, S., Tsai, C.T., Roth, D.: Named entity recognition with partially annotated training data. arXiv preprint arXiv:1909.09270 (2019)
16. Chang, M.W., Ratinov, L., Roth, D.: Guiding semi-supervision with constraint-driven learning. In: Proceedings of the 45th Annual Meeting of the Association of Computational Linguistics, pp. 280–287 (2007)

17. Lee, W.S., Liu, B.: Learning with positive and unlabeled examples using weighted logistic regression. In: ICML, vol. 3, pp. 448–455 (2003)
18. Clark, K., Luong, M.T., Manning, C.D., Le, Q.V.: Semi-supervised sequence modeling with cross-view training. arXiv preprint arXiv:1809.08370 (2018)
19. Cao, Y., Hu, Z., Chua, T.S., Liu, Z., Ji, H.: Low-resource name tagging learned with weakly labeled data. arXiv preprint arXiv:1908.09659 (2019)
20. Liu, B., Lee, W.S., Yu, P.S., Li, X.: Partially supervised classification of text documents. In: ICML, vol. 2, pp. 387–394. Citeseer (2002)
21. Liu, B., Dai, Y., Li, X., Lee, W.S., Yu, P.S.: Building text classifiers using positive and unlabeled examples. In: Third IEEE International Conference on Data Mining, pp. 179–186. IEEE (2003)
22. Elkan, C., Noto, K.: Learning classifiers from only positive and unlabeled data. In: Proceedings of the 14th ACM SIGKDD international conference on Knowledge discovery and data mining, pp. 213–220 (2008)
23. Peng, M., Xing, X., Zhang, Q., Fu, J., Huang, X.: Distantly supervised named entity recognition using positive-unlabeled learning. arXiv preprint arXiv:1906.01378 (2019)
24. Ratinov, L., Roth, D.: Design challenges and misconceptions in named entity recognition. In: Proceedings of the Thirteenth Conference on Computational Natural Language Learning (CoNLL 2009), (2009)
25. Hochreiter, S., Schmidhuber, J.: Long short-term memory. Neural Comput. **9**(8), 1735–1780 (1997)
26. Huang, Z., Xu, W., Yu, K.: Bidirectional LSTM-CRF models for sequence tagging. arXiv preprint arXiv:1508.01991 (2015)
27. Qiao, Y., Xiong, C., Liu, Z., Liu, Z.: Understanding the behaviors of BERT in ranking. arXiv preprint arXiv:1904.07531 (2019)
28. Wei, J., Zou, K.: EDA: easy data augmentation techniques for boosting performance on text classification tasks. arXiv preprint arXiv:1901.11196 (2019)
29. Vaswani, A., et al.: Attention is all you need. In: Advances in Neural Information Processing Systems, pp. 5998–6008 (2017)
30. Jiao, Z., Sun, S., Sun, K.: Chinese lexical analysis with deep Bi-GRU-CRF network. arXiv preprint arXiv:1807.01882 (2018)
31. Cui, Y., et al.: Pre-training with whole word masking for Chinese BERT. arXiv preprint arXiv:1906.08101 (2019)

Named Entity Recognition Using
a Semi-supervised Model Based on BERT
and Bootstrapping

Yuanyuan Liu[1,2(✉)], Xiang Li[3], Jiaxin Shi[1,2], Lei Zhang[3], and Juanzi Li[1,2]

[1] Department of Computer Science and Technology, BNRist, Beijing, China
lijuanzi@tsinghua.edu.cn
[2] KIRC, Institute for Artificial Intelligence, Tsinghua University, Beijing 100084, China
[3] State Key Laboratory of Intelligent Technology and Systems, Tsinghua University, Beijing 100084, China

Abstract. Named Entity Recognition (NER) aims to recognize entities from texts, which is important for lots of natural language processing tasks, including knowledge graph and so on. Supervised NER method requires expensive datasets. In order to reduce expensive labor costs, we propose a semi-supervised model based on BERT and Bootstrapping to recognize more entities with a small amount of labeled data and less human labor. The method works in the following steps. Firstly, we build a small entity dictionary, and tag the collected texts using this dictionary. Secondly, we learn a model based on the small dataset. Thirdly, recognize entities that are not in dictionary and add the high-quality entities to the dictionary. Then, we repeat these steps until no new high-quality entities are found. Experiments show that the proposed model performs better than baselines on the Financial Entity Recognition task in 2019 CCF BDCI.

Keywords: BERT · Bootstrapping · Named Entity Recognition

1 Introduction

The information on the Internet is exploding, and there is a large quantity of unstructured texts, which are difficult for computers to understand and learn. How to mine and utilize massive information has become an urgent problem to be solved. Information extraction technology aims to extract structured information such as entities, entity relationships, and event information from massive unstructured texts. It is a fundamental part of natural language processing tasks. Among them, Named Entity Recognition (NER) aims to extract specific types of entities such as person names, place names, and organization names, which is of great significance for the further research of natural language processing [1].

In reality, usually there is insufficient labeled data in professional fields such as finance, medical care and computer. Tagging a large number of labels requires a lot of experts' time and labor. The research on Named Entity Recognition using a small amount

H. Chen et al. (Eds.): CCKS 2020, CCIS 1356, pp. 54–63, 2021.
https://doi.org/10.1007/978-981-16-1964-9_5

of labeled data has important practical significance and application value. The existing Named Entity Recognition models, such as BERT BiLSTM-CRF [2] are usually used in scenarios with a large amount of training data, but they cannot handle small sample scenarios well. For small sample scenarios, existing methods expand the training data by the way of Bootstrapping. However, it is difficult for the encoder to capture semantic information due to insufficient training data. This paper uses pre-trained BERT to solve the problem of semantic information, and uses Bootstrapping to solve the problem of insufficient labeled samples. This paper proposes a semi-supervised Named Entity Recognition method based on BERT and Bootstrapping.

2   Related Works

The methods of Named Entity Recognition are divided into rule-based methods, statistical methods, and deep learning methods. The rule-based methods cost a lot of human efforts but have bad portability. In statistical methods, the Named Entity Recognition task is regarded as a sequence labeling task. The models include Hidden Markov Models (HMM) [3] and Maximum Entropy Markov Models (MEMM), Conditional Random Field (CRF) [4]. Compared with traditional statistical machine learning methods, methods based on deep learning do not require artificial features and obtain better results. Recurrent neural network RNN [5] can make full use of context dependence, but it has a gradient dispersion problem, the sequence data is too long. Literature [6] uses Long Short-Term Memory (LSTM) to solve the gradient dispersion problem, but the one-way LSTM can only obtain the information before the sentence and lose the semantic information after the sentence. Literature [7] uses BiLSTM to model the sentence in two directions, it could learn the context information of the sentence, but could not learn the dependence between the output annotations. Literature [8] adds a CRF layer after BiLSTM, taking the dependence between tags into account, and effectively improves the performance.

Most deep learning models use Word2Vec to train word vectors, they consider more the contextual co-occurrence, rather than the word order. In 2018, Google proposed Bidirectional Encoder Representation from Transformers (BERT) [9] based on Transformer [10], which can extract semantic information of text in a deeper level. Literature [11, 12] proposes BiLSTM-CRF model based on BERT, which further improves the performance.

Neural network models such as CRF and BiLSTM-CRF can achieve good performance, but require a lot of labeled data. In current situation, it is difficult to obtain a large number of labeled samples in professional fields like finance and medicine. The labeled samples require professional knowledge, and the labeling costs a lot of labor and time. Therefore, how to use limited training data and human efforts to train a better Named Entity Recognition model, is a practical problem to be solved.

Active Learning [13] is designed for insufficient labeled data. Active learning delivers the most valuable unlabeled samples to oracle for labeling, and trains the model with the samples labeled by the oracle. Through the interaction between the model and the oracle, the amount of labeling data required for model learning is reduced. The model starts learning with a small labeled training set L, selects the sample U with the largest information, and delivers U to the oracle for labeling. Then it incorporates the newly obtained labeled data to the original labeled training set L, continues training the model. Repeat these processes until the model performance meets the requirement.

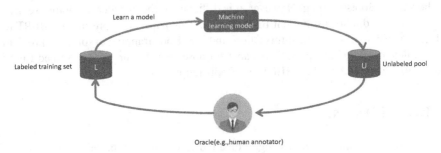

Fig. 1. Active learning

Literature [14–17] use Bootstrapping to recognize new entities. They use a small number of entities as the initial seed set, learn the model, use the model to recognize new entities, and use new entities to learn the model after manual filtering, iterate until the performance meets the requirement. However, the existing methods use core models such as SVM and LSTM-CRF. Due to the lack of training data, they could not obtain deep semantic information well. Therefore, this paper proposes a Named Entity Recognition model based on pre-trained BERT and Bootstrapping.

3 Model

Choosing BERT based NER models such as BERT, BERT BiLSTM, BERT CRF and BERT BiLSTM-CRF, as the core model, we propose a model based on Bootstrapping.
1) We build and start from a small entity dictionary.
2) We learn a model using the entities from the dictionary.
3) We recognize entities that are not in the dictionary using the learned model on another part of data.
4) We manually choose high-quality entities and add them to the entity dictionary.
5) Repeat step2–4, until no more new entities are recognized or the model meets our requirement.

3.1 BERT Based NER Models

We choose BERT based NER models such as BERT, BERT BiLSTM, BERT CRF and BERT BiLSTM-CRF, as the core model. Take BERT BiLSTM-CRF for example, the model has three layers: BERT layer, BiLSTM layer and CRF layer, as Fig. 2 shows:

BERT
The input of BERT is obtained by summing the word vector Token Embeddings, sentence vector Segment Embeddings and position vector Position Embeddings.

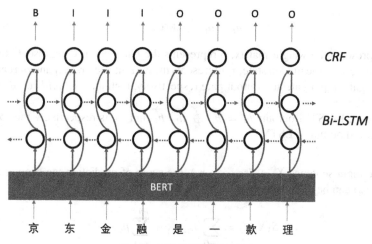

Fig. 2. BERT BiLSTM-CRF

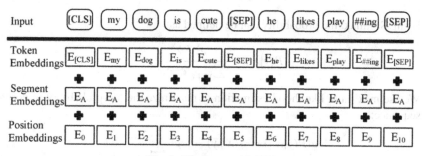

Fig. 3. The Input of BERT

This article uses the BERT-Base-Chinese model, the Transformer layer has 12 layers, the hidden layer has 768 dimensions, the number of attention head is 12, and the total parameter is 110M.

BiLSTM

LSTM unit is composed of input gate, forget gate, output gate and cell state.

The LSTM process at time t:

$$f_t = sigmoid(W_f[h_{t-1}, x_t] + b_f)$$

$$i_t = sigmoid(W_i[h_{t-1}, x_t] + b_i)$$

$$\widehat{C}_t = \tanh(W_c[h_{t-1}, x_t] + b_c)$$

$$C_t = f_t \bullet C_{t-1} + i_t \bullet \widehat{C}_t$$

$$o_t = sigmoid(W_o[h_{t-1}, x_t] + b_o)$$

$$h_t = o_t \bullet \tanh(C_t)$$

W represents the weight matrix, b represents the bias vector, C represents the cell state, h_t represents the hidden state, x_t represents the current vector, f_t, i_t and o_t represent the forget gate, input gate and output gate respectively, tanh and *sigmoid* represent the activation function.

At time t, BiLSTM outputs $h_t = \overrightarrow{h_t} \oplus \overleftarrow{h_t}$, $\overleftarrow{h_t}$ and $\overrightarrow{h_t}$ is respectively the forward and backward output of LSTM.

CRF

Given the input sentence $X = (x_1, \cdots, x_n)$, the score of prediction tag sequence $y = (y_1, \cdots, y_n)$ can be defined as:

$$S(X, y) = \sum_{i=0}^{n} a_{y_i, y_{i+1}} + \sum_{i=1}^{n} p_{i, y_i}$$

$a_{y_i, y_{i+1}}$ measures the score from tag y_i to tag y_{i+1}, p_{i, y_i} measures the score of x_i tagging to y_i.

Using Softmax, the probability can be defined as:

$$P(y|X) = \frac{e^{S(X, y)}}{\sum_{\tilde{y} \in Y_X} e^{S(X, \tilde{y})}}$$

We use maximum likelihood estimation to maximize the log-probability of correct tag sequence:

$$\log(P(y|X)) = S(X, y) - \log(\sum_{\tilde{y} \in Y_X} e^{S(X, \tilde{y})})$$

Using Viterbi Algorithm, we choose the tag sequence y* with maximum score as the prediction result:

$$y^* = argmax_{\tilde{y} \in Y_X} S(X, \tilde{y})$$

3.2 A Named Entity Recognition Method Based on BERT and Bootstrapping

Given unlabeled dataset and an entity dictionary, the algorithm works as follows:

1. First we divide the unlabeled dataset into N parts: SET[0], SET[1]......SET[N].
2. Initialize the step to 0, and repeat the following steps utill no more new entities are found or the performance meets our requirement.
1) Automatic annotation: Use the current Entity Dictionary to label the entities in SET[n%N] through automatic annotation. In order to discover more new entities, every step when data is loaded, data augmentation is performed by randomly replacing entities and adding noise. Specifically, we randomly select an entity in each sentence, randomly replace it with another entity in the entity dictionary, and randomly select a character in this entity to be replaced with its homophone.

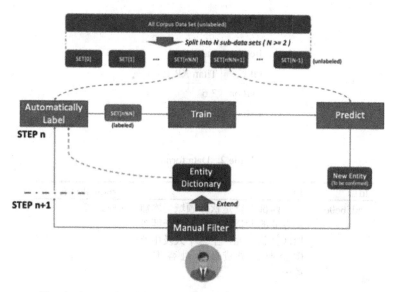

Fig. 4. A named entity recognition method based on bootstrapping.

2) Model learning: Entity Dictionary + SET[n%N] → model. That is, we use Entity Dictionary to automatically annotate SET[n%N] to train the model. The core model is chosen form BERT based NER models.

3) Dictionary expansion: Model + data SET[n%N + 1] → New Entity, Entity Dictionary + New Entity → Entity Dictionary. That is, we use the model obtained in the training stage to recognize entities that are not in the dictionary from SET[n%N + 1] to obtain a new entity set New Entity. Manually filter out high-quality entities from New Entity and add them to Entity Dictionary for the next step of automatic annotation.

4) n = n + 1

At last, we can obtain a high-quality entity dictionary and a learned model with good performance.

4 Experiments

4.1 Environments and Dataset

Dataset: Dataset of 2019 Financial Entities Recognition Contest dataset of 2019 CCF Big Data & Computing Intelligence Contest.

The training data is given in csv format. Each piece of data includes an identification number (id), text content (text), and a list of unknown entities (unknownEntities).

The format is shown as follows:

The purpose is to identify unknown financial entities from the text, including financial platform names, company names, project names, and product names.

Table 1. Environments

Item	Value
GPU	Titan XP(12G)
Python	3.6
Pytorch	1.2.0

Table 2. Data format.

Id	Text	unknownEntites
1ad5be0d	······西部助贷是伙伴领域资本旗下专业的助贷平台，专注于西部地区贷款研究，主营个人及中小微企业融资贷款金融咨询服务······	西部助贷
6dd28e9b	······揭秘趣步骗局，趣步是什么，趣步是怎么赚钱的？趣步公司可靠吗？趣步合法吗 ······	趣步
f3b61b38	······度小满金融在首年国民投资理财情绪指数报告》中提到······	度小满金融
84b12bae	······纷享车链首款官方定制的可信设备——七巧智盒3in1已于8月初已经正式面市······	七巧智盒3in1

4.2 Metrics

The evaluation metrics used in Experiment 2 are: Precision, Recall and F1 value.

The calculation formula is:

Accuracy P = number of correct entities recognized / number of entities recognized,
Recall R = number of correct entities recognized / number of entities in the texts,
F1 = (2*P*R)/(P + R).

4.3 Settings

Experiment 1

There are 2500 pieces of data and 1416 labeled entities in the dataset. 50% of the labeled entities are hidden and don't participate in training. We use the remaining 50% of entities (708) to train the model, and test the model's performance to find the hidden 708 entities.

Experiment 2

There are 2500 pieces of data and 1416 labeled entities in the dataset. 50% of the labeled entities are hidden and don't participate in training. We use the remaining 50% of entities

(708) to train the model, and test the model's performance to recognize new entities on test dataset containing 2000 pieces of data.

4.4 Results

Experiment 1

Table 3. Experiment 1.

Model	Num of new entities
BiLSTM-CRF	83
BERT	170
BERT BiLSTM	170
BERT CRF	175
BERT BiLSTM-CRF	173
Bootstrapping BERT	**242**
Bootstrapping BERT BiLSTM	240
Bootstrapping BERT CRF	222
Bootstrapping BERT BiLSTM-CRF	218

Experiment 2

Table 4. Experiment 2.

Model	Precision	Recall	F1
BiLSTM-CRF	0.3640	0.0999	0.1568
BERT	0.5000	0.1785	0.2630
BERT BiLSTM	0.4246	0.2076	0.2788
BERT CRF	0.4367	0.2173	0.2902
BERT BiLSTM-CRF	0.3524	0.2085	0.2620
Bootstrapping BERT	0.3360	0.2464	0.2843
Bootstrapping BERT BiLSTM	0.3394	0.2881	**0.3116**
Bootstrapping BERT CRF	0.3423	0.2706	0.3023
Bootstrapping BERT BiLSTM-CRF	0.3377	0.2503	0.2875

Experiment 1 shows that, on the dataset with insufficient labels, Bootstrapping can find more hidden entities in the dataset. There is little difference between the performance of BERT NER models, but no matter which BERT NER model is chosen, Bootstrapping improves the performance than the corresponding model.

Experiment 2 shows that when the training dataset is insufficiently labeled, Bootstrapping can better discover new entities in test dataset than the corresponding model. Among them, in BERT BiLSTM, Bootstrapping improves F1 score by 0.0328, which is most than the corresponding model, which is 0.0328. Especially, Bootstrapping improves the recall in a big degree because it helps improve the entity dictionary, which helps improve the recognition of new entities.

The experiment also finds that as the number of steps increases, the performance of the model continues to increase, and the workload of manual filtering gradually decreases.

5   Conclusions

On the problem of Named Entity Recognition with insufficient annotated data, this paper proposes a semi-supervised Named Entity Recognition algorithm based on BERT and Bootstrapping. Starting from a small known entity dictionary, it iteratively repeats the process of automatic annotation, model learning, dictionary expansion, and achieves the purpose of recognizing more new entities. Experiments show that the algorithm both saves the cost of labeling and ensures the performance of the model.

Future work will test the effects of different forms and intensities of noises on the performance, and add a dictionary to reduce the workload of manual filtering.

Acknowledgments. This work is supported by the Key-Area Research and Development Program of Guangdong Province (2019B010153002) and NSFC Key Projects (U1736204, 61533018).

References

1. Sang, E.F., De Meulder, F.: Introduction to the CoNLL-2003 shared task: language-independent named entity recognition. arXiv preprint cs/0306050 (2003)
2. 王子牛, 姜猛, 高建瓴, 等. 基于 BERT 的中文命名实体识别方法. 计算机科学, **46**(11A), 138–142 (2019)
3. Rabiner, L.R.: A tutorial on hidden Markov models and selected applications in speech recognition. Proc. IEEE **77**(2), 257–286 (1989)
4. Learning, M.: An introduction to conditional random fields. Mach. Learn. **4**(4), 267–373 (2011)
5. Elman, J.L.: Finding structure in time. Cogn. Sci. **14**(2), 179–211 (1990)
6. Hochreiter, S., Schmidhuber, J.: Long short-term memory. Neural Comput. **9**(8), 1735–1780 (1997)
7. Chen, X., Qiu, X., Zhu, C., et al.: Long short-term memory neural networks for chinese word segmentation. In: Proceedings of the 2015 Conference on Empirical Methods in Natural Language Processing, pp. 1197-1206 (2015)
8. Huang, Z., Xu, W., Yu, K.: Bidirectional LSTM-CRF models for sequence tagging. arXiv preprint arXiv:1508.01991 (2015)
9. Vaswani, A., Shazeer, N., Parmar, N., et al.: Attention is all you need. In: Advances in Neural Information Processing Systems, pp. 5998–6008 (2017)
10. Devlin, J., Chang, M.W., Lee, K., et al.: Bert: pre-training of deep bidirectional transformers for language understanding. arXiv preprint arXiv:1810.04805 (2018)

11. 孔玲玲. 面向少量标注数据的中文命名实体识别技术研究. 浙江大学 (2019)
12. 吴俊, 程垚, 郝瀚, 等. 基于 BERT 嵌入 BiLSTM-CRF 模型的中文专业术语抽取研究[J]. 情报学报, **39**(4), 409–418 (2020)
13. Settles, B.: Active learning literature survey. University of Wisconsin-Madison Department of Computer Sciences (2009)
14. Tran, Q.T., Pham, T.X.T., Ngo, Q.H., et al.: Named entity recognition in Vietnamese documents. Progress Inform. J. **5**, 14–17 (2007)
15. Krishnanand, K.N., Ghose, D.: Glowworm swarm optimization for simultaneous capture of multiple local optima of multimodal functions. Swarm Intell. **3**(2), 87–124 (2009)
16. Thenmalar, S., Balaji, J., Geetha, T.V.: Semi-supervised bootstrapping approach for named entity recognition. arXiv preprint arXiv:1511.06833 (2015)
17. Kim, J., Ko, Y., Seo, J.: A bootstrapping approach with CRF and deep learning models for improving the biomedical named entity recognition in multi-domains. IEEE Access **7**, 70308–70318 (2019)

Reinforcement Learning for Clue Selection in Web-Based Entity Translation Mining

Lingyong Yan[1,3(✉)], Xianpei Han[1,2], and Le Sun[1,2]

[1] Chinese Information Processing Laboratory, Beijing, China
{lingyong2014,xianpei,sunle}@iscas.ac.cn
[2] State Key Laboratory of Computer Science,
Institute of Software, Chinese Academy of Sciences, Beijing, China
[3] University of Chinese Academy of Sciences, Beijing, China

Abstract. Web-based mining is a promising approach for entity translation. Traditional web-based mining methods construct spotting queries in a heuristic, one-step paradigm, which cannot resolve the diversity of entity names and is incapable of leveraging the information in the returned bilingual pages for iterative spotting query refinement. To resolve the above drawbacks, this paper proposes a reinforcement learning-based method, which models web-based entity translation mining as a Markov Decision Process (MDP). Specifically, we regard the query construction as a multi-turn, state-to-action mapping procedure, and learn a Dueling Deep Q-network based clue selection agent which can adaptively select spotting clues based on both short-term and long-term benefits. Experiments verified the effectiveness of our approach.

1 Introduction

Web-based entity translation mining aims to mine entity translations from web pages. Web-based mining is usually regarded as a promising approach for translating named entities rather than the conventional machine translation techniques [2,11,29]. Because entity names are commonly not translated literally (e.g., the Chinese translation of the film *La La Land* is "爱乐之城 [the city which likes music]"), conventional machine translation techniques often fail to work, while the vast majority of entity translations can be found and mined from the timely-updated web pages. To translate the named entities, web-based mining methods often perform the following steps: *bilingual page spotting*, which usually takes the concatenation of original entity mention and spotting clues (i.e., additional foreign keywords) as the spotting queries (e.g., "*La La Land* 电影 [film]") and feeds them into search engines to spot web pages; *translation extraction*, which extracts translation candidates from the spotted pages; *translation ranking*, which ranks the candidates by summarizing evidence (e.g., co-occurrence statistics) and returns the top-ranked one as the final translation.

H. Chen et al. (Eds.): CCKS 2020, CCIS 1356, pp. 64–77, 2021.
https://doi.org/10.1007/978-981-16-1964-9_6

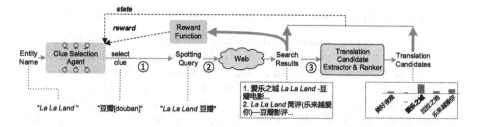

Fig. 1. The architecture of our web-based entity translation mining method.

Among the above steps, it is critical to select effective clues to construct spotting queries [4,7,23], because an effective clue can help to spot the high-quality bilingual pages containing the correct translations. To select effective spotting clues, previous studies commonly use one-step heuristic rules, such as using the partial translations of the original entity [7,24] or selecting top-ranked foreign words from initial search results [2,11], to construct spotting queries for different entity types. However, the above approaches, despite the promising results, could suffer from two problems. First, it is hard for the heuristic rules to adaptively select effective clues for different entities. For example, the partial literal transliterations can be good spotting clues for person names (e.g., the "琼恩 [Jon]" for "Jon Snow"), but the partial translations are not favorable for film names (e.g., the Chinese translation of *La La Land* is "爱乐之城 [the city which likes music]"). Second, the one-step paradigm cannot take the returned bilingual pages into consideration, which often contain effective evidence. Therefore, to generate effective spotting queries for better bilingual page spotting, a system should select the most suitable clues in a multi-turn paradigm, i.e., selecting the clues based on both entity mentions and previously returned bilingual pages.

To this end, this paper proposes a novel web-based entity translation mining method, which adaptively selects effective spotting clues to construct queries and conduct entity translation mining in a multi-turn paradigm (see Fig. 1). Specifically, we model the web-based entity translation mining as a Markov Decision Process (MDP) and learn to adaptively construct spotting clues via a Dueling Deep Q-network (Dueling DQN) [21] based clue selection agent. Given an entity name in a source language, the information from previous search results and translation candidates is summarized as the MDP state, and the next selection of keyword clues is used as the MDP action. Then, the Dueling DQN-based clue selection agent takes the MDP state as the input and outputs the next action (i.e., selecting a spotting clue to construct the query). After search results are returned, the pre-learned translation candidate extractor and ranker are used to extract and rank translation candidates from the return pages. To effectively learn the Dueling DQN-based clue selection agent, we train it via the reinforcement learning algorithm and summarize both the short-term and long-term search result quality as the reinforcement learning reward.

Compared with previous methods, our method has the following advantages:

1) We mine the entity translation in a multi-turn paradigm by modeling web-based entity translation mining as an MDP. To our best knowledge, this work is the first to model translation mining as a Markov Decision Process.
2) We propose a neural network-based clue selection agent and learn it via the reinforcement learning algorithm. By taking the short-term and long-term search result quality as the reward, the agent can learn to adaptively select effective spotting clues for different entities.

For evaluation, we construct a new dataset for named entity translation, which contains three types of entities, i.e., person, company, and film. Experiments show that our method outperforms traditional heuristic methods on all three entity types.

2   Web-Based Entity Translation Mining

Generally, a typical web-based entity translation mining system contains the following three components:

Clue Selection Agent. Given an entity name in a source language, the clue selection agent selects effective spotting clues (e.g., "电影 [film]") and constructs spotting queries by concatenating spotting clues with the original entity name (e.g., "*La La Land* 电影 [film]"). Then the spotting queries are sent to search engines for retrieving bilingual web pages. In the next section, we propose to select clues and generate spotting queries in a multi-turn paradigm (i.e., an MDP) to take previously applied queries and returned bilingual pages into consideration.

Translation Candidate Extractor. Specifically, this paper uses a pre-trained LSTM-CRF model as the translation candidate extractor which extracts translation candidates by tagging tokens in the returned snippets with BIO tags[1]. Different from previous work that requires handcraft annotated data for training, this paper uses a distant supervision-based strategy to generate training data, leading to improved scalability of our approach.

Translation Candidate Ranker. Given the extracted translation candidates, the translation candidate ranker ranks all translation candidates and return the top-ranked one as the final translation. This paper follows [29] to use a pre-trained Ranking SVM model as the ranker, which uses multiple features to capture the association between entities and translation candidates.

The next two sections describe how we mine entity translations from web pages in a multi-turn paradigm, where we model the web-based entity translation mining as an MDP and learn the neural network-based clue selection agent via

---

[1] The BIO tags are widely used in the sequence labeling tasks, where B, I, and O refer to the beginning, inside and outside token of the chunks in target types.

Table 1. Example of the MDP state–candidate feature table (a) and the clue feature table (b),when mining the translation of the film entity *La La Land*.

Candidate	Score
爱乐之城	0.633
乐来越爱你	0.159
美好夜晚	0.097
...	...

(a)

Clue	Len	Freq	Perc	CoStat with Entity	CoStat with Candidates		CoStat with Clues	
					爱乐之城	...	电影	...
电影[film]	2	20	.82	(.2, .5, .6, 0, 3.7)	(.6, .8, .6, 0, 3.5) ...		*	...
导演[director]	2	5	.34	(.3, .6, .8, 0, 9.3)	(.3, .2, .9, 0, 9.2) ...		(.2, .5, .3, 0, 6.5) ...	
...	...	...	...	...	...	...	...	...

(b)

reinforcement learning. Besides, because the translation candidate extractor and ranker are not our main contributions, this paper does not detail them, but pre-trains them in advance (see details in Sect. 5.1) and directly uses them.

3 Web-Based Entity Translation Mining via Reinforcement Learning

To mine entity translations, this paper proposes a multi-turn entity translation mining approach by modeling the entire process as a Markov Decision Process (MDP) instead of regarding it as a pipeline. The MDP for web-based entity translation mining can be formalized as a tuple $<S, A, T, R>$, where:

- $\mathbf{S} = \{\mathbf{s}\}$ are the states used to capture information from both previous translation candidates and the returned bilingual pages from search engines.
- $\mathbf{A} = \{\mathbf{a}\}$ are the actions used to indicate the selected next clues for spotting query construction.
- $\mathbf{T}(\mathbf{s}\prime|\mathbf{s}, \mathbf{a})$ is the state transition function, which is related to the state update.
- $\mathbf{R}(\mathbf{s}, \mathbf{a})$ is the reward function for state-action pair (s, a), which models the likelihood that the correct translation is found by performing action a at state s. We describe a specifically-designed reward function in Sect. 4.2.

Based on the above MDP, a Dueling Deep Q-network based clue selection agent is designed to map states to actions, i.e., adaptively construct spotting queries for different entity names based on the previously returned search results and translation candidates.

3.1 MDP State

The state of our MDP summarizes information from previously returned search results and translation candidates. To ensure the adaptivity of the clue selection model, we use type-independent features to model the state. Specifically, each state is represented by a candidate feature table and a clue feature table:

Candidate Feature Table. (see Table 1a for example). This table contains top k translation candidates with their confidence scores from our translation candidate ranker after taking the last action, which is used to indicate the updated translation confidence.

Clue Feature Table. (see Table 1b for example). This table captures information about all potential clues from previously returned bilingual pages, which are also the candidate actions for the next step. Following previous methods [2,11], every candidate clue is a frequent foreign word in returned bilingual pages. For efficiency, we select top k foreign words in returned snippets to form a fixed-sized clue pool based on their TF-IDF scores among all returned pages.

Intuitively, an efficient spotting clue should: 1) frequently co-occur with both original entity names and good translation candidates; 2) less frequently co-occur with previously used clues (since they tend to provide little new information). Based on the above observations, we use the following features for each clue: 1) the clue's length (*Len*); 2) its occurrence frequency in the returned snippets (*Freq*); 3) the percentage of the snippets containing the clue (*Perc*); 4) the co-occurrence statistics (*CoStat*) with the original entity name; 5) the co-occurrence statistics with translation candidates; 6) the co-occurrence statistics with other clues. For each co-occurrence feature, we use a tuple of following five statistics:

- **Co-occurrence ratio**: the ratio of snippets containing both entities.
- **Pointwise mutual information** (PMI): $PMI(e_1, e_2) = \log \frac{N \times a}{(a+b) \times (a+c)}$, where a is the number of snippets containing both e_1 and e_2, b is the number of snippets containing e_1 but not e_2, c is the number of snippets containing e_2 but not e_1, N is the number of snippets.
- **Context similarity**: cosine similarity of vectors $v_{e_1} = (t_1^1, ..., t_1^i, ...)$ and $v_{e_2} = (t_2^1, ..., t_2^i, ...)$, where t_1^i and t_2^i are the numbers of e_1 and e_2 in the ith snippet.
- **Minimum distance**: the minimum distance between two entities in returned snippets.
- **Average distance**: the average distance between two entities in all returned snippets.

3.2 MDP Action

The action in our MDP corresponds to a clue selection decision. For example, selecting "电影 [film]" as the next clue of film entity *La La Land* is an action. Once a clue is selected, a spotting query is constructed and sent to the search engine for retrieving new search results.

As we mentioned above, all clues in the current clue feature table is a candidate action. Besides, we also introduce a STOP action to the action set, which indicates to stop the translation mining process. Intuitively, we want our clue selection agent to select the STOP action when: 1) the best translation candidate has high confidence; 2) all new clues are highly correlated with previously used clues, indicating that they are unlikely to provide new information.

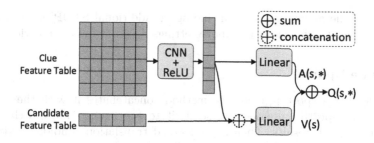

Fig. 2. Dueling DQN-based clue selection agent.

3.3 State-to-Action Mapping via Dueling DQN-Based Clue Selection Agent

At each state s, the clue selection agent $Q(s, a)$, corresponding to the Q-function in MDP, is used to decide which clue to select based on the expected future reward of action a at current state s. Accordingly, the clue with the highest Q-function value is selected at each step.

In this paper, we use the Dueling DQN, a variant of DQN [13], as the Q-function approximator $Q(s, a) \approx Q(s, a|\theta)$, which has been shown to be effective when the state space is continuous. To make the learning process stable, the Dueling DQN decomposes Q-function $Q(s, a)$ into two parts: 1) a value function $V(s; \theta, \alpha)$ which serves as the basic reward of the current state and is independent of actions; 2) an advantage function $A(s, a; \theta, \beta)$, which models the advantage of an action compared with other potential actions. Formally, we adopt the alternative form for Dueling DQN [21] as follows:

$$Q(s, a) = V(s; \theta, \alpha) + (A(s, a; \theta, \beta) - \mathbf{B})$$
$$B = \frac{1}{\mathcal{A}} \sum_{a'} A(s, a'; \theta, \beta) \tag{1}$$

where $\mathcal{A}$ is the action set, θ, α and β are parameters.

The architecture of our clue selection agent is shown in Fig. 2. Given the clue feature table and the candidate feature table of the current state, the clue feature table is mapped to a hidden representation by several convolution filters along with ReLU activation. Then a fully connected layer is applied to obtain the output $A(s, *; \theta, \beta)$, i.e., each dimension of the output corresponds to a candidate clue in the clue feature table as well as one additional dimension corresponding to the STOP action. Furthermore, the hidden representation, together with the candidate feature table, is simultaneously connected to another fully connected layer, whose output is $V(s; \theta, \alpha)$. Then $V(s; \theta, \alpha)$ is summed up every dimension of $A(s, *; \theta, \beta)$ to obtain the final Q-values for all state-action pairs. We describe how to learn it via a reinforcement learning-based algorithm in Sect. 4.

For efficiency, in our experiments, we only retain the top 30 clues with the highest TF-IDF scores and the top 10 translation candidates. Therefore, the final size of the action set is 31, including an additional STOP action. In our pilot experiments, the use of the above settings does not influence performance.

3.4   State Update

After selecting a spotting clue, our method concatenates it with the original entity as the spotting queries and sends it to the background search engine to get new search results. Then a pre-learned translation candidate extractor is applied to extract new translation candidates, and a pre-learned translation candidate ranker subsequently assigns new confidence scores to all candidates.

Previously returned search results and all translation candidates with new confidence scores are then used to construct a new MDP state as in Sect. 3.1. If our clue selection agent selects the STOP action, the search is stopped, and the translation candidate with the highest confidence score is returned as the final entity translation.

4   Learning Clue Selection Agent via Reinforcement Learning

As mentioned above, learning an effective clue selection agent is the key step for the proposed method. Unfortunately, as we discussed above, entity translation mining is a multi-step, iterative process, and no direct supervision signal can be provided for the clue selection agent. Therefore, we design a targeted reward function for learning the clue selection agent under the Q-learning framework, which can consider both short-term and long-term benefits of actions.

4.1   Learning Object

Our clue selection agent learns an optimal Q-function by maximizing total reward over all timesteps. Generally, the optimal Q-function $Q^*(s,a)$ can also be written as the following Bellman equation [20]:

$$Q^*(s,a) = \mathbb{E}_{s'}[R(s,a) + \gamma \max_{a'} Q^*(s',a')] \qquad (2)$$

where $R(s,a)$ is the instant reward after action a is performed at state s, γ is discount factor, and s' is the next state after performing action a.

To train our Dueling DQN, we use off-policy learning strategy—Q-learning [22] with experience replay to increase data efficiency and to reduce variance. The γ in the Bellman equation is set to 0.8. We adopt the ε-greedy strategy to explore at the beginning of training and stochastic gradient descent with the Adam algorithm to learn optimal parameters θ, α, and β in the Dueling DQN model.

4.2   Reward Function

The reward function is essential in learning the clue selection agent. Intuitively, if a clue matches many bilingual web pages containing the correct translation, we should assign a high, positive reward to that clue. Based on this intuition, we design a reward function based on the percentage of returned snippets containing the correct translation after taking action a, which is denoted as p_a. Specifically, given an action a at state s, the reward function is defined as:

$$R(s,a) = \begin{cases} 1, & p_a \geq M(p_{a^-}) \\ p_a - A(p_{a^-}), & p_a < M(p_{a^-}) \\ -1, & \text{using } a \text{ twice} \end{cases} \quad (3)$$

where a^- is the set of actions done before action a, $M(p_{a^-})$ is the max percentage of returned snippets by a^-, and $A(p_{a^-})$ is the average percentage of returned snippets by a^-. Besides, we penalize the mining process stopping too early by set the reward -1 if the STOP action is selected in the first two steps.

5   Experiments

5.1   Experiment Settings

Dataset. In this paper, we construct a new English-to-Chinese entity translation dataset containing three entity types: person, company, film[2]. For each entity type, we manually collect 500 entity translation pairs as the training set, 100 as the development set, and 100 as the test set.[3]

Metrics. We evaluate our method using search quality and translation quality. To evaluate the search quality, we use the following information retrieval metrics: P@n (precision of top n snippets containing the correct translation), MRR (mean reciprocal rank of the correct translation appearing in returned snippets), MAP (mean average precision of returned snippets), NDCG@n (normalized discounted cumulative gain in top n snippets), and Recall(ratio of entities whose correct translations appear in retrieved snippets). To measure the translation quality, we use P@n (the precision of top n translation candidates) and Recall (ratio of entities whose translations are in candidate list).

Baselines. We use 5 clue selection methods as baselines:

---

[2] The translation pairs in the LDC datasets can be easily fetched from a specific website (http://www.ichacha.com), leading to almost static search results with varied queries. We therefore construct a new dataset. The person, company, and film names are correspondingly collected from Wikidata (https://www.wikidata.org), Forbes (https://www.forbes.com), and IMDB (http://www.imdb.com).

[3] Dataset and source code are available at https://github.com/lingyongyan/entity_translation.

Table 2. Overall search quality performance.

Clue type	P@1	P@10	P@50	MAP	MRR	NDCG@10	NDCG@50	Recall
Correct translation	**0.640**	**0.630**	**0.606**	**0.690**	**0.696**	**0.667**	**0.781**	**0.893**
No clue	0.023	0.031	0.029	0.055	0.068	0.532	0.726	0.263
Partial translation	0.267	0.250	0.244	0.307	0.344	0.525	0.675	0.590
Related word	0.257	0.218	0.215	0.282	0.351	0.541	0.722	0.633
Intent word	0.180	0.184	0.161	0.247	0.299	0.575	0.696	0.740
Our method	**0.417**	**0.350**	**0.296**	**0.383**	**0.507**	**0.557**	**0.786**	**0.797**

Table 3. Search quality of different methods on different entity types.

Entity type	Clue type	P@1	P@10	MAP	MRR	NDCG@50	Recall
Person	Partial translation	0.310	0.288	0.326	0.364	0.683	0.560
	Related word	0.220	0.174	0.201	0.269	0.694	0.460
	Our method	**0.430**	**0.359**	**0.381**	**0.496**	**0.835**	**0.730**
Company	Partial translation	0.310	**0.320**	**0.365**	0.420	**0.752**	0.730
	Related Word	0.250	0.227	0.305	0.370	0.736	0.750
	Our method	**0.380**	0.314	0.347	**0.505**	0.746	**0.870**
Film	Partial translation	0.180	0.143	0.231	0.249	0.589	0.480
	Related word	0.300	0.254	0.340	0.413	0.736	0.690
	Our method	**0.440**	**0.376**	**0.423**	**0.521**	**0.777**	**0.790**

1) **No Clue**: which only uses original entity names as spotting queries.
2) **Partial Translation**:this baseline is similar to [18], which uses partial translations of an original entity as spotting clues (e.g., using "土地 [land]" as a clue of the entity *La La Land*). In our experiments, we try different partial translations of an entity and report the highest score.
3) **Related Word**: this baseline is similar to [2], which uses the foreign word with the highest TF-IDF value in initial search results as the spotting clue.
4) **Intent Word**: which uses some common intent words (e.g., "翻译 [translation]", "中文 [Chinese]", and "名字 [name]") as spotting clues and report the highest score.
5) **Correct Translation**: we use the correct translation as the golden clue. This is used to assess the gap between the competing methods and this upper limit.

The Pre-training of Translation Candidate Extractor and Ranker. We pre-train the translation candidate extractor and ranker using distant supervision methods. That is, we distantly annotated search snippets of another 10,000 entity translation pairs from Wikidata. For each entity translation pair, we concatenate the original entity mention and its translation as the search query to retrieve the bilingual snippets in the Web. And the returned snippets are labeled using these translation pairs. Then, the translation candidate extractor is trained using these labeled data. After that, the same search snippets of these translation

Table 4. Translation quality performance.   **Table 5.** Search quality of different search steps.

Clue type	P@1	P@3	Recall
Correct translation	**0.563**	**0.720**	**0.730**
No clue	0.113	0.167	0.173
Partial translation	0.273	0.440	0.473
Related word	0.343	0.527	0.590
Intent word	0.210	0.370	0.417
Our method	**0.357**	**0.543**	**0.693**

Step	P@10	MAP	MRR	Recall
0	0.134	0.207	0.254	0.543
1	0.327	0.386	0.492	0.770
3	0.350	0.383	0.507	0.797

pairs are used for training the translation candidate extractor, where the correct translations are used as positive samples, and the wrong candidates returned by the translation candidate extractor are used as negative samples.

5.2 Experimental Results

Overall Results. Table 2 shows the search quality using different clue selection methods. Table 3 shows the search quality for different entity types. Table 4 shows the translation quality performance. From Table 2–4, we can see that:

1) **Compared with all heuristic methods, our method achieved the best spotting clue selection performance.** The golden baseline—Correct Translation can find 89.3% entity translations, while our method can find 79.7% entity translations, i.e., our method can find 89.2% entity translations if they exist in the Web. Furthermore, our method achieves a higher MAP than other heuristic methods, indicating that our method can select higher-quality clues.

2) **Our method can adaptively select spotting clues for different entity types.** In Table 3, we can see that our method achieves considerable performance improvement over Partial Translation and Related Word on all three types of entities. A likely cause for the better performance of our method is that both Partial Translation and Related Word are specialized to entity types: Partial Translation is better at mining person and company names, and Related Word is better at mining films. In contrast, our method can adaptively select optimal clues for different entities.

3) **Our method can achieve competitive translation mining performance.** Table 4 shows the translation quality by running the same ranking model using different spotting clues. We can see that, among all entity translations, 73% of them can be found using the Correct Translations as clues, and our method can find 94.9% of entity translations of them. In contrast, the best baseline (Related Word) can only find 80.8% of them. Furthermore, most correct translations are ranked within the top 3 by our method, which supports our intuition that the high-quality clues can be selected by considering the returned bilingual pages in the multi-turn paradigm.

Effect of Search Steps. For detailed analysis, Table 5 shows the clue quality of our method at different search steps. We can see that our method can enhance the clue quality when more information is returned: more search steps lead to better search quality and therefore better translation ranking performance. Indeed, as shown by Table 5,the initial clue only retrieves 54.3% entity translations, while the clue after 1 or 3 searching step(s) improves Recall by 41.8% or 46.7%, respectively. Besides, comparing to the Related Word, which also selects clues after initial searching, our method still outperforms it at step 1.

Table 6. Search quality on the unseen entities (i.e., the chemical entities) of different methods.

Clue type	P@1	P@10	MAP	MRR	NDCG@50	Recall
Correct translation	**0.810**	**0.709**	**0.755**	**0.827**	**0.832**	**0.920**
No clue	0.020	0.063	0.112	0.122	0.616	0.810
Partial translation	0.190	0.173	0.228	0.275	0.601	0.750
Related word	0.350	0.301	0.407	0.446	0.684	**0.850**
Intent word	0.240	0.284	0.326	0.430	0.688	0.790
Our method	**0.400**	**0.396**	**0.422**	**0.541**	**0.755**	0.830

Adaptation to Unseen Entity Type. To further verify that our method can adaptively select clues for different entity types, we additionally experiment on 100 entities of the unseen type—*chemical entities* using the clue selection agent learned on other types of entities, i.e., person, company, and film entities. From Table 6, we can see that our model still outperforms other heuristic methods on the unseen entity type. We believe this is because our model uses the entity-type-independent abstract information to represent the search state (as shown in Sect. 3.2), which makes the agent be adapted to different entity types.

6   Related Work

Most early studies on entity translation mining focused on extracting entity translations from parallel and comparable corpus by aligning entities and their translations [3,5,8,10,25,26].

Because parallel and comparable corpus usually have low entity translation coverage, a promising approach is to mine entity translations from the Web [17, 27]. A simple approach to spotting bilingual web pages is directly using original entity names as spotting queries [28,29]. Many other studies exploit different methods to select spotting clues, such as partial transliterations of original entity name [7,23], chunk translations of original entity name [1,18,24], translations of related words in retrieved web pages [6,27], top-5 foreign nouns using TF-IDF metrics [2,4,11]. Besides, [19] generate entity translation candidates iteratively. As described above, these heuristic methods do not consider the diversity of entity names and are incapable of leveraging the information in the returned pages for better spotting query construction.

Reinforcement learning has shown beneficial in various NLP tasks. For instance, the DQN model has been used to incorporate external evidence for improving information extraction [14]. Besides, the advanced actor-critic algorithm [12] is also used to learn neural machine translation with simulated human feedback [15]. Recently, reinforcement learning is used to learn the curriculum for presenting training samples to an NMT system [9] and learns to remove wrongly labeled sentences in distantly supervised relation extraction [16].

7   Conclusion

This paper proposes a novel reinforcement learning-based approach for web-based entity translation mining, which can automatically construct the most effective spotting queries and conducting entity translation mining in a multi-turn paradigm. Specifically, we use a neural Dueling Deep Q-Network as the clue selection agent and learn effective clue selection policy via Q-learning. Experiments show that our clue selection policy significantly outperforms traditional heuristic methods. For future work, because multi-step, iterative mining is a common technique for knowledge acquisition tasks, we will apply them on related tasks, such as entity set expansion and relation extraction.

Acknowledge. This research work is supported by the National Key Research and Development Program of China under Grant No. 2017YFB1002104, the National Natural Science Foundation of China under Grants no. U1936207, Beijing Academy of Artificial Intelligence (BAAI2019QN0502), and in part by the Youth Innovation Promotion Association CAS (2018141).

References

1. Fang, G., Yu, H., Nishino, F.: Chinese-English term translation mining based on semantic prediction. In: Proceedings of the COLING/ACL 2006 Main Conference Poster Sessions, pp. 199–206. Association for Computational Linguistics (2006)
2. Ge, Y.D., Hong, Yu., Yao, J.M., Zhu, Q.M.: Improving web-based OOV translation mining for query translation. In: Cheng, P.-J., Kan, M.-Y., Lam, W., Nakov, P. (eds.) AIRS 2010. LNCS, vol. 6458, pp. 576–587. Springer, Heidelberg (2010). https://doi.org/10.1007/978-3-642-17187-1_54
3. Hassan, A., Fahmy, H., Hassan, H.: Improving named entity translation by exploiting comparable and parallel corpora. In: Proceedings of the Conference on Recent Advances in Natural Language Processing, Borovets, Bulgaria, pp. 1–6 (2007)
4. Hsu, C.C., Chen, C.H.: Mining synonymous transliterations from the World Wide Web. ACM Trans. Asian Lang. Inf. Process. 9(1), 1 (2010)
5. Huang, F., Vogel, S.: Improved named entity translation and bilingual named entity extraction. In: Proceedings of the 4th IEEE International Conference on Multimodal Interfaces, pp. 253–258 (2002)
6. Huang, F., Zhang, Y., Vogel, S.: Mining key phrase translations from web corpora. In: Proceedings of the conference on Human Language Technology and Empirical Methods in Natural Language Processing, pp. 483–490 (2005)

7. Jiang, L., Zhou, M., Chien, L.F., Niu, C.: Named entity translation with web mining and transliteration. In: Proceedings of the 20th international Joint Conference on Artificial Intelligence, vol. 7, pp. 1629–1634 (2007)

8. Kim, J., Jiang, L., Hwang, S.W., Song, Y.I., Zhou, M.: Mining entity translations from comparable corpora: a holistic graph mapping approach. In: Proceedings of the 20th ACM International Conference on Information and Knowledge Management, pp. 1295–1304. ACM (2011)

9. Kumar, G., Foster, G., Cherry, C., Krikun, M.: Reinforcement learning based curriculum optimization for neural machine translation. arXiv:1903.00041 (2019)

10. Lee, T., Hwang, S.W.: Bootstrapping entity translation on weakly comparable corpora. In: Proceedings of the 51st Annual Meeting of the Association for Computational Linguistics, pp. 631–640 (2013)

11. Liu, L., Ge, Y.D., Yan, Z.X., Yao, J.M.: A CLIR-oriented OOV translation mining method from bilingual webpages. In: Proceedings of the 2011 International Conference on Machine Learning and Cybernetics, pp. 1872–1877. IEEE (2011)

12. Mnih, V., Badia, A.P., Mirza, M., Graves, A., Lillicrap, T., Harley, T., Silver, D., Kavukcuoglu, K.: Asynchronous methods for deep reinforcement learning. In: International Conference on Machine Learning, pp. 1928–1937 (2016)

13. Mnih, V., et al.: Human-level control through deep reinforcement learning. Nature **518**(7540), 529–533 (2015)

14. Narasimhan, K., Yala, A., Barzilay, R.: Improving information extraction by acquiring external evidence with reinforcement learning. In: Proceedings of the Conference on the Empirical Methods in Natural Language Processing, pp. 2355–2365 (2016)

15. Nguyen, K., Daumé III, H., Boyd-Graber, J.: Reinforcement learning for bandit neural machine translation with simulated human feedback. In: Proceedings of the Conference on the Empirical Methods in Natural Language Processing, pp. 1464–1474 (2017)

16. Qin, P., XU, W., Wang, W.Y.: Robust distant supervision relation extraction via deep reinforcement learning. In: Proceedings of the 56th Annual Meeting of the Association for Computational Linguistics, pp. 2137–2147 (2018)

17. Qu, J., Nguyen, L.M., Shimazu, A.: Cross-language information extraction and auto evaluation for OOV term translations. China Commun. **13**(12), 277–296 (2016)

18. Ren, F.: A practical Chinese-English ON translation method based on ON's distribution characteristics on the web. In: Proceedings of COLING 2012: Demonstration Papers, pp. 377–384 (2012)

19. Ren, F., Zhu, J., Wang, H.: Translate Chinese organization names using examples and web. In: Natural Language Processing and Knowledge Engineering, pp. 1–7. IEEE (2009)

20. Sutton, R.S., Barto, A.G.: Reinforcement Learning: An Introduction. MIT Press, Cambridge (1998)

21. Wang, Z., Schaul, T., Hessel, M., Van Hasselt, H., Lanctot, M., De Freitas, N.: Dueling network architectures for deep reinforcement learning. In: Proceedings of the 33rd International Conference on International Conference on Machine Learning, pp. 1995–2003 (2016)

22. Watkins, C.J.C.H., Dayan, P.: Q-learning. Mach. Learn. **8**(3), 279–292 (1992). https://doi.org/10.1007/BF00992698

23. Wu, J.C., Chang, J.S.: Learning to find English to Chinese transliterations on the web. In: Proceedings of the 2007 Joint Conference on Empirical Methods in Natural Language Processing and Computational Natural Language Learning, pp. 996–1004 (2007)

24. Yang, F., Zhao, J., Liu, K.: A Chinese-English organization name translation system using heuristic web mining and asymmetric alignment. In: Proceedings of the Joint Conference of the 47th Annual Meeting of the ACL and the 4th International Joint Conference on Natural Language Processing of the AFNLP, pp. 387–395 (2009)

25. You, G.w., Cha, Y.r., Kim, J., Hwang, S.w.: Enriching entity translation discovery using selective temporality. In: Proceedings of the 51st Annual Meeting of the Association for Computational Linguistics, Sofia, Bulgaria, pp. 201–205. Association for Computational Linguistics, August 2013

26. You, G.w., Hwang, S.w., Song, Y.I., Jiang, L., Nie, Z.: Mining name translations from entity graph mapping. In: Proceedings of the 2010 Conference on Empirical Methods in Natural Language Processing, pp. 430–439 (2010)

27. Zhang, Y., Huang, F., Vogel, S.: Mining translations of OOV terms from the web through cross-lingual query expansion. In: Proceedings of the 28th Annual International ACM SIGIR Conference on Research and development in Information Retrieval, pp. 669–670. ACM (2005)

28. Zhang, Y., Su, Y., Jin, C., Zhang, T.: Multi-feature representation for web-based English-Chinese OOV term translation. In: International Conference on Machine Learning and Cybernetics, ICMLC 2011, Proceedings, pp. 1515–1519. IEEE (2011)

29. Zhao, Y., Zhu, Q., Jin, C., Zhang, Y., Huang, X., Zhang, T.: Chinese-English OOV term translation with web mining, multiple feature fusion and supervised learning. In: Sun, M., Liu, Y., Zhao, J. (eds.) CCL/NLP-NABD-2014. LNCS (LNAI), vol. 8801, pp. 234–246. Springer, Cham (2014). https://doi.org/10.1007/978-3-319-12277-9_21

Neural Fusion Model for Chinese Semantic Matching

Renshou Wu, Qingliang Miao, Miaoyuan Shi, Min Chu, and Kai Yu[✉]

AI Speech Co., Ltd., Suzhou, China
{renshou.wu,qingliang.miao,miaoyuan.shi,min.chu,kai.yu}@aispeech.com

Abstract. Deep neural networks are the most popular choices for semantic matching nowadays. In this paper, we propose a novel Chinese semantic matching approach, namely Neural Fusion Model, which consists of char-word fusion encoder and encoding-interaction fusion classifier. The char-word fusion encoder models the char and word sequences separately, and then uses a char-word fusion units to fuse them together. The encoding-interaction fusion classifier jointly learns from three simple classifiers. There are a encoding based classifier, a interaction-based classifier and a fusion classifier. Among them, the fusion classifier combines encoding-based and interaction-based representation from multiple-perspective. These three classifiers share the two-layer feed-forward network for prediction. Empirical studies demonstrate the effectiveness of the proposed model for Chinese semantic matching, and it achieves the best results among non-BERT models. In addition, our BERT-equipped model obtains new state-of-the-art results on the Chinese semantic matching benchmark corpus: LCQMC and BQ.

Keywords: Semantic matching · Neural networks · BERT.

1 Introduction

Semantic matching task is a fundamental task of natural language processing (NLP). It has been used in many fields such as information retrieval, question answering and dialogue systems. In Retrieval-based Question Answering (RBQA) system, two questions are defined as semantically equivalent if they convey the same intent, so they could be answered by the same answer. In recent years, neural network models have achieved great success in semantic matching. Various neural network models for semantic matching have been proposed, these models are usually composed of two parts: (1) encoder, which transforms the input sentence into a fixed-length sentence representation vector, and (2) classifier, which extracts semantic-level features from the sentence representation vector and predicts probability for each class.

According to whether to perform Chinese word segmentation, encoder can be classified into two kinds: character-based (char-based) encoder and word-based encoder. In the context of deep learning, segmented words are usually treated

© Springer Nature Singapore Pte Ltd. 2021
H. Chen et al. (Eds.): CCKS 2020, CCIS 1356, pp. 78–90, 2021.
https://doi.org/10.1007/978-981-16-1964-9_7

as the basic units for operations. However, word-based models have a few disadvantages. For example, word-level data sparsity inevitably leads to over-fitting, and the ubiquity of out-of-vocabulary (OOV) words limits the model's learning capacity. Besides, Chinese word segmentation performance is far from perfect, which may propagate errors to downstream NLP tasks. Meng et al. [10] benchmark neural word-based models against neural char-based models in four end-to-end NLP benchmark tasks. The results show that char-based models consistently outperform word-based models. However, in Chinese natural language processing, words are the smallest meaningful language components capable of independent activity. Word segmentation can alleviate polysemy to a certain extent. For example, panda (熊猫), without Chinese word segmentation, are two animals: bear (熊) and cat (猫).

Depending on whether to use cross sentence features, classifier for semantic matching can be classified roughly into two types: (1) encoding-based classifier [15], which computes sentences similarity directly from sentence vectors obtained by encoding. The model of this type has advantages in simplicity of the network design and generalization to other NLP tasks. (2) interaction-based classifier [17], which takes word alignment and interactions between sentence pairs into account and often shows better performance. However, in order to obtain better performance, the model usually contains multiple alignment layers to maintain its intermediate states and gradually refines its predictions [18], these deeper architectures are harder to train.

In this paper, we investigate how to maximize the utilization of char and word information to improve the encoder performance, and explore how to use encoding-based and interaction-based approach to build a simple, high-performance classification for semantic matching. The main contributions of this work are summarized as follows: **Firstly**, we propose a char-word fusion encoder, which includes a hierarchical CNN-based encoder (HierCNN encoder) to incorporate character (word) context and sequence order into modeling for better vector representation, followed by a char-word fusion units to compare char and word representations and then fuse them together. **Secondly**, we propose an encoding-interaction fusion classifier. It is composed of a simple encoding-based classifier, a simple interaction-based classifier and a simple fusion classifier. Although they are three simple classifiers, they can achieve good performance after joint learning. **Thirdly**, our proposed Neural Fusion Model, whether non-BERT or BERT-based, achieves the state-of-the-art on two Chinese semantic matching benchmark corpus, namely LCQMC [11] and BQ [2]. We also conduct ablation studies to further illustrate the effectiveness of the char-word fusion encoder and encoding-interaction fusion classifier.

2   Related Works

Semantic matching is a widely used NLP technology which aims to model the semantic relationship between two texts, and it is often regarded as a binary classification problem. Recent researches show that deep neural network based semantic matching algorithms outperform traditional algorithms using on designing hand-craft features to capture n-gram overlapping, word reordering and syntactic alignments phenomena [5]. In the context of deep learning, segmented words are usually treated as the basic units for operation. However, word-based models have a few disadvantages, such as OOV. Unlike the previous char-based [12] and char-word hybrid models [17], which simply concatenated characters and words representation, our proposed char-word fusion encoder performs interactive aggregation of characters and words information. Depending on whether to use cross sentence features or not, two types of deep neural network models have been proposed for semantic matching. The first kind is encoding-based models, where each sentence is encoded into a fixed-length vector and then computes sentence similarity directly, such as Deep Structured Semantic Models (DSSM) [6], Shortcut-Stacked Sentence Encoders (SSE) [15], or Multi-task Sentence Encoding Model (MSEM) [7]. The other ones are called interaction-based models, which take word alignment and interactions between the sentence pairs into account and often show better performance. The main representative work is Enhanced Sequential Inference Model (ESIM) [3], Bilateral Multi-Perspective Matching Model (BiMPM) [17]. More effective models can be built if inter-sequence matching is allowed to be performed more than once. In RE2 [18], tokens in each sequence are first embedded by the embedding layer and then processed consecutively by N same-structured blocks with independent parameters. Different from previous studies, we propose the encoding-interaction fusion classifier, which jointly learns from three simple classifiers, and achieves good performance.

3 Approach

In this section, we first give a general definition of the semantic matching task and introduce the notations used in our approach. Then, we present the proposed Neural Fusion Model in detail.

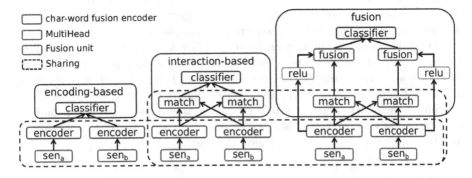

Fig. 1. Overview of the Neural Fusion Model.

Assuming we have two sentences $sen_a = (t_1^a, t_2^a, ..., t_{l_a}^a)$, $sen_b = (t_1^b, t_2^b, ..., t_{l_b}^b)$ which consist of a sequence of tokens, where each t_i^a or t_i^b comes from a fixed vocabulary V, and l_a and l_b are the length of the corresponding sentence. The goal of semantic matching is to predict a label $y \in \{0,1\}$ that indicates the similarity between sen_a and sen_b.

Figure 1 illustrates the proposed Neural Fusion Model, which has two components: (1) char-word fusion encoder, and (2) encoding-interaction fusion classifier. The details are introduced as follows.

3.1   The Char-Word Fusion Encoder

Unlike the previous char-word hybrid models, only by simply concatenating characters and words[11]. Our proposed char-word fusion encoder performs interactive aggregation of characters and words information, which is composed of a HierCNN encoder and a char-word fusion unit. The HierCNN encoder incorporates characters (words) context and sequence order into modeling for better vector representation. The char-word fusion unit compares char and word representations and then fuses them together.

HierCNN Encoder. Given a sentence sen with words $w_i, i \in [0, l_s]$, we first convert the words w_i to their respective embeddings e^w, where $e^w \in \mathbb{R}^k$ is an embedding of k-dimensional vector, which can be initialized with some pre-trained word (or character) embeddings. Similarly, we get character-level embeddings e^c, which have been shown to be helpful to deal with out-of-vocab tokens.

Next with the Embedding layer, we apply the hierarchical CNN to capture the implicit semantic information in the input sequence and generate a fixed-length vector. Similar to image, language also contains local correlation, such as the internal correlation of phrase structure[12]. The parameter sharing of the convolutional kernels enables the model to extract these common features in the sentence and indicate the correlation among the source annotations.

Figure 2 shows the detailed structure. Convolution units with kernel size of 1 in the first CNN layer is used to capture the information of a single token and increase non-linearity. In the second CNN layer, we use three different kernels where $k_1 = 3$, $k_2 = 5$ and $k_3 = 7$ to extract certain robust and abstract features, specifically n-gram (3, 5 or 7 adjacent words) features, such as keywords and phrases information. Patterns like "I don't hate", "I don't like" could be expressions and therefore CNNs can identify them in the sentence regardless of their position. The details of convolution block is described below:

$$h_k(x_i) = relu(W_k[x_{i-\lfloor k/2 \rfloor}, ..., x_{i+\lfloor k/2 \rfloor}] + b_k), k \in \{1, 3, 5, 7\}, \qquad (1)$$

where $relu(\cdot)$ refers to the non-linear activation function Rectified Linear Unit.

To provide richer features for encoding processes, a particular fusion unit has been designed to fuse the outputs of different CNN layers. The simplest way of fusion is a concatenation, addition, difference or element-wise product of two representation vectors [13], followed by some linear or non-linear transformation. In the above method, both vectors to be fused are treated equally. It is assumed that there is no preference of a particular vector during fusion. However, in

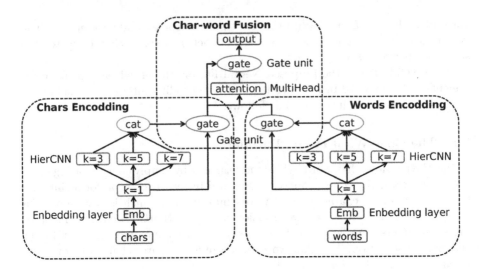

Fig. 2. Structure of char-word fusion encoder.

most real world cases, the vectors to be fused do not contribute equally to the final semantic matching. Usually one of them contains more information, so it should be used as the primary contributor for fusion, and the other vector with less information should act as a supplementary role. For example, the output of different CNN layers in the HierCNN is different, and the output of the second layer (v_2) contains features that are significantly better than the first layer (v_1).

$$v_1 = h_1(e^t), t \in \{c, w\} \tag{2}$$

$$v_2 = [h_3(v_1); h_5(v_1); h_7(v_1)] \tag{3}$$

To address the above limitations, we propose a **Imbalance Fusion Unit**, which discriminates the importance of vectors, filters unimportant information, then compares and combines different representations based on the more important vector. The details are described below:

$$o(x_1, x_2) = g(x_1) \circ m(x_2) + (1 - g(x_1)) \circ x_1 \tag{4}$$

$$m(x) = relu(W_m x + b_m) \tag{5}$$

$$g(x) = sigmoid(W_g x + b_g), \tag{6}$$

where x_1 and x_2 should be padded to the same sequence length. $\circ$ denotes the element-wise product, $W_m \in \mathbb{R}^{d_o \times d_m}$, b_m, $W_g \in \mathbb{R}^{d_o \times d_g}$ and b_g are trainable parameters. d_o, d_m and d_g respectively represent the output dimensions of $o(\cdot)$, $m(\cdot)$ and $g(\cdot)$. By encoding characters and words separately, we get the character-level and word-level representations r^c and r^w derived as below.

$$r^t = o(v_2, v_1), t \in \{c, w\}, \tag{7}$$

where trainable parameters in $o(\cdot)$ is shared for sentence sen_a and sen_b.

Char-Word Fusion Units. We further implement multihead-attention units [16] upon HierCNN encoder so as to dig out the correlations of characters and words, and then fuse them together by imbalance fusion unit proposed above. To be specific, we refer Q and V to the word-level representations r^w, while $K = r^c$.

$$MultiHead(Q, K, V) = Concat(head_1, ..., head_h)W_O \qquad (8)$$

$$head_i = Attention(QW_i^Q, KW_i^K, VW_i^V), \qquad (9)$$

where $Attention(\cdot)$ is Scaled Dot-Product Attention [16], W_O is trainable parameters. In this scenario, the output of multihead-attention units is regarded as the important part.

$$r^{cw} = o(MultiHead(r^c, r^w, r^w), r^c) \qquad (10)$$

3.2 The BERT-based Char-Word Fusion Encoder

The BERT [4] model and char-word fusion encoder can be easily fused using the imbalance fusion unit proposed in the previous section.

First, we process the sentence pair into a specific form and use it as the input to the BERT. The output of the last layer in BERT can be simply used as the BERT-based representation of sentence pairs. Specifically, we insert a [CLS] token at the first position of a sentence pair, insert a [SEP] token after sentence A, and then insert sentence B. For example, a pair of sentence are cut by BERT-Tokenizer: $sen_a = (t_1^a, t_2^a, t_3^a)$ and $sen_b = (t_1^b, t_2^b, t_3^b, t_4^b)$:

$$Input = [CLS], t_1^a, t_2^a, t_3^a, [PAD], [SEP], t_1^b, t_2^b, t_3^b, t_4^b, [SEP] \qquad (11)$$

$$r^{bert} = BERT\_sequence\_output(Input) \qquad (12)$$

After obtaining the BERT-based representation r^{bert}, we divide it into separate representation corresponding to sen_a and sen_b. Then, we can use the imbalance fusion unit to fused r^{bert} and r^{cw} together as below.

$$r^{cw\_bert} = o(r^{cw}, r^{bert}) \qquad (13)$$

3.3 The Encoding-Interaction Fusion Classifier

Our proposed encoding-interaction fusion classifier is composed of a simple encoding-based classifier, a simple interaction-based classifier and a simple fusion classifier. They share a two-layer feed-forward network for prediction conducting joint learning. As different classifiers have different noise patterns, the encoding-interaction fusion classifier that learns three classifier simultaneously is able to learn a more general representations, and helps the model to focus on important features [1]. The detail will be described below.

Encoding-Based Classifier. The Encoding-based classifier directly uses the encoded vectors r_a^{cw} and r_b^{cw} to compute the semantic similarity through the two-layer feed-forward network. A heuristic matching trick with difference and element-wise product are employed, which are found effective in combining different representations.

$$d_1(x_1, x_2) = tanh(W_{d_1}[x_1; x_2; x_1 - x_2; x_1 \circ x_2] + b_{d_1}) \tag{14}$$

$$\tilde{y} = softmax(W_y x + b_y), \tag{15}$$

where W_{d_1}, b_{d_1}, W_y and b_y are trainable parameters.

Interaction-Based Classifier. It calculates sentence pair interactions using the outputs of the char-word fusion encoder at first. We use a multihead-attention units to simulate the chunks alignment between two sentences. The alignment information is useful for sentence pair modeling because the semantic relation between two sentences depends largely on the relation of aligned chunks. In addition, since the vector at position i in the encoded sentence vector already contains context information, and the work [18] shows that independent word representation is important for alignment, we therefore splice the output of the convolution $kernel = 1$ in the encoder and the outputs of the char-word fusion encoder as input of the multihead-attention mechanism.

$$p_{ab} = MultiHead([r_a^{cw}; h_1(e_a^c)], r_b^{cw}, r_b^{cw}) \tag{16}$$

$$p_{ba} = MultiHead([r_b^{cw}; h_1(e_b^c)], r_a^{cw}, r_a^{cw}) \tag{17}$$

After obtaining the interaction representation p_{ab} and p_{ba}, we can compute the semantic similarity through the sharing two-layer feed-forward network.

Fusion Classifier. The imbalance fusion unit proposed in 3.1 is employed to compare encoding-based and interaction-based representations from multiple-perspective, and then combine them together. In this scenario, we consider interaction-based representations as the important part.

$$f_a = o([r_a^{cw}; p_{ab}; r_a^{cw} - p_{ab}; r_a^{cw} \circ p_{ab}], p_{ab}) \tag{18}$$

$$f_b = o([r_b^{cw}; p_{ba}; r_b^{cw} - p_{ba}; r_b^{cw} \circ p_{ba}], p_{ba}) \tag{19}$$

After obtaining the fusion representation f_a and f_b, the semantic similarity can be computed via the shared two-layer feed-forward networks.

3.4   Training and Inference

The proposed model can also be jointly trained end-to-end. The loss function consists of three parts: the loss computed by encoding-based classifier, interaction-based classifier and fusion classifier respectively.

$$Loss = \lambda_1 L_{encoding} + \lambda_2 L_{inter} + L_{fusion}, \tag{20}$$

where L is negative conditional log likelihood loss, λ_1 and λ_2 are hyper-parameters to tune the impacts of the encoding-based classifier and interaction-based classifier.

4   Experiment

4.1   Dataset

LCQMC Corpus. A Chinese question matching corpus which constructed from Baidu Knows and is manually annotated. Different from the word-level paraphrase corpus, the LCQMC is mainly focused on the variance of expressions for the same intent, rather than the variance of vocabularies for the same meaning. The distribution of LCQMC corpus is given in Table 1.

BQ Corpus. A Chinese corpus for question semantic equivalence identification in the bank domain, which constructed from real bank customer service logs. Compared with existing corpora, it is of high quality and challenging. The distribution of BQ corpus is given in Table 1.

Table 1. The distribution of different data sets in LCQMC and BQ corpus.

Data	LCQMC			BQ		
	Total	Positive	Negative	Total	Positive	Positive
Train	238766	138574	100192	100000	50000	50000
Vaild	8802	4402	4400	10000	5000	5000
Test	12500	6250	6250	10000	5000	5000

4.2   Experiment Settings

Hyper-parameters. The hyper-parameters values in the model are tuned in the development set. The hidden state size of first layer CNN cell ($kernel = 1$) is 300. The hidden state size of second layer CNN cell ($kernel = 3, 5, 7$) is 100. The BERT version is Google BERT-Base, Chinese, which is pre-trained on Chinese Simplified and Traditional. An Adam optimizer [9] is used to optimize all the trainable weights. After a series of experiments, the λ_1 and λ_2 in loss function are both set 0.1 can get better performance.

Bert-Based Model Training Strategy. We first fine-tune the BERT-classifier, which consists of the BERT model followed by a two-layer feed-forward network (described in Sect. 3.2) for prediction. Then freeze BERT-classifier to fine-tune the Neural Fusion Model, share the two-layer feed-forward network for prediction. Finally jointly tune both modules until convergence.

4.3   Baselines Approaches

To demonstrate the effect of our approach, in this section, we provide some approaches for thorough comparison as follow:

Text-CNN. A typical CNN model [8] for sentence classification.

BiLSTM. Two sentences go through the same BiLSTM unit, then concatenate the representations of both sentences to make a classification [14].

Table 2. Experimental results on LCQMC and BQ corpus

Methods	LCQMC				BQ			
	P	R	F1	Acc	P	R	F1	Acc
Text-CNN	68.4	84.6	75.7	72.8	67.8	70.6	69.2	68.5
BiLSTM	70.6	89.3	78.9	76.1	75.0	70.5	72.9	73.5
BiMPM	77.7	93.5	84.9	83.3	82.3	81.2	81.7	81.9
ESIM	76.5	93.6	84.2	–	81.9	81.8	81.9	–
SSE	78.2	93.6	85.2	–	80.2	80.3	80.2	–
MSEM	78.9	93.7	85.7	–	82.9	84.4	83.6	–
Glyce	80.4	93.4	86.4	85.3	81.9	85.5	83.7	83.3
RE2	79.6	95.1	86.6	85.4	83.8	82.8	83.3	83.4
Our	**82.8**	91.7	**87.0**	**86.2**	82.3	**86.0**	**84.1**	**83.8**
BERT	83.2	94.2	88.2	87.5	83.5	85.7	84.6	84.8
Glyce+BERT	86.8	91.2	88.8	88.7	84.2	86.9	85.5	85.8
Our+BERT	**88.1**	90.7	**89.4**	**89.2**	**85.0**	**87.5**	**86.2**	**86.1**

BiMPM. An interaction-based model which uses four different types of multi-perspective matching layers to match two sentences [17].

ESIM. An interaction-based model, which encode sentence contexts and uses the attention mechanism to calculate the information between two sentences [3].

SSE. An encoding based sentence-matching model, which enhances multi-layer BiLSTM with short-cut connections [15].

MSEM. A semantic retrieval framework that integrates the encoding-based matching model with the approximate nearest neighbor search technology [7].

Glyce. Glyph-vectors for Chinese character representations [12], which use historical Chinese scripts to enrich the pictographic evidence in characters, and use image-classification as an auxiliary.

RE2. An interaction-based model [18], which use an embedding layer first embeds discrete tokens, followed by several same-structured blocks consisting of encoding, alignment and fusion layers then process the sequences consecutively.

4.4 Results

Table 2 shows the overall and performances of all approaches to semantic matching. From this table, we can see that our approach significantly outperforms the Text-CNN and BiLSTM baselines by a large margin, and achieves the best results among non-BERT models on all of the two benchmarks (p-value<0.05 via t-test). Compared with the current non-BERT SOTA model RE2, especially in LCQMC corpus, our model has 0.4 and 0.8 relative gain in F1 and Acc respectively, which proves the effectiveness of our proposed Neural Fusion Model. In addition, BERT outperforms all non-BERT models, and Our+BERT performs the best, setting new SOTA results on all of the two benchmarks. Compared with the BERT model, Our+BERT model owns an advantage of $F1$ score 1.2 and 1.6 on the LCQMC and BQ respectively.

Table 3. Influence in encoder

Methods	P	R	F1	Acc
Word-based	81.1	91.4	85.9	85.1
Char-based	82.3	90.1	86.0	85.4
Char-word fusion	**82.8**	**91.7**	**87.0**	**86.2**

Table 4. Influence in classifier

Methods	P	R	F1	Acc
Encoding-based	82.1	91.0	86.3	85.6
Interaction-based	82.4	91.3	86.7	85.8
Fusion	81.7	92.6	86.8	85.9
Encoding-interaction fusion	**82.8**	**91.7**	**87.0**	**86.2**

4.5 Ablations

In order to evaluate the individual contribution of each model component, we run an ablation study in LCQMC corpus.

In Table 3, the encoding approaches to query matching are implemented with different ways. The word(char)-base encoder is the implementation of HierCNN encoder, which only uses the words (characters) as the input. The results show that only using characters or words as the input of encoder, the performance degrades significantly, especially only using the words. Compared to the char-base encoder, the model equipped with the char-word fusion encoder performs better. This indicates that char and word information are quite complementary and the char-word fusion encoder is the key component in the whole model.

In Table 4, the classifiers are implemented with four different ways. Among them, encoding-based, interaction-based and fusion classifier are single models, encoding-interaction fusion classifier is the joint model proposed in Sect. 3.3. The experimental results are shown in Table 4. In the single model, fusion classifier is better than encoding-based and interaction-based classifier, indicating that the fusion unit we proposed is effective. At the same time, compared to the single-task model, the performance of joint model has been improved, achieving the best performance, indicating that joint learning is effective, which helps the model learn a more general representations and focus on important features.

4.6 Analysis and Discussion

In addition, we present some typical examples for case study, as shown in Table 5. By analyzing the test results of each model, we can observe that: (1) The ubiquity of OOV words limits the word-based model's learning capacity. In example 1, Jieba[1] has tokenized unsuitable word segmentation "心各有" and "心有", and these words are not in the vocabulary. Both queries are expressed as "<unk> 所属 是 什么 意思 ?", which leads the error of word-based model. (2) Compared with characters, words are more diverse. Therefore, in similar sentence pairs, characters overlap is higher than word overlap, which leads the char-based model to predict highly overlapped sentences as similar. In example 1, although 95.2% lexical overlap between two sentences, they are not similar. (3) Encoding-based approach can capture relatively complete semantic information, but it can't distinguish key points between two sentences. Therefore, when most of the semantic information among sentence pair is similar, the encoding-based model tends to give a high similarity score, as shown in example 2. (4) In interactive-based approach, each word only cares about the information that related to itself in another sentence. Therefore, when the relevant information cannot be found, the interactive-based model emphasise differences between two sentences. For example 3, "我这是什么眼?" has no matching information in Query 1, interactive module gives a low similarity score. As shown in former two case studies, word diversity leads OOV issue, but avoids the over dependency on character overlap. Though char-based model avoids OOV issue, it leads the model over influenced by character overlap. Therefore, character and word information are complementary. Example 1 shows that the proposed char-word fusion encoder can fuse both character and word information effectively and alleviate the OOV and dependence on character overlap finely. Similarly, case study (3) and (4) shows that encoding-based representation and interactive-based representation are also complementary. Encoding-based representation can capture relatively complete information, and interactive-based representation can emphasise key

---

[1] https://github.com/fxsjy/jieba.

Table 5. Some typical examples. Where en-inter fusion denotes the encoding-interaction fusion

Id	Example	Methods	Label
1	Query 1:心各有所属是什么意思?	manual	0
	What does it mean that each heart belongs to each other?	word-based	1
	Query 2:心有所属是什么意思?	char-based	1
	What does it mean that someone has someone in mind?	char-word fusion	0
2	Query 1:哪里是一个自讨苦吃的地方	manual	0
	Where is a place to ask for trouble	encoding-based	1
	Query 2:一个自讨苦吃的地方在哪里。是什么生肖。	interaction-based	0
	Where's a place to ask for trouble. What is the zodiac.	en-inter fusion	0
3	Query 1:我是桃花眼吗?	manual	1
	Am I amorous eyes ?	encoding-based	1
	Query 2:我这是什么眼? 是桃花眼吗?	interaction-based	0
	What kind of eye am I? Is it amorous eye?	en-inter fusion	1

points between two sentences. Using encoding-interaction fusion classifier can effectively distinguish different parts between sentences and determine whether different and similar parts are intention related. As shown in example 2 and 3, the sentence pairs in example 3 are matching because of the different part "这是什么眼?" is intention related, while the different part "是什么生肖" in example 2 is not intention related.

5 Conclusion

In this paper, we propose a novel approach to Chinese semantic matching, namely Neural Fusion Model. We adopt char-word fusion encoder to enhance the representation of sentences. Besides, we utilize a fusion classifier to semantic matching and joint learning, that consists of an encoding-based classifier, an interaction-based classifier and a fusion classifier. In addition, our model can be fused with BERT simply by the fusion unit to enhance the semantic representation of sentence pairs. Empirical studies show that the proposed approach performs significantly better than several strong baseline approaches.

References

1. Bingel, J., Søgaard, A.: Identifying beneficial task relations for multi-task learning in deep neural networks. In: Proceedings of the EACL, vol. 2, pp. 164–169 (2017)
2. Chen, J., Chen, Q., Liu, X., Yang, H., Lu, D., Tang, B.: The BQ corpus: a large-scale domain-specific Chinese corpus for sentence semantic equivalence identification. In: Proceedings of EMNLP, pp. 4946–4951 (2018)
3. Chen, Q., Zhu, X., Ling, Z.H., Wei, S., Jiang, H., Inkpen, D.: Enhanced LSTM for natural language inference. In: Proceedings of ACL, vol. 1, pp. 1657–1668 (2017)

4. Devlin, J., Chang, M.W., Lee, K., Toutanova, K.: BERT: pre-training of deep bidirectional transformers for language understanding. In: Proceedings of NAACL, vol. 1, pp. 4171–4186 (2019)

5. Heilman, M., Smith, N.A.: Tree edit models for recognizing textual entailments, paraphrases, and answers to questions. In: Proceedings of NAACL, pp. 1011–1019. Association for Computational Linguistics (2010)

6. Huang, P.S., He, X., Gao, J., Deng, L., Acero, A., Heck, L.: Learning deep structured semantic models for web search using clickthrough data. In: Proceedings of CIKM, pp. 2333–2338 (2013)

7. Huang, Q., Bu, J., Xie, W., Yang, S., Wu, W., Liu, L.: Multi-task sentence encoding model for semantic retrieval in question answering systems. In: Proceedings of IJCNN, pp. 1–8. IEEE (2019)

8. Kim, Y.: Convolutional neural networks for sentence classification. In: Proceedings of EMNLP, pp. 1746–1751 (2014)

9. Kingma, D.P., Ba, J.: Adam: a method for stochastic optimization. arXiv preprint arXiv:1412.6980 (2014)

10. Li, X., Meng, Y., Sun, X., Han, Q., Yuan, A., Li, J.: Is word segmentation necessary for deep learning of Chinese representations? In: Proceedings of ACL, pp. 3242–3252 (2019)

11. Liu, X., et al.: LCQMC: a large-scale Chinese question matching corpus. In: Proceedings of COLING, pp. 1952–1962 (2018)

12. Meng, Y., et al.: Glyce: glyph-vectors for Chinese character representations. In: Proceedings of NIPS, pp. 2742–2753 (2019)

13. Mou, L., et al.: Natural language inference by tree-based convolution and heuristic matching. In: Proceedings of ACL, vol. 2, pp. 130–136 (2016)

14. Mueller, J., Thyagarajan, A.: Siamese recurrent architectures for learning sentence similarity. In: Proceedings of AAAI (2016)

15. Nie, Y., Bansal, M.: Shortcut-stacked sentence encoders for multi-domain inference. arXiv preprint arXiv:1708.02312 (2017)

16. Vaswani, A., et al.: Attention is all you need. In: Proceedings of NIPS, pp. 5998–6008 (2017)

17. Wang, Z., Hamza, W., Florian, R.: Bilateral multi-perspective matching for natural language sentences. In: Proceedings of IJCAI, pp. 4144–4150 (2017)

18. Yang, R., Zhang, J., Gao, X., Ji, F., Chen, H.: Simple and effective text matching with richer alignment features. In: Proceedings of ACL, pp. 4699–4709 (2019)

Knowledge Extraction: Relation

Improving Relation Extraction Using Semantic Role and Multi-task Learning

Zhandong Zhu, Jindian Su$^{(\boxtimes)}$, and Xiaobin Hong$^{(\boxtimes)}$

College of Computer Science and Engineering, South China University of Technology, Guangzhou 510640, China
{sujd,mexbhong}@scut.edu.cn

Abstract. Relation extraction (RE) aims at identifying the relationship between two given entities and plays an essential role in natural language processing (NLP). Most of existing relation extraction models use convolutional or recurrent neural network and fail to capture the in-depth semantic features from the entities. These models also only focus on the training data and ignore external knowledge. In this paper, we propose a relation extraction model that makes use of external knowledge and the semantic roles of entities. In our model, we first adopt RoBERTa to make use of the knowledge learned from the unsupervised pretraining corpus. Then we obtain the semantic role embeddings and propose an entity attention network to select important words for relation extraction. We also offer the multi-task learning module and further improve our model by learning from auxiliary tasks. Our model obtains a Macro-F1 score of 89.96% on the benchmark dataset, outperforming most of the existing methods. More ablation experiments on two different datasets show that semantic role information and multi-task learning can help improve the relation extraction.

Keywords: Relation extraction · Semantic role · Multi-task learning

1 Introduction

Relation extraction (RE) is the task of extracting a relational fact between a pair of entities in a given sentence. For example, given a sentence "Russian President [Vladimir Putin] will travel to [Paris] next Friday." with two annotated entities e_1 = "Vladimir Putin" and e_2 = "Paris", a relation extraction model can extract a relation of "Entity-Destination($e_1 - e_2$)" between the two entities. Relation extraction is an essential task in Natural Language Processing (NLP) which is an intermediate step in many higher-level NLP tasks such as question answering [1].

Early relation extraction models can be divided into feature-based models and kernel-based models. A feature-based model converts the input into a feature vector and trains a classifier, which is depended strongly on feature engineering. In contrast, a kernel model uses a string-kernel function to compute the similarity between the given sentence and the example of a particular relation. Although both kinds of models have achieved good results, it is burdensome to select helpful features or kernel functions.

© Springer Nature Singapore Pte Ltd. 2021
H. Chen et al. (Eds.): CCKS 2020, CCIS 1356, pp. 93–105, 2021.
https://doi.org/10.1007/978-981-16-1964-9_8

With the growing popularity of deep learning, recent relation extraction models mainly focus on using deep neural networks such as convolutional neural networks (CNN) or recursive neural network to encode the sentences and extract features automatically [2, 3]. However, these models usually only use static word embedding methods like Word2Vec [4], making the models hard to encode the in-depth contextual semantic information of the entities in the sentences. Also, such models only consider the training samples, and the useful knowledge contains in external corpora is neglected.

Recently, pretrained language models have been proven to be very successful in many natural language processing tasks. The pretrained language model BERT [5] proposed by Google has been applied to many natural language processing tasks and achieves state-of-the-art on most of the tasks. After BERT, many pretrained language models, such as RoBERTa [6] and XLNet [7], are proposed and also have important implications. However, very few researches apply a pretrained language model to relation extraction, which relies on the two entities as well as the sentence itself.

Our work is based on the following observations. First, many pretrained language models, such as BERT and RoBERTa, are off-the-shelf. These language models are pretrained on large unsupervised corpora and contain useful knowledge. Through transfer learning, the downstream task would be able to use the knowledge of the semantics of language acquired in pretraining. This helps alleviate the problem that most of the downstream tasks have rather small datasets. Second, the semantic roles of the two entities can provide helpful information in relation extraction. For example, in the sentence "Russian President [Vladimir Putin] will travel to [Paris] next Friday," "Vladimir Putin" has the semantic role of Arg0 and "Paris" has the semantic role of Arg1 with the verb "travel." The semantic information of the main verb "travel" is more related to the relation "Entity-Destination". Also, the semantic roles of the two target entities help judge the direction of the relation. Since "Vladimir Putin" has the role of Arg0 and "Paris" has the role of Arg1, their relation would be "Entity-Destination($e_1 - e_2$)" rather than "Entity-Destination($e_2 - e_1$)." Third, other supervised datasets also contain a large amount of knowledge. MT-DNN [8] proves that through multi-task learning, models can learn from different supervised datasets. With multi-task learning, a relation extraction model can learn from other tasks and gets more robust and generalized.

The main contributions of our paper are as follow:

(1) This paper proposes a novel network architecture for relation extraction that can make use of the external knowledge from auxiliary tasks and semantic roles of the target entities at the same time. Both explicit contextual semantics and implicit external knowledge can be integrated with pretraining language model RoBERTa and improve performance.

(2) Our model obtains a Macro-F1 of 89.96% on the benchmark dataset, outper-forming most of the existing methods. More ablation experiments on two different datasets show that semantic role information and multi-task learning can both help improve the relation extraction.

The rest of the paper is structured as follows. In Sect. 2, we will review related work about relation extraction. Then Sect. 3 describes our proposed model in detail. Section 4 presents the experiments and their results. Finally, we conclude our work with a summary and some directions for future work in Sect. 5.

2 Related Work

As one of the essential tasks in NLP, many efforts have been devoted to relation extraction. Early methods include feature-based models [9] and kernel-based models [10]. These models are depended highly on the features or the kernel functions and have limited performances.

Recent relation extraction methods mainly focus on deep learning. Zeng et al. [2] first introduce CNN in relation extraction. They use CNN to extract features automatically and applied a softmax layer for prediction, which outperforms previous methods. Santos et al. [11] propose CR-CNN, which uses a loss function based on pairwise ranking to reduce the noise in the examples of no relation. Socher et al. [3] develop the MVRNN model, which encodes the syntactic structure of a sentence through recursive neural network. Lee et al. [12] propose a recurrent neural network model which uses attention mechanism to give different words appropriate weights for relation extraction. Mandya et al. [13] encodes the sentence into subgraphs and uses graph neural network (GNN) for relation extraction.

With the success of the pretrained language model BERT, many researchers introduce BERT in their models on many tasks, such as text classification and sequence labeling. Recently, Wu et al. [14] employ the pretrained language model BERT in relation extraction and achieve the state-of-the-art.

All the above methods are of supervised paradigm and require lots of annotated data. To address this problem, some researchers focus on distantly supervised relation extraction [15, 16]. However, a distantly supervised dataset uses a knowledge graph to annotate sentences automatically and contains much noise. In this paper, we focus on supervised relation extraction, which contains little noise and plays a more critical role in relation extraction.

3 Our Approach

In this section, we will present our model in detail. Our model can be divided into two parts: the RoBERTa and semantic role based single-task model and the multi-task learning module.

3.1 RoBERTa and Semantic Role Based Single-Task Model

As illustrated in Fig. 1, the RoBERTa and semantic role based single-task model contains four parts: sentence encoder, semantic role encoder, entity attention layer, and output layer.

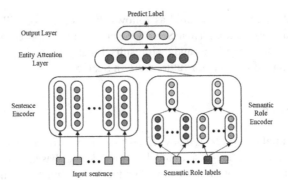

Fig. 1. The architecture of the RoBERTa and semantic role based single-task model

Sentence Encode

Given a sentence with two entities, the sentence encoder use the pretrained model RoBERTa to encode the sentence. To make the model pay more attention to the two entities, we use unique tokens as entity markers. Specifically, we insert a special token '{' at both sides of the head entity and a special token '}' at both sides of the tail entity. We also add '<s>' at the beginning and '</s>' at the end of the sentence.

For example, after we insert the special tokens, the example sentence in Sect. 1 will turn into: "<s> Russian President {Vladimir Putin {will travel to} Paris} next Friday </s>".

After the insertion, we use RoBERTa to encode the sentence and get the output of the final hidden state $H = [h_1, h_2, \ldots, h_n]$. $H \in \mathbb{R}^{n*d}$ where n is the length of input and d indicates the dimensions of hidden state vectors. Since RoBERTa may turn a word into subwords while tokenizing and one entity may contain two or more words, several vectors may correspond to an entity. Suppose h_{start} to h_{end} are the final hidden state of an entity e_i, we use the average of those vectors as a contextual vector representation of the entity:

$$\bar{h}_i = \frac{1}{end - start + 1} \sum\nolimits_{k=start}^{end} h_k \tag{1}$$

where $i = 1, 2$ denotes the two target entities.

Semantic Role Encoder

In our paper, NLP tool allennlp [17] is used get the semantic role label. Semantic role encoder maps a semantic role label as a vector representation using an embedding matrix $W_{srl} \in \mathbb{R}^{|SRL|*d_{srl}}$, where $|SRL|$ is the size of semantic role labels and d_{srl} indicates the dimensions of semantic role vectors. There may be a variety of semantic role labeling results in one sentence, which can be viewed as the semantic information corresponding to different perspectives of the sentence. Therefore, one entity may correspond to several semantic role vectors. We also apply the average operation to get the semantic vector representation s of an entity:

$$s_i = \sum\nolimits_{j=1}^{n} W_{srl}\left(r_i^j\right) \tag{2}$$

where $i = 1, 2$ denotes target entities, r_i^j is the j_{th} semantic role of the i_{th} entity.

Entity Attention Layer

Given a contextual vector representation $\bar{h}_i$ and a semantic vector representation s_i, we apply a tanh activation function and a fully connected layer to obtain the final vector representation m_i of an entity:

$$m_i = W_e \left\{ tanh \left[concat \left(\bar{h}_i, s_i \right) \right] \right\} + b_e \tag{3}$$

where $W_e \in \mathbb{R}^{d*(d+d_{srl})}$.

Considering that different words in the sentence contribute to the relation differentially and H cannot discriminate the contributions of different words in the sentence, we use attention based on entity representation to weight words differentially. We use the average of the representation of the two entities as the query vector, and the process can be formalized as follow:

$$q = \frac{m_1 + m_2}{2} \tag{4}$$

$$\alpha_i = \frac{\exp(h_i \cdot q)}{\sum_j \exp(h_j \cdot q)} \tag{5}$$

where α_i is the importance weight of h_i. To prevent giving the two entities so many weights that other words have weights close to zero, we mask the two entities while calculating the attention scores. Given the attention scores, the contextual representation of the sentence calculated as follow:

$$m_H = \sum_{i=1}^{n} \alpha_i h_i \tag{6}$$

Finally, we concatenate m_H, m_1, m_2 to get the final representation of an input sentence:

$$M = concat(m_H, m_1, m_2) \tag{7}$$

The illustration of entity attention layer is shown in Fig. 2.

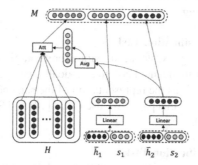

Fig. 2. The illustration of entity attention layer

Output Layer

In the output layer, softmax classifier is used to obtain the conditional probability for each relation:

$$y = softmax[W_o(M) + b_o] \tag{8}$$

where $W_o \in \mathbb{R}^{3d*|REL|}$, $b_o \in \mathbb{R}^{|REL|}$, $|REL|$ is the number of different relations in the dataset. Our model regard the relation with the highest probability value as the final prediction.

We use the cross-entropy loss as our loss function:

$$loss(y, label) = -\sum_{i=1}^{|y|} \widehat{y_i} \log(y_i) \tag{9}$$

where y_i is the i_{th} element of y, indicating the conditional probability for the i_{th} relation. $\widehat{y_i}$ is the indicator variable:

$$\widehat{y_i} = \begin{cases} 1 \ if \ label = i \\ 0 \ else \end{cases} \tag{10}$$

3.2 Multi-task Learning Module

Section 3.1 presents the single-task model, which can perform well in relation extraction. To further utilize the knowledge in other tasks and improve the performance of our model, we also propose a multi-task learning module.

In multi-task learning module we adopt auxiliary tasks and share parameters between models while training. The auxiliary tasks can be divided into the related task, the single sentence classification task, and the sentence-pair classification task. Relation extraction can be viewed as a special kind of sentence classification task so we believe there is a correlation between relation extraction and sentence classification.

The Related Task

Our model focus on relation extraction. While there is more than one relation extraction task, we can regard one dataset as the main task and others as the related tasks. The models of the main task and related tasks share the parameters in all layers except the output layer, as shown in Fig. 3.

The Single Sentence Classification Task

As shown in Fig. 4, in a single sentence classification task, given a sentence, we use a RoBERTa model shared with the main task to encode the sentence. Then we use the last hidden state vector of '<s>' as the representation of the sentence and apply a softmax classifier for classification. In this paper, we use CoLa [18] and SST-2 [19] as the single sentence classification tasks.

The Sentence-Pair Classification Task

As shown in Fig. 5, the sentence-pair classification model concatenate the two sentences

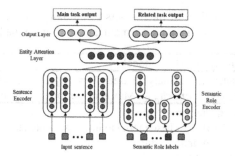

Fig. 3. Multi-task learning with the related task

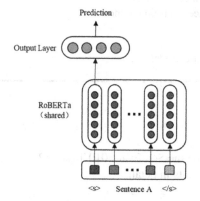

Fig. 4. The single sentence classification model

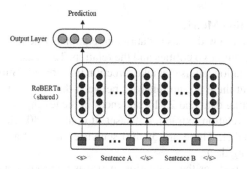

Fig. 5. The sentence-pair classification model

and use RoBERTa and a softmax classifier for classification. In our paper, QNLI [20], MRPC [21], and MNLI [22] are used as the sentence-pair classification tasks.

The Training Procedure

The training procedure for the whole model is shown in Algorithm 1. In each epoch, we sample the same amount of training instances from every task and use them for training the model so the different sizes of the datasets would not influence the model. After a few epochs, the model is finetuned on the main task to achieve the best result.

Algorithm 1: Training the whole model.

Input: Main dataset:D_{re}, Auxiliary datasets:$D = [D_1, D_2, ..., D_T]$,
Multi-task learning epoch:$epoch_{mtl}$, Max epoch:$epoch_{max}$, Sampling size:n

---

Initialize model parameters randomly.
Use the pretrained RoBERTa-base model to initialize sentence encoder
for $epoch$ in 1, 2, ..., $epoch_{max}$ **do:**
    if $epoch < epoch_{mtl}$ **then:**
        for D_t in $[D_1, D_2, ..., D_T] \cup D_{re}$ **do:**
            Sample n instances from D_t
            Use the instances to train the model
        end
    else:
        Use D_{re} to train the model.
end

---

4   Experiments

We conduct experiments on the SemEval-2010 Task 8 [23] dataset and the Wiki80 [24] dataset and analyze the results in detail.

4.1   Experimental Settings

Dataset and Evaluation Metric
SemEval-2010 Task 8 is a widely used dataset in relation extraction. All sentences in the dataset are annotated with two entities and the relation. The dataset contains 19 relation labels include nine relation labels that take direction into consideration and a special one "Other," denoting that there are no relations between the two entities. There are 8,000 sentences in the training data and 2,717 sentences in the testing data.

Wiki80 is a new relation extraction dataset. It contains 80 relation labels and 700 sentences for each label. There are 50,400 sentences in the training data and 5,600 sentences in the testing data.

We adopt the macro-f1 as the evaluation metric as previous works do.

Parameter Settings.
Table 1 shows the hyper-parameters we set in our model.

Table 1. Hyper-parameters

Parameters	Description	Value
d	Dimension of RoBERTa	768
d_{srl}	Dimension of semantic role embedding	100
lr_1	Learning rate of RoBERTa	5×10^{-5}
lr_2	Learning rate of other layers	5×10^{-3}
$dropout$	Dropout rate	0.1
$batch\_size$	Batch size	32

4.2 Performance Comparison

To evaluate the effect of our method, we compare its result with various methods on SemEval-2010 Task 8. The results of the models are demonstrated in Table 2.

Table 2. Comparison with different models

Model	Macro-F1
Rank et al. (2010) [9]	82.20
Socher et al. (2012) [3]	82.40
Zeng et al. (2014) [2]	82.70
Lee et al. (2019) [12]	85.20
Wu et al. (2019) [14]	89.25
Tao et al. (2019) [25]	**90.36**
Tao et al. (w/o IS)	89.30
Mandya et al. (2020) [13]	86.30
Our model	89.94

From the results in Table 2, we can see that:

(1) Our approach achieves a macro-f1 score of 89.94% on the SemEval-2010 Task 8 dataset, which is second only to Tao et al., outperforming most of the previous models.

(2) Tao et al. achieve 90.36% on the SemEval-2010 Task 8 dataset, which is the current state-of-the-art result. Their model uses an indicator sequence that helps to distinguish relation types. However, the extraction of the indicator sequence needs the guidance of syntactic knowledge and is based on complex rules. According to their paper, the score of their model drops to 89.30% when not using indicator sequence. Our model is only based on neural networks, which avoid the effort to design rules manually. Also, our model can utilize external knowledge through

multi-task learning, which helps improve performance even when there are not enough training data.

4.3 Ablation Studies

To demonstrate the contributions of different components in our model, we conduct experiments with different configurations on both SemEval-2010 Task 8 and Wiki80:

- **RoBERTa base**: This model removes the semantic role encoder and the multi-task learning module, which only use RoBERTa to encode the sentence for relation extraction.
- **RoBERTa+SRL**: The RoBERTa and semantic role based single-task model introduced in Sect. 3.1.
- **RoBERTa+MTL**: This model removes the semantic role encoder in the proposed model

As shown in Table 3, we can see that:

(1) Compared with RoBERTa base, using the semantic role encoder helps improve the performance. The macro-f1 is increased from 89.50% to 89.70% on SemEval-2010 Task 8 and from 85.18% to 85.97% on Wiki80. It can be said that the semantic role of the entities contains rich information for extracting relations.
(2) With multi-task learning, our model can utilize external knowledge in other datasets, which helps improve the generalization of the sentence encoder. The result is increased from 89.50% to 89.74% on SemEval-2010 Task 8 and from 85.18% to 85.77% on Wiki80.
(3) Our model can get the best result using both the semantic role encoder and the multi-task learning module, which means that these two components improve relation extraction in a different way. The semantic role encoder uses the semantic information of the entities, and the multi-task learning utilize external knowledge.

Table 3. The score of different configurations

Model	SemEval-2010 Task 8	Wiki80
Our model	**89.94**	**86.34**
RoBERTa+MTL	89.74	85.77
RoBERTa+SRL	89.70	85.97
RoBERTa base	89.50	85.18

4.4 Effect of Different Auxiliary Tasks

In this section, we further explore the effect of different auxiliary tasks in multi-task learning module. We use RoBERTa base as the basic model and conduct experiments with different auxiliary tasks. From Table 4 we can see that:

(1) Whether using SemEval-2010 Task 8 or Wiki80 as the main task, the introduction of each other as auxiliary task resulted in greater improvement than the introduction of other auxiliary tasks. It can be said that both tasks aim to extract the relations between entities and have strong correlations. Multi-task learning with both tasks makes it easier for the model to capture the semantic features associated with the relations.

(2) With all auxiliary tasks, the model can achieve the best result. Although other tasks have different goals, some of the semantic features they need are related. Adding all auxiliary tasks can make our relation extraction model learn from other tasks and get a better result. It is also convenient to add more auxiliary tasks without changing the architecture of the single task model.

Table 4. Macro-F1 with different auxiliary tasks

Auxiliary task(s)	Main Task	
	SemEval-2010 Task 8	Wiki80
RoBERTa base	89.50	85.18
+SemEval-2010 Task 8	–	85.60
+Wiki80	89.63	–
+single sentence classification	89.27	85.20
+sentence-pair classification	89.51	85.24
+all auxiliary tasks	**89.74**	**85.77**

5 Conclusion

In this paper, we propose a novel relation extraction approach that can make use of the external knowledge and semantic roles of the target entities. In our method, we first use RoBERTa to encode the sentence to make use of the knowledge learned from the unsupervised pretraining corpus. Then we get the semantic role embeddings and propose an entity attention network to select important words for relation extraction. We also suggest the multi-task learning module and further improve our model by learning from auxiliary tasks. Experiments show that our model can improve the performance of relation extraction with the semantic role and multi-task learning.

Acknowledgments. This work was supported in part by Research and Development Program in Key Areas of Guangdong Province under Grant 2018B010109004, in part by the National Natural Science Foundation of China under Grant 61936003, and in part by the Applied Scientific and Technology Special Project of Department of Science and Technology of Guangdong Province under Grant 20168010124010.

References

1. Abujabal, A., Yahya, M., Riedewald, M., Weikum, G.: Automated template generation for question answering over knowledge graphs. In: WWW, pp. 1191–1200 (2017)
2. Zeng, D., Liu, K., Lai, S.: Relation classification via convolutional deep neural network. In: COLING 2014: Technical Papers, pp. 2335–2344. ACL, Stroudsburg (2014)
3. Socher, R., Huval, B., Manning, C.D.: Semantic compositionality through recursive matrix-vector spaces. In: EMNLP-CoNLL, pp. 1201–1211. ACL, Stroudsburg (2012)
4. Mikolov, T., Sutskever, I., Chen, K.: Distributed representations of words and phrases and their compositionality. In: NIPS, pp. 3111–3119. MIT Press, Cambridge (2013)
5. Devlin, J., Chang, M.W., Lee, K., et al.: BERT: pre-training of deep bidirectional transformers for language understanding. arXiv preprint arXiv:1810.04805 (2018)
6. Liu, Y., Ott, M., Goyal, N., et al.: RoBERTa: a robustly optimized bert pretraining approach. arXiv preprint arXiv:1907.11692 (2019)
7. Yang, Z., Dai, Z., Yang, Y., et al.: XLNet: generalized autoregressive pretraining for language understanding. In: NIPS. Cambridge, MIT Press, pp. 5754–5764 (2019)
8. Liu, X., He, P., Chen, W., et al.: Multi-task deep neural networks for natural language understanding. arXiv preprint arXiv:1901.11504 (2019)
9. Rink, B., Harabagiu, S.: UTD: classifying semantic relations by combining lexical and semantic resources. In: SemEval, pp. 256–259. ACL, Stroudsburg (2010)
10. Collins, M., Duffy, N. :Convolution kernels for natural language. In: Advances in Neural Information Processing Systems, pp. 625–632. MIT Press, Cambridge (2002)
11. Santos, C.N., Xiang, B., Zhou, B.: Classifying relations by ranking with convolutional neural networks. arXiv preprint arXiv:1504.06580 (2015)
12. Lee, J., Seo, S., Choi, Y.S.: Semantic relation classification via bidirectional LSTM networks with entity-aware attention using latent entity typing. Symmetry **11**(6), 785 (2019)
13. Mandya, A., Bollegala, D., Coenen, F.: Contextualised graph attention for improved relation extraction. arXiv preprint arXiv:2004.10624 (2020)
14. Wu, S., He, Y.: Enriching pre-trained language model with entity information for relation classification. In: CIKM, pp. 2361–2364. ACM, New York (2019)
15. Hoffmann, R., Zhang, C., Ling, X.: Knowledge-based weak supervision for information extraction of overlapping relations. In: ACL HLT, vol. 1, pp. 541–550 (2011)
16. Surdeanu, M., Tibshirani, J., Nallapati, R.: Multi-instance multi-label learning for relation extraction. In: Empirical Methods in Natural Language Processing, pp. 455–465 (2012)
17. Gardner, M., Grus, J., Neumann, M., et al.: AllenNLP: a deep semantic natural language processing platform. arXiv preprint arXiv:1803.07640 (2018)
18. Warstadt, A., Singh, A., Bowman, S.R.: Neural network acceptability judgments. Trans. Assoc. Comput. Linguist. **7**, 625–641 (2019)
19. Socher, R., Perelygin, A., Wu, J., et al.: Recursive deep models for semantic compositionality over a sentiment treebank. In: EMNLP, pp. 1631–1642. ACL, Stroudsburg (2013)
20. Rajpurkar, P., Zhang, J., Lopyrev, K., et al.: Squad: 100,000+ questions for machine comprehension of text. arXiv preprint arXiv:1606.05250 (2016)

21. Conneau, A., Kiela, D.: SentEval: an evaluation toolkit for universal sentence representations. arXiv preprint arXiv:1803.05449 (2018)
22. Williams, A., Nangia, N., Bowman, S.R.: A broad-coverage challenge corpus for sentence understanding through inference. arXiv preprint arXiv:1704.05426 (2017)
23. Hendrickx, I., Kim, S.N., Kozareva, Z.: SemEval-2010 task 8: multi-way classification of semantic relations between pairs of nominals. In: SemEval, Stroudsburg, PA, pp. 94–99 (2009)
24. Han, X., Gao, T., Yao, Y., et al.: OpenNRE: an open and extensible toolkit for neural relation extraction. arXiv preprint arXiv:1909.13078 (2019)
25. Tao, Q., Luo, X., Wang, H., et al.: Enhancing relation extraction using syntactic indicators and sentential contexts. In: ICTAI, Piscataway, NJ, pp. 574–1580. IEEE (2019)

Learning Global Representations for Document-Level Biomedical Relation Extraction

Lishuang Li[(✉)], Hongbin Lu, Shuang Qian, Shiyi Zhao, and Yifan Zhu

Dalian University of Technology, Dalian 116024, China

Abstract. Document-level relation extraction is in the face of two challenges: (1) The task contains a large number of cross-sentence relations. (2) The same entity may appear more than once in a document, called mentions of the entity. Most existing systems merely rely on the local context surrounding the mentions (e.g. in a sentence) to classify each mention pair respectively and merging the predictions of all mention pairs to extract the relation between entities. These methods have no advantages in learning semantic information from long-distance context and the approaches of merging multiple entity mentions may lead to information loss. In this paper, we present a novel model based on multi-attention mechanism to learn the global representation for document-level relation extraction. The Global Context-aware Attention (GCA) is proposed to obtain the global semantic information in a document. Meanwhile, the Global Entity-aware Attention (GEA) is employed to integrate the semantic information of all mention pairs. Our model is validated on the BioCreative V Chemical Disease Relation (CDR) dataset which is widely used in the domain of biomedical document-level relation extraction. Experimental results demonstrate that our model is competitive compared to the existing methods.

Keywords: Document-level relation extraction · Global representations · Attention mechanism · Chemical Disease Relation

1 Introduction

Relation Extraction (RE) plays an important role for the construction of knowledge graph and question answering tasks, aiming at extracting semantic relations between the named entities. Compared with the traditional intra-sentence RE, there are two difficulties in the document-level RE. Firstly, document-level RE task exists a large number of relations across sentence boundaries, involved in the problem of long-distance dependency. Secondly, the target entity is usually mentioned many times in a document, which means how to integrate the information of all mentions should be concerned. As shown in the Fig. 1 (The text is from BioCreative V Chemical Disease Relation (CDR) [1], which is a representative task for document-level RE), there is a Chemical-Induced Disease (CID) relation between the chemical entity *Streptomycin* and the disease entity *deafness*. Each entity may appear several times in the text. For example, the chemical entity *Streptomycin* appears three times in the text and each is called entity mention.

© Springer Nature Singapore Pte Ltd. 2021
H. Chen et al. (Eds.): CCKS 2020, CCIS 1356, pp. 106–118, 2021.
https://doi.org/10.1007/978-981-16-1964-9_9

Age-dependent sensitivity of the rat to neurotoxic effects of streptomycin. Streptomycin sulfate was injected for various periods into preweanling rats and for 3 weeks into weanling rats. Beginning at 8 days of age, body movement and hearing were examined for 6 and up to 17 weeks, respectively. Abnormal movements and deafness occurred only in rats treated during the preweanling period; within this period the greatest sensitivities for these abnormalities occurred from 2 to 11-17 and 5 to 11 days of age, respectively, indicating that the cochlea is more sensitive to strep than the site responsible for the dyskinesias.

Fig. 1. The CID relation between *Streptomycin* and *deafness* in an abstract

Most of previous studies [2–6] make isolated predictions for each pair of mentions depending on the local context around mentions, such as a sentence or several sentences. Then the final prediction of each entity pair can be obtained by merging the predictions of the corresponding mention pairs. In the above example, the aforementioned methods will predict each mention pairs through learning their local surroundings and then merge the results of the three mention pairs according to the designed rule. The rule assumes that if there is at least one pair of the mentions could support the relation, and it is deemed that there is a true relation in the corresponding entity pair. Unfortunately, the relation between *Streptomycin* and *deafness* will not be extracted by these methods since only the local contextual semantic information (e.g. intra-sentence) is considered. Intuitively, learning the long-distance semantic information throughout the entire document is crucial. In addition, predicting each mention pair separately and then merging them by the rule will split the semantic information of the entire document, resulting in information loss.

At present, some researchers [7–10] are working on long-distance dependency. For example, Convolutional Neural Networks (CNN), Transformer and Graph Neural Networks (GNN) are adopted to learn long-distance semantic information. However, not all the words in the document are helpful for the target relation extraction. The above methods ignore the processing of these irrelevant words when capturing long-distance information. At the aspect of multiple mentions indicating the same entity in the document, the above methods employ the multi-instance learning (MIL) fusing the mention pairs to classify the relation of entities in the document. However, the method of MIL relies on the semantic information of a few mention pairs, which may result in losing the information of the other mention pairs. Obviously, the global semantic information of all mention pairs in the document should be considered for extracting the relation between the corresponding entity pair.

For the two challenges of document-level RE, we propose a novel model that learns the global semantic representations for document-level RE. The main contributions of our work can be summarized as follows:

(1) The global context-aware attention (GCA) is proposed to obtain the global semantic representations. The method can learn semantic information throughout the entire article and filter the redundant information, which alleviates the problem derived from the long-distance dependency of inter-sentence relations.

(2) The global entity-aware attention (GEA) is employed to learn the global entity representations, which is able to integrate the semantic information of all mentions through measuring the correlation of each mention pair involved in the target entities.

2 Related Work

In general, CDR task requires to judge whether a given pair of chemical entity and disease entity is asserted with the induction relation in an article. Early researches [11–13] are mainly committed to extract the relation of a single mention pair in a single sentence for intra-sentence RE task. And the document-level relations are split into intra-sentence relations and the inter-sentence relations, but only the intra-sentence relations are considered. For example, Lowe et al. [2] develop a simple pattern-based system to find relations within the same sentence in CDR dataset. Actually, approximately 30% of relations in the BioCreative V CDR dataset are expressed across sentence boundaries [1]. Therefore, a large amount of indispensable inter-sentence relation information will be lost.

In recent years, inter-sentence relation extraction is getting more and more attention. Most works separately train two independent classifiers for inter-sentence and intra-sentence relations, and then extract relations of entities depending on the rule aforementioned. For example, Gu et al. [3] train a Maximum Entropy (ME) model and a CNN model to extract inter-sentence and intra-sentence relations, respectively. Xu et al. [4] separate relations at the sentence level and the document-level and employ support vector machines (SVMs) with artificial features and domain knowledge. Even though the above methods take the inter-sentence relation into account, learning the semantic information of long-distance context is still insufficient, which hinders the performance of the inter-sentence classifier.

Some studies discard breaking down the relations into intra-sentence and inter-sentence, and they learn the information of all mentions to directly predict the relation between the entities. For example, Zheng et al. [7] integrate CNN network and multilayer RNN network to learn the global and local information of the article, and generate the training instances by marking entities in the document. Since there is too much information in the document, this model may bring in redundant information. Verga et al. [8] form pairwise predictions over the entire document using the self-attention encoder. Meanwhile, they adopt Multi-Instance Learning (MIL) to treat multiple mentions of target entities in a document. MIL method relies on a few mention pairs features and ignores the contribution of other mention pairs to the relation extraction, which may result in information loss. Compared with the above methods, we propose GCA to select the key information related to entities in the whole article and GEA to obtain entity representations by learning the information from all mention pairs.

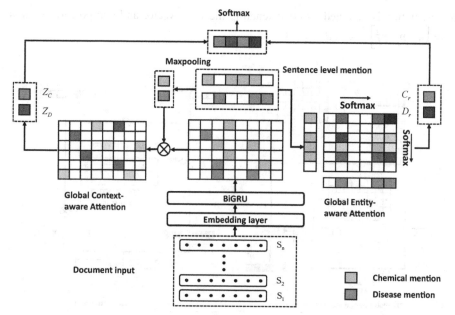

Fig. 2. Architecture of the document-level relation extraction model

3   Model

In this work, we design a novel model integrating GCA and GEA to extract the relation of the entity pair in a document directly. Figure 2 depicts the overall architecture of our model. The model consists of five modules: input embedding layer, encoder layer, GCA layer, GEA layer and output layer. Each module will be described in the following sections.

3.1   Inputs

The input to the model is a document $DOC = \{S_1, \ldots, S_i, \ldots, S_n\}$, $i \in \{1, 2, \ldots, n\}$, n is the number of sentences in the document. And the input sequence of the i-th sentence is $S_i = \{x_1, \ldots, x_j, \ldots, x_m\}$, $j \in \{1, 2, \ldots, m\}$, m is the number of words in the sentence. Each word representation x_j is the concatenation of the word embedding and its position vectors.

Specifically, we employ the pre-trained word vector to obtain the fixed word embedding of each word. We denote the embedding look-up matrix as $L_w \in \mathbb{R}^{d_v \times |V|}$ and word embedding $w_j \in \mathbb{R}^{d_v}$, where d_v is the dimensionality of the word vector and $|V|$ is the vocabulary size. The position vector is calculated according to the Euclidean distance from the closest mention of an entity to the current word. And the position vectors (p_j^1 and p_j^2) of the current word are related to the chemical entity and the disease entity, respectively. We learn the position embedding matrix $L_p \in \mathbb{R}^{d_p \times |P|}$ initialized randomly, which contains a separate $p_1, p_2 \in \mathbb{R}^{d_p}$, where d_p is the dimensionality of embedding for each position and $|P|$ is the vocabulary size. The final representation is obtained by concatenating the word vector and two position vectors $x_j = \left[w_j; p_j^1; p_j^2 \right] \in \mathbb{R}^d, d = d_v + 2 \times d_p.$

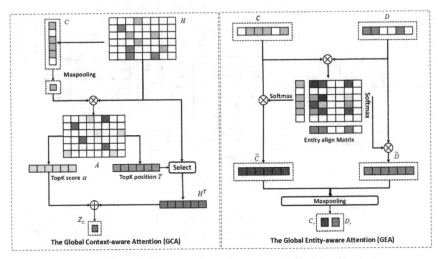

Fig. 3. Architecture of GCA and GEA

3.2 Encoder

We utilize the Bidirectional Gated Recurrent Unit (BGRU) [14] to encode the sequences and capture the contextual information of the words from the forward and backward directions in each sentence sequence. Specifically, given the input word embedding $x_t \in \mathbb{R}^d$, the hidden output of the forward GRU is formalized as Eq. (1):

$$h_t = GRU(x_t), h_t \in \mathbb{R}^l, \tag{1}$$

where l is the dimensionality of the hidden units. The hidden states can also be obtained by the backward GRU in a similar way. We concatenate the two hidden outputs as the final representation of the input word. Therefore, for a document $DOC \in \mathbb{R}^{n \times m \times d}$, the output sequence encoded by BGRU is $H \in \mathbb{R}^{n \times m \times 2l}$.

3.3 Global Context-Aware Attention (GCA)

By the BGRU encoder, we can obtain the contextual representation of each word in the entire document. However, some of the representation information is redundant as not all words can imply the relation between entities. Therefore, we propose the global context-aware attention (GCA) to extract the significant contextual representations related to the target entities. Specifically, the model picks out important context information related to the chemical entity and the disease entity (two target entities in the CID relation) from the whole context, respectively.

To obtain the entity-related global contextual representations, we extract the mention representations of the target entities from the encoded sequence H for each sentence. Concretely, we extract the entity mentions' representation according to their locations annotated in the dataset, obtaining the corresponding contextual information of the mentions. If there are multiple mentions in a sentence, the max pooling operation is used to get the overall representation. Finally, we gain the chemical mention sequence C and the disease mention sequence D in the entire document as follows:

$$C = \{C_1, C_2, \ldots C_i \ldots C_n\} \in \mathbb{R}^{n \times 2l}, \tag{2}$$

$$D = \{D_1, D_2, \ldots D_j \ldots D_n\} \in \mathbb{R}^{n \times 2l}, \tag{3}$$

where C_i is the representation of the chemical mention in the i-th sentence and D_j is the representation of the disease mention in the j-th sentence.

In the following, taking the chemical entity as an example, we will describe how to obtain the entity-related context representation with GCA. The left part of Fig. 3 depicts the architecture of the GCA module. We firstly get the representation V_C of the chemical entity from the chemical mention sequence $C \in \mathbb{R}^{n \times 2l}$ through the max pooling operation as Eq. (4).

$$V_c = maxpool(C), V_c \in \mathbb{R}^{2l}. \tag{4}$$

V_C can represent the global information of the chemical entity because it is induced from all mentions of the target chemical entity in the entire document.

Then, to measure the importance of context to the chemical entity representation V_C, we compute the global score matrix A by scoring each word of the entire document H with the entity vector V_C as Eq. (5). $\otimes$ denotes multiplication, $A_i, i \in \{1, 2, \ldots, n \times m\}$ represents the relevance between word x_i and the chemical entity V_C.

$$A = H \otimes V_c, A \in \mathbb{R}^{n \times m}. \tag{5}$$

Subsequently, we pick out the top K (K is hyper-parameter) highest scores g and their corresponding positions T in matrix A. The corresponding fragment H^T that is important for the global chemical entity V_C is selected from the contextual representation H, as Eq. (6)–(7):

$$g, T = top_{k(A,K)}, g, T \in \mathbb{R}^K, \tag{6}$$

$$H^T = index(H, T), H^T \in \mathbb{R}^{k \times 2l}. \tag{7}$$

Finally, the attention weights of the selected context representations are obtained by SoftMax. We can gain the contextual information $Z_C \in \mathbb{R}^{2l}$ that is related to the chemical entity by calculating the weighted sum of the selected representations H^T. The process can be summarized as Eq. (8)–(9):

$$\alpha = softmax(g), \alpha \in \mathbb{R}^K, \tag{8}$$

$$Z_c = \sum_i^K h_i^T \cdot \alpha_i. \tag{9}$$

Similarly, the contextual information $Z_D \in \mathbb{R}^{2l}$ related to the disease entity can also be captured by the GCA module. Therefore, Z_C and Z_D represent the global context-aware information relevant to the target entities of the entire document respectively, which is crucial to the final classification.

3.4 Global Entity-Aware Attention (GEA)

In document-level relation extraction, one entity often involves more than one mention, hence integrating the information of all mention pairs in the document is important for classifying the relations. Therefore, we propose the global entity-aware attention (GEA) module to capture the global entity information, as illustrated in the right part of Fig. 3.

Firstly, inspired by Chen et al. [15], we compute the correlation between each chemical mention and disease mention with Eq. (10):

$$Matrix = C \otimes D, Matrix \in \mathbb{R}^{n \times n}. \tag{10}$$

$Matrix_{ij}$ is the correlation between the i-th chemical mention and the j-th disease mention, which means that this matrix measures the impact of each pair of the mentions on the target relation in the entire document.

The global entity-aware information is determined by the correlation matrix. For a chemical mention C_i in sequence C, its semantic information related to all disease mentions is obtained using $Matrix_{ij}$, denoted as $\tilde{C}_i$, more specifically with Eq. (11). The same is performed for each mention representation in sequence D with Eq. (12).

$$\tilde{C}_i = \sum_{j=1}^n \frac{exp(Matrix_{ij})}{\sum_{k=1}^n exp(Matrix_{ik})} C_i, \forall i \in \{1, 2, \ldots, n\}, \tag{11}$$

$$\tilde{D}_j = \sum_{i=1}^n \frac{exp(Matrix_{ij})}{\sum_{k=1}^n exp(Matrix_{kj})} D_j, \forall j \in \{1, 2, \ldots, n\}. \tag{12}$$

After the above operations, the new chemical and disease mentions' representations are $\tilde{C} \in \mathbb{R}^{n \times 2l}$ and $\tilde{D} \in \mathbb{R}^{n \times 2l}$, which contain the global entity-aware interaction information in each target entity pair of the entire document.

Finally, we use max pooling to get two global entity representations with their corresponding mention sequences, named $C_r \in \mathbb{R}^{2l}$, $D_r \in \mathbb{R}^{2l}$, as Eq. (13)–(14):

$$C_r = maxpool\left(\tilde{C}\right), \tag{13}$$

$$D_r = maxpool\left(\tilde{D}\right), \tag{14}$$

3.5   Classification

At last, we concatenate the above two global semantic vectors and two global entity representation vectors as the final representation $\gamma \in \mathbb{R}^{8l}$. It is fed to a softmax classifier for identifying the relation between a pair of chemical and disease entities. As shown in Eq. (15)–(16), [;] denotes concatenation, $P \in \mathbb{R}^{|Y|}$ is the probability distribution for the relations between the pairs of entities, $W_\gamma \in \mathbb{R}^{|Y| \times 8l}$ is a weight matrix, $b \in \mathbb{R}^{|Y|}$ is a bias vector and $Y = \{0, 1\}$ is the label of a category (0 and 1 denote negative and positive, respectively).

$$\gamma = [Z_C; Z_D; C_r; D_r], \tag{15}$$

$$P = softmax\left(W_\gamma \cdot \gamma + b\right). \tag{16}$$

4   Experiments and Results

4.1   Dataset and Experiment Setting

We evaluate our model on the BioCreative V chemical disease relation (CDR) dataset [1]. Table 1 lists the statistics of the corpus. The CDR corpus consists of training, development and test, each of which includes 500 Medline articles. Most articles are selected from an existing CTD-Pfizer (Comparative Toxicogenomics Database (CTD) [16]) collaboration related dataset. Since each entity is annotated with Mesh ID, but an entity involves multiple mentions, we list the number of mentions and Mesh IDs in Table 1. The "CIDs" denotes the number of the true relations.

Table 1. The statistics of CDR dataset

Dataset	Chemical		Disease		CIDs
	Mention	Mesh ID	Mention	Mesh ID	
Training	5203	1467	4182	1965	1038
Development	5347	1507	4244	1865	1012
Test	5385	1435	4424	1988	1066

Following Xu et al. [4], we merge the original training and development sets, 90% of which are used for training and the rest for validating. The reported results are based on the test set. According to the statistics of corpus, the number of sentences in each document is set to 23 and each sentence has a maximum of 200 words. The dimension of the word and position embedding vectors are set as 300 and 30, respectively. The look-up table is initialized by GloVe [17], pre-trained on 10G background text from PubMed. The size of GRU units is set to 256, and the Dropout rate is set to 0.3. The value of K on GCA is 400. The batch size is set to 16 for all experiments. We employ RMSprop optimization, and set the learning rate $lr = 0.001$ and the momentum item parameter $rho = 0.9$. The performance of our system is evaluated by the standard evaluation measures metrics: Precision (P), Recall (R) and F-score (F).

4.2 Effects of GCA and GEA Modules

To explore the effects of the proposed method, we construct experiments with different modules. The performance comparisons can be seen in Table 2.

BGRU: We take Bidirectional GRU (BGRU) network as our baseline, which can encode each word with its sentence-level context from both forward and backward directions for the entire document. The representations obtained from the hidden states after max pooling operation are used for prediction.

Table 2. The performance comparisons with various modules of our model

Model	P (%)	R (%)	F (%)
BGRU	54.0	58.9	56.3
BGRU + GCA	50.8	65.9	57.3
BGRU + GEA	54.3	63.4	58.5
BGRU + GEA + GCA	60.4	59.8	**60.1**

GCA: As described in Sect. 3.3, the GCA module can focus on the contextual semantic information related to the target entities in the entire document. Table 2 shows that the GCA module achieves an improvement of 1.0% on F-score based on BGRU, which demonstrates the global information captured by GCA is effective to the relation extraction.

GEA: Using the GEA module, the interactive information among the mention pairs can be captured. As shown in Table 2, the GEA module obtains an F-score of 58.5%, which is 2.2% higher than that of BGRU. The result indicates that constructing the relevance among all mention pairs is effective for document-level RE.

GEA + GCA: As shown in Table 2, the approach combining GCA and GEA can achieve an F-score of 60.1%, which is higher than adopting GEA or GCA solely. This also shows that considering the global context information and global entity interactive information is helpful for model learning. The features derived from the combination are more comprehensive, which help to achieve a better precision, and correspondingly the recall score is decreased compared with a single module (**GEA** or **GCA**).

4.3 The Analysis of Intra- and Inter-sentence Classification

In order to demonstrate the capability to capture long-distance dependency of our model, we analyze the results of intra- and inter-sentence RE. In the document-level relation extraction, the relations between entities are often expressed across sentence boundaries (inter-sentence relations). We address this problem by capturing global information by GCA and GEA, and the detail results can be seen in Table 3, which are promising.

Table 3. The performance comparison of the intra- and inter-sentence CIDs on test set

Model	Level	P (%)	R (%)	F (%)
BGRU	Intra-sentence	55.6	70.5	62.2
	Inter-sentence	45.6	29.6	35.9
	Document	54.0	58.9	56.3
Our model	Intra-sentence	62.3	69.0	65.5
	Inter-sentence	52.7	36.2	42.9
	Document	60.4	59.8	60.1

From Table 3, it can be seen that the performance of intra- and inter-sentence relation extraction is both improved, achieving the F-scores of 65.5% and 42.9%, respectively. Compared with the BGRU model, the F-score of intra-sentence relation of our model is improved by 3.3%. Moreover, our model improves the F-score of the inter-sentence relation extraction by 7.0%. And the overall F-score at document-level of our model is promoted by 3.8%. Experimental results demonstrate that our approach can significantly improve the performance of inter-sentence relation extraction. This is mainly because our model can capture the global information by integrating the GCA module and the GEA module. By GCA, we pick the relevant contextual information related to the target entities throughout the entire document. Meanwhile, the global entity representations containing interactive information among all mention pairs are obtained with GEA. Therefore, our method is effective for inter-sentence RE.

4.4 Comparison with the Existing Methods

We compare our model with the existing methods on BioCreative V CDR corpus. Table 4 shows the performance compared against the previous systems on the test set without post-processing or external resources.

The early researches mainly deal with intra-sentence relations, but ignore the relationships across sentence boundaries, so the ability of these relation extraction models is limited. Lowe et al. [2] develop a pattern-based system to find the relations within the sentence. Zhou et al. [5] exploit syntactic and semantics information to extract relations. However, the neglected inter-sentence relations limit the overall performance of these methods. In order to extract document-level relations, many studies employ two independent classifiers for intra- and inter-sentence relations respectively. Xu et al. [4] utilize different features for intra- and inter-sentence relations extraction with the support vector machine (SVM) classifier, achieving an F-score of 50.7%. Gu et al. [3] build a maximum entropy (ME) model and a CNN model for inter- and intra-sentence level RE respectively, achieving a competitive F-score of 60.2%.

Instead of the separate classifiers aforementioned, some methods extract document-level relations uniformly. Li et al. [6] employ recurrent piecewise CNN to train the document-level classifier by the combination of piecewise strategy, attention mechanism and multi-instance learning (MIL), reaching an F-score of 59.1%. Zheng et al.

Table 4. The comparison with other existing methods

Model	P (%)	R (%)	F (%)
Lowe et al. [2]	52.6	51.8	52.2
Xu et al. [4]	59.6	44.0	50.7
Zhou et al. [5]	64.9	49.3	56.0
Gu et al. [3]	60.9	59.5	**60.2**
Li et al. [6]	55.1	63.6	59.1
Zheng et al. [7]	54.3	65.9	59.5
Verga et al. [8]	49.9	63.8	55.5
Sahu et al. [9]	52.8	66.0	58.6
Our model	60.4	59.8	**60.1**

[7] integrate CNN and RNN to learn semantic representations of documents, achieving the similar results. But they are still insufficient to capture long-distance information. Various methods are adopted to capture the long-distance information. Sahu et al. [9] use GCN to train document graph and apply MIL to extract document-level relations and obtain an F-score of 58.6%. The graph network method depends on the construction of the document graph, which needs to design the dependency syntax and article structure artificially. Verga et al. [8] employ multi-layer Transformer to capture the information of mentions from the article, achieving the F-score of 55.5%. Transformer is used to alleviate the problem of the long-distance dependency in their model. However, these models disregard the redundant information in the documents.

In this work, we apply GCA to capture long-distance information and simultaneously skip the irrelevant information. Our model learns document-level semantic representation without the preliminary design, such as external features and graph construction. Besides, compared with the Transformer encoder, there are fewer parameters in our model. The important distant information of the context can be induced by GCA based on the BGRU encoder. Additionally, instead of MIL that may lead to information loss, the GEA module can integrate the semantic information of all mention pairs and gain the global entity representations. By combining GCA and GEA, we achieve an F-score of 60.1%, which is promising and suggests that our model is effective for document-level relation extraction.

Actually, it can be seen in Table 4, Gu et al. [3] report a slightly better result than that of our model (60.2% *vs.* 60.1%). Different from our model, they build two separate models for intra- and inter-sentence relations, and make more efforts in extracting intra-sentence relations compared with inter-sentence. With linguistic and dependency features, their model achieves the overall F-score of 60.2%. However, the result of their inter-sentence classifier is lower than that of our method (11.7% *vs.* 42.9%). In this paper, our model extracts document-level relations with a single model directly and achieves a promising result.

5 Conclusion

Document-level relation extraction has always been a difficult problem in biomedical relation extraction. Learning semantic representation of long-distance context and merging all the mention pairs to extract the relation between two entities are crucial for document-level relation extraction. In this paper, we propose a novel architecture that can learn the global representations with GCA and GEA. GCA is employed to explore the document-level semantics through selecting the most significant information from the entire document. Meanwhile, the GEA is adopted to measure the correlation between each mention pair for the integration of all mentions. Experimental results demonstrate the effectiveness of our approach on BioCreative V CDR dataset.

Acknowledgment. This work is supported by grant from the Science and Technology Innovation Foundation of Dalian (2020JJ26GX035) and the National Natural Science Foundation of China (No. 62076048). The Titan Xp GPU used for this research was donated by the NVIDIA Corporation.

References

1. Li, J., et al.: Annotating chemicals, diseases, and their interactions in biomedical literature. In: Proceedings of the Fifth BioCreative Challenge Evaluation Workshop, Seville, Spain, pp. 173–182 (2015)
2. Lowe, D., O'Boyle, N., Sayle, R.: LeadMine: disease identification and concept mapping using Wikipedia. In: Proceedings of the Fifth BioCreative Challenge Evaluation Workshop, Spain, pp. 240–246 (2015)
3. Gu, J., Sun, F., Qian, L., Zhou, G.: Chemical-induced disease relation extraction via convolutional neural network. Database **2017**, 1–10 (2017)
4. Xu, J., Wu, Y., Zhang, Y., Wang, J., Lee, H.J., Xu, H.: CD-REST: a system for extracting chemical-induced disease relation in literature. Database **2016**, 1–9 (2016)
5. Zhou, H., Deng, H., Chen, L., Yang, Y., Jia, C., Huang, D.: Exploiting syntactic and semantics information for chemical–disease relation extraction. Database **2016**, 1–10 (2016)
6. Li, H., Yang, M., Chen, Q., Tang, B., Wang, X., Yan, J.: Chemical-induced disease extraction via recurrent piecewise convolutional neural networks. BMC Med. Inf. Decis. Making **18**(2), 45–51 (2018)
7. Zheng, W., et al.: An effective neural model extracting document-level chemical-induced disease relations from biomedical literature. J. Biomed. Inf. **83**, 1–9 (2018)
8. Verga, P., Strubell, E., McCallum, A.: Simultaneously self-attending to all mentions for full-abstract biological relation extraction. In: Proceedings of the 2018 Conference of the North American Chapter of the Association for Computational Linguistics: Human Language Technologies, New Orleans, Louisiana, pp. 872–884 (2018)
9. Sahu, S.K., Christopoulou, F., Miwa, M., Ananiadou, S.: Inter-sentence relation extraction with document-level graph convolutional neural network. In: Proceedings of the 57th Annual Meeting of the Association for Computational Linguistics, Florence, Italy, pp. 4309–4316 (2019)
10. Christopoulou, F., Miwa, M., Ananiadou, S.: Connecting the dots: document-level neural relation extraction with edge-oriented graphs. In: Conference on Empirical Methods in Natural Language Processing, Hong Kong, China, pp. 4924–4935 (2019)

11. Kim, S., Liu, H., Yeganova, L., Wilbur, W.J.: Extracting drug-drug interactions from literature using a rich feature-based linear kernel approach. J. Biomed. Inf. **55**, 23–30 (2015)
12. Liu, S., Tang, B., Chen, Q., Wang, X.: Drug-Drug interaction extraction via convolutional neural networks. Comput. Math. Methods Med. **2016**, 6918381–6918388 (2016)
13. Masaki, A., Miwa, M., Sasaki, Y.: Extracting drug-drug interactions with attention CNNs. In: Proceedings of the BioNLP workshop, Vancouver, Canada, pp. 9–18 (2017)
14. Cho, K., Merïenboer, V., Gulcehre, C., Bougares, F., Bougares, H., Bengio, Y.: Learning phrase representations using RNN encoder–decoder for statistical machine translation. In: Proceedings of the 2014 Conference on Empirical Methods in Natural Language Processing, Doha, Qatar, pp. 1724–1734 (2014)
15. Chen, Q., Zhu, X., Ling, Z., Wei, S., Jiang, H., Inkpen, D.: Enhanced LSTM for natural language inference. In: Proceedings of the 55th Annual Meeting of the Association for Computational Linguistics (Volume 1: Long Papers), Vancouver, Canada, pp. 1657–1668 (2017)
16. Davis, A.P., Murphy, C.G., Saracenirichards, C.A., Rosenstein, M.C., Wiegers, T.C., Mattingly, C.J.: Comparative toxicogenomics database: a knowledgebase and discovery tool for chemical–gene–disease networks. Nucleic Acids Res. **27**, 786–792 (2009)
17. Pennington, J., Socher, R., Manning, C.D.: GloVe: global vectors for word representation. In: Proceedings of the 2014 Conference on Empirical Methods in Natural Language Processing, Doha, Qatar, pp. 1532–1543 (2014)

Type-Aware Open Information Extraction via Graph Augmentation Model

Qinghua Wen[1,2], Yunzhe Tian[3], Xiaohui Zhang[4], Ruoyun Hu[5], Jinsong Wang[6], Lei Hou[1,2(✉)], and Juanzi Li[1,2]

[1] Department of Computer Science and Technology, BNRist, Beijing, China
{houlei,lijuanzi}@tsinghua.edu.cn
[2] KIRC, Institute for Artificial Intelligence, Tsinghua University,
Beijing 100084, China
[3] Beijing Key Laboratory of Security and Privacy in Intelligent Transportation,
Beijing Jiaotong University, Beijing, China
[4] State Grid Customer Service Center, Tianjin, China
[5] State Grid Zhejiang Marketing Service Center, Hangzhou, China
Hu_ruoyun@sgcc.zj.com.cn
[6] Case Western Reserve University, Cleveland, OH, USA
jxw741@case.edu

Abstract. Open information extraction (IE) can support knowledge graph enrichment. Open IE systems are capable of extracting relational tuples from texts without the need for a pre-specified vocabulary. There have been more researches on open IE in English than in Chinese, and most of them rely on word segmentation and syntactic analysis tools, which have a great influence on the results. Besides, the lack of annotated Chinese corpus also makes it difficult to classify triples in a supervised manner. To address the problems, we propose an unsupervised Chinese open IE model, named graph augmentation model (GAM). It first uses the knowledge graph to obtain linked entities and types of entities, where the linked entities can benefit the word segmentation accuracy and the entity types can help obtain the domain and range of relations for knowledge graph schema completion. Then it uses manually set rules to obtain candidate triples and uses a designed graph-based algorithm to iteratively calculate the importance and accuracy of triples. Experiments demonstrate that our method outperforms existing baseline methods. Specifically, GAM is proved to effectively extract domain and range of relations that other methods cannot. GAM achieves high accuracy of triples above a certain threshold, and the triples obtained show benefits in enriching a knowledge graph without the need for data annotation.

Keywords: Open information extraction · Knowledge graph · Knowledge graph completion · Entity linking

1 Introduction

With the development of artificial intelligence (AI) technology, the importance of knowledge graphs is also increasing. Although there have been several knowledge

© Springer Nature Singapore Pte Ltd. 2021
H. Chen et al. (Eds.): CCKS 2020, CCIS 1356, pp. 119–131, 2021.
https://doi.org/10.1007/978-981-16-1964-9_10

graphs in the general field, many challenges still exist in constructing knowledge graphs in specific domains, and one of the challenges is the lack of domain knowledge. To address this challenge, open information extraction (IE) is emphasized to expand domain knowledge, which has attracted much attention from both academia and industry.

Traditional IE methods mainly aim to extract entities, relations, and events with limited types from small-scale texts [1]. With the number of data on the Internet increasing, traditional IE methods that require manual annotations are no longer efficient. Under such circumstances, open IE is proposed in order to efficiently extract named entities, semantic relations, and fact event of unlimited types from large-scale web pages, most of which contain noisy and inconsistent information.

Numerous methods have been proposed since open IE was first introduced by Banko et al. Most existing open IE methods are English-oriented, such as TextRunner (Banko et al. 2007) [2], ReVerb (Fader et al. 2011; Etzioni et al. 2011) [3], and OLLIE (Schmitz et al. 2012) [4]. In the field of Chinese open IE, Tseng et al. (2014) [6] proposed CORE based on syntactic analysis, and Qiu et al. (2014) [7] proposed ZORE based on syntax and relational propagation algorithms. The above methods have achieved relatively good results, but there are still several limitations. (1) When only sentence segmentation is considered, the accuracy of word segmentation and part-of-speech (POS) tagging will be low. (2) Existing methods heavily rely on large amount of annotation data for training, such that triple results could be classified and filtered in a supervised way. (3) Existing methods seldom use the knowledge graph to improve the quality of extractions and lack the scoring evaluation of results.

To address these problems, we propose a Graph Augmentation Model (**GAM for short**) for knowledge graph enrichment. It initially uses external link entities from the knowledge graph to improve word segmentation quality, and we manually make a set of rules to get candidate triples. Moreover, we design a novel graph model to iteratively calculate the importance scores of each element in the triples, and obtain the domain and range distribution of relations which can increase the accuracy of extracted triples, finally we filter out the required triples. The model does not require a large amount of annotation data for training. Additionally, our method can use the domain and range of relations to further complete knowledge graph schema. Experimental results prove that the proposed method can obtain triples with higher accuracy, and can benefit more for enriching knowledge graph in a specific domain.

2   Related Works

Banko et al. (2007) [2] proposed the first learning-based open IE system: TextRunner, a self-supervised learning approach. It uses a dependency parser to identify a set of extractions as positive and negative training examples from a small sample of sentences, this data is then used as input to a Naive Bayes classifier. The extractor then generates candidate tuples by first identifying pairs of noun phrase arguments and then designating each word in between as part of a relation phrase or not. After that each candidate extraction is presented to the classifier. Finally, a redundancy-based assessor assigns a probability to each retained tuple based on the number of sentences, thus exploiting the redundancy of information in Web text and assigning higher confidence to extractions that occur multiple times. Fader et al. (2011) [3] proposed the rule-based system: ReVerb. It proposes two kinds of constraints, respectively (1) syntactic constraint for filtering out incoherent extraction and incomplete extraction, and (2) lexical constraint for filtering out redundant extractions. For the first type, the relational phrase must match the given POS pattern. For the second type, an effective relationship should appear multiple times in a large-scale corpus with different head and tail entities. Schmitz et al. (2012) [4] proposed the OLLIE. It first uses ReVerb to generate high-precision seed triples and uses bootstrap to obtain a larger training set. Then it learns a new syntax template on the training set, and matches the template with the dependency syntax to identify the triples. Next it synthesizes the content of the context and expands the triples with two additional fields. Finally it uses a classifier to filter out the triples with lower confidence. There are also other methods like ClausIE (Del Corro and Gemulla, 2013) [5] is presented, which exploits linguistic knowledge about the grammar of the English language to map the dependency relations of an input sentence to clause constituents, and then get a better result.

Chinese open IE is more difficult due to the complex and changeable nature of Chinese languages. Besides, the lack of Chinese training corpus makes it even harder. Tseng et al. (2014) [6] proposed the CORE, which is the first attempt in the field of Chinese open IE. Given a Chinese text as input, CORE uses word segmentation, POS tagging, and syntactic analysis to automatically tag Chinese sentences to complete the extraction of entity-relational triples. Qin et al. (2015) [11] proposed an unsupervised model UnCORE. It generates candidate triples through the distance limitation between entities and the relation position limitation, then uses global relations and type relations to obtain a typical relation indicator list for different entity type pairs. While the relation of a candidate triple is not in the relation vocabulary of the head and tail entity pair type of the candidate triple, it is filtered as an error. Jia et al. (2018) [8] proposed an open IE model DSNFs based on the syntactic dependency tree with the Chinese linguistic features, and it uses 7 features to match the sentences and then get the triples from them. At present, the DFNFs achieved the most competitive results.

To sum up, most of the open IE methods use supervised methods and only use NLP tools for named entity recognition and syntactic dependency analysis. The named entity recognition of these tools has problems such as low accuracy and sparse results. In the Chinese open IE field, due to the lack of annotated data, supervised methods are severely limited. To solve these problems, we propose the unsupervised open IE method in this article. Compared with the previous methods, our method is very unique.

3   Method

3.1   Problem Formulation

For our open IE task, given a document set $D = \{D_1, D_2, ..., D_n\}$, we get sentence set $S = \{S_1, S_2, ..., S_n\}$ from each document, then we convert each sentence $S_i = (w_1, w_2, ..., w_n)$ into structured representation $S_i' = (\{e_1, ..., e_n\}, \{r_1, ..., r_m\})$ (e as entity, r as relation), and record the position of entities and relations, then we obtain the structured document which is composed of S_i'. Our task is to obtain triples from the structured document and get the domain and range distribution of each relation. The triple is represented as (e_i_head, r, e_j_tail). Where e_i_head and e_j_tail represent the head entity and the tail entity in the triple respectively, the relation is represented as r.

3.2   Method Framework

Figure 1 illustrates the overall framework. Our framework mainly consists of three parts: *data processing, graph augmentation model, triple scoring and filtering*. Data processing turns raw Baidu Baike data into structured representation containing entities and relations and gets the candidate triples in it. Graph augmentation model which we called GAM uses the triples to construct a graph and get domain and range distribution corresponding to each relation node and compute scores of the nodes. Triple scoring and filtering get the final result of triples based on a formula. We will describe each part of the method in the following passages.

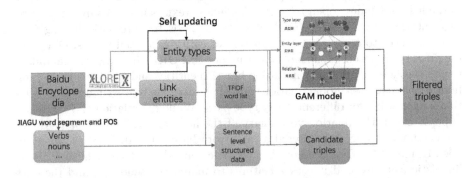

Fig. 1. The method framework shows the process of our model.

3.3   Data Processing

Since Baidu Baike is the largest data source in Chinese and has the most comprehensive information, we crawled 1,218 Beijing tourist attractions web pages in Chinese as the data for our work. We introduce the linked entities owned by Baidu Baike texts and use the XLORE [9] to link entity mentions in the texts. Proper noun phrases, phrases with special symbols, and file names are also considered, then we generate an entity dictionary with nearly 13,000 words.

We used the *jiagu* NLP tool which loads the entity dictionary above in the experiment. We record the position of each entity in the segmented sentence and keep the position of comma for subsequent triple processing operations. Finally, we generate a structured representation list with named entities, noun phrases and verbs. Moreover, we enrich the representation list by converting the short words into long geographical noun phrases and adding subjects for clauses.

Using the structured data above, we adopt a rule-based candidate triple extraction method. To meet the needs of the task of enriching the tourism graph, we mainly consider using verbs as relation words to obtain candidate triples. 1) We get the verbs in each sentence as relation words. 2)According to the work of Qin et al. [11], we get the head entities and tail entities according to their relative position to the relation words in the sentence. 3) We manually set rules to filter the head entities and tail entities, and finally get the triples. We set different priorities for different entities. For the head entity, the priority is (*subject/entity* > *linked/entity* > *nearest/noun*).For the tail entity, the priority is (*added/entity* > *linked/entity* > *nearest/noun*). If a relation word matches both the head entity and the tail entity, the three are combined into a triple which stored in a triple list.

3.4 Graph Construction

Inspired by the work about the joint task of expanding entity set and attribute extraction [10], we similarly propose a graph representation, which can capture the relationship among entities, entity types, and relations. Moreover, the relationship between similar types or similar relations is represented in the graph. These nodes are modeled as a tripartite graph as illustrated in Fig. 2. Each entity (and relation or type) is represented as a node in the graph. The relationship between e_i and r_j is represented as a weighted edge. The W_{ij} is a positive real number, indicating the strength of the relatedness between e_i and r_j. The tripartite graph consists of a relation layer, entity layer, and type layer. Here, the process of each layer's construction is described in detail as follows:

Relation Layer Construction. The construction of relation layer is based on all relation words in each document, which are currently verbs. As similar relation words often have the same domain and range distribution, we connect an edge between the nodes when their vector similarity is greater than a threshold $\sigma(\sigma = 0.7)$, we use a word vector file which is trained by Wikipedia and Baidu Baike to calculate the similarity. We select the relation words whose length is greater than 1, and exclude some words with incorrect POS tagging and nonsense, so as to improve the quality of relation layer. We initialize the importance score of each relation to 1.

Entity Layer Construction. For each relation node in relation layer, we search the corpus and obtain entities that co-occur with the relations in one sentence. To distinguish head entities and tail entities, we denote entities appearing in front of the relations as x_head (e.g., 慈禧 (cixi)_head) and entities appearing after the relations as x_tail (e.g., 宫殿 (palace)_tail). In order to improve the entity node

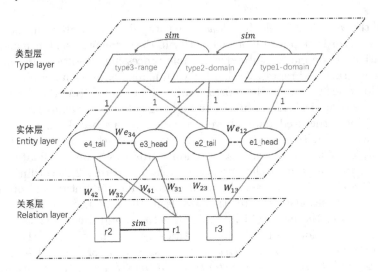

Fig. 2. An tripartite graph representation for types and entities and relations, where t_i and e_i and r_i are type and entity and relation respectively, and W_{ij} denote the weight of edges between the entities and relations, and We_{ij} denote the weight of edges between the head entities and tail entities in a triple. *sim* denote the weight of edges between similar types (and relations).

and quality, we calculate the TFIDF value of each entity in each document and set the threshold $\eta(\eta > 0.005)$ to obtain the list of important entities in each text. We initialize the importance score of each entity to 1. Then, we add an edge between the relation node and its co-occurred entity node with the frequency of co-occurrence as the weight.

Type Layer Construction. We get the type list of each entity from the XLORE [9]. There are 12751 different types. However, some fine-grained types are subsets of coarse types. If all types are used as type layer nodes, it will be too large for calculation, thus we selected 50 coarse-grained types from all types, which covers entities as much as possible. We count the frequency of each type and select 50 high frequency and representative types. As an entity is likely to belong to multiple types, in order to confirm one type for an entity, we manually set weights for these types. The weight reflects the domain of our document which is in the tourism field. Then each entity will select the type with the largest weight from its type list as the entity type. Table 1 shows the top 7 types from 50 types we selected.

For each type in our type set, we generate two type nodes in the type layer, one is domain type node (e.g., 地点 (location)_domain) and the other is range type node (e.g., 地点 (location)_range). Then we connect entity nodes and their corresponding type nodes by edges whose weight is 1. Note that, head entity nodes are only connected to the domain type nodes and tail entity nodes are only connected to the range type nodes. We also add an edge between two domain type nodes or two range type nodes with PMI (Point-wise Mutual

Information) [12], which measures the compatibility between types as its weight, and we only keep the edges with a weight greater than a threshold $\xi(\xi = 1.3)$. The initial importance score of a type node is the sum of the initial score of all entity nodes connected to it.

$$PMI(t_1, t_2) = log(p(t_1, t_2)/(p(t_1) \times p(t_2))) \tag{1}$$

where t_1 and t_2 denote two entity types, $p(t_1)$ and $p(t_2)$ denote the probability that an entity belongs to type t_1 and t_2 respectively, and $p(t_1, t_2)$ denotes the probability that an entity belongs to both type t_1 and t_2. The greater the value, the two types are more compatible.

Table 1. The list of top 7 types and their frequencies and weights

类型 Types	景观景点 Attraction	地点 Location	地理 Geography	建筑 Building	行政区划 State	人物 Person	组织机构 Organization
Frequency of types	1096	4204	4390	1652	1163	8181	1961
Weight of types	130	120	110	110	101	100	100

3.5 Importance Propagation

We propose an algorithm inspired by this work [10]. Graph augmentation algorithm propagates the importance of nodes among different layers through iteration. We get the importance score of each node through an iterative calculation of entities, relationships, and types in the text, which can judge the quality of the generated triples. Two assumptions are used in our method:

Hypothesis 1: The entities linked by many important relations and many important types tend to be important, the relations linked by many important entities tend to be important, and the types linked by many important entities tend to be important.

Based on the above hypothesis, we weight the importance of nodes in a recursion way. Because the entity layer is a bridge with type layer and relation layer, both type nodes and relation nodes will propagate their importance scores to entities nodes. The importance scores of entity e_i, relation r_j, and type t_i at the $(k + 1)$-th iteration are calculated as:

$$s_1(e_i)^{k+1} = \sum_{\forall m:r_m-e_i} s(r_m)^k \frac{w(r_m - e_i)}{\sum_{\forall n:r_m-e_n} w(r_m - e_n)} + \sum_{\forall m:t_m-e_i} s(t_m)^k \frac{w(t_m - e_i)}{\sum_{\forall n:t_m-e_n} w(t_m - e_n)} \tag{2}$$

$$s_1(r_j)^{k+1} = \sum_{\forall i:r_j-e_i} s(e_i)^k \frac{w(e_i - r_j)}{\sum_{\forall n:e_i-r_n} w(e_i - r_n)} \tag{3}$$

$$s_1(t_i)^{k+1} = \sum_{\forall m:e_m-t_i} s(e_m)^k \frac{w(e_m - t_i)}{\sum_{\forall n:e_m-t_n} w(e_m - t_n)} \tag{4}$$

where $s_1(\cdot)$ denotes that it is computed based on *Hypothesis 1*. $s(e_i)^k$ is the importance score of entity e_i at the k-th iteration, $s(r_m)^k$ is the importance score of relation r_m at the k-th iteration, and $s(t_m)^k$ is the importance score of type t_m at the k-th iteration. $r_m - e_i$ is the edge between relation r_m and entity e_i, and $w(r_m - e_i)$ is the weight of the edge.

Due to data sparseness, *Hypothesis 1* may underestimate the importance of long-tail entities, relations, and types. In order to the alleviate negative impact of data sparseness, we additionally propagate the importance score of one node to its neighbors at the same layer based on the following hypothesis. Here is the hypothesis:

Hypothesis 2: The relations linked by many important relations tend to be important, and the types linked by many important types tend to be important.

According to *Hypothesis 2*, the importance scores of relation r_i and type t_i at the (k+1)-th iteration are calculated as follows:

$$s_2(r_i)^{k+1} = (1 - \beta) * s(r_i)^k + \beta * \sum_{\forall j: r_j - r_i} s(r_j)^k \frac{w(r_i - r_j)}{\sum_{\forall n: r_j - r_n} w(r_j - r_n)} \tag{5}$$

$$s_2(t_i)^{k+1} = (1 - \beta) * s(t_i)^k + \beta * \sum_{\forall j: t_j - t_i} s(t_j)^k \frac{w(t_i - t_j)}{\sum_{\forall n: t_j - t_n} w(t_j - t_n)} \tag{6}$$

The above equations are consist of two items. The first item is the importance score from the node itself, ensuring its importance score does not deviate too much from the previous iteration. The second item is the incoming importance score from neighboring nodes at the same layer. $\beta \in [0,1]$ is an interpolation parameter to control the importance of the two items. Because there are too many entities and the computational complexity is too high, we only consider the relations and types in *Hypothesis 2*.

Each node has two importance scores $s_1(\cdot)$ and $s_2(\cdot)$ based on *Hypothesis 1* and *Hypothesis 2* respectively. Finally, we weight each node by combining the above two hypotheses using a weighted mean method:

$$s(e_i) = s_1(e_i) \tag{7}$$

$$s(r_i) = \alpha * s_1(r_i) + (1 - \alpha) * s_2(r_i) \tag{8}$$

$$s(t_i) = \alpha * s_1(t_i) + (1 - \alpha) * s_2(t_i) \tag{9}$$

where $\alpha \in [0,1]$ is an interpolation parameter.

We update the importance score of entities, relations, and types using equations until the number of iterations exceeds a predefined threshold. After testing, we find that the change of scores tends to be stable when the number of iterations is in $[1, 10]$.

3.6   Triples Scoring

After the above operations, we can generate candidate triples, construct a graph, and calculate the score of each node. Here we propose the model running process. 1) Get the domain and range distribution of each relation node. For each relation node, we obtain all the head entity nodes and tail entity nodes connecting with it, and by matching the entity types, we obtain the distribution of the head entity types and tail entity types, which is the domain and range distributions. The format of distribution is: $(r : [< type_i - domain, score_i >, ...], [< type_j - range, score_j >, ...])$, where r represents relation, $type_i - domain$ and $score_i$ denote domain $type_i$ and score of it respectively, $type_j - range$ and $score_j$ denote range $type_j$ and score of it respectively. 2) Get the graph of each document. We construct a graph for each document, then we obtain the importance score of the entities and relations by iterative calculation. 3) Design formula to get scores of triples. For each triple *(head entity, relation, tail entity)*, we get the score of the head entity, tail entity, and relation. we get the domain score and range score from 1). Then we use a formula to combine these scores, we finally obtain the importance score of the triples. The formula is as follow:

$$S_{triple} = log(\lambda * (S_{domain} * S_{range}) + (1 - \lambda) * \mu * (S_{head} + S_r + S_{tail}) + \theta) \quad (10)$$

where the triple is $(head, r, tail)$. the S_{domain} and the S_{range} are the score of a head entity type and the score of a tail entity type in the distribution of relation r, the S_{head} is the score of a head entity and S_{tail} is the score of a tail entity, S_r is the score of a relation. $\lambda \in [0, 1]$ is a weight parameter. μ is the parameter used to control the influence of elements of the triple. θ is a constant used to prevent formula errors.

4   Experiments

4.1   Experimental Settings

Dataset. We obtain 1,218 web pages related to Beijing attractions from Baidu Baike. We obtain 91,649 sentences and extract 12,932 entities from them. To match the types of entities, we acquire 12,915 entities with types from XLORE [9] and we choose 50 most frequent types as the nodes. TFIDF algorithm is used to filter entities and noun phrases, and we get 36,353 words from all documents for filtering.

Parameter Setting. The parameters α and β from the *Importance Propagation* section are empirically set to 0.2 and 0.1, and the number of iterations is set to 5. In the last formula, λ, μ, and θ are respectively set to 0.8, 0.0001, and 0.5.

Baseline. Due to the differences between Chinese and English, the method of English domain use directly in Chinese corpus will bring great uncertainty. In the field of Chinese open IE, UnCORE [11] and DSNFs [8] are the recent method and achieve competitive results. We use them as baselines to compare with our model.

4.2 Evaluation Metrics

Precision: We extracted 70,108 triples from our data set using our proposed method. It is impossible to label all extractions. Thus, we adopt the idea of multiple sampling. For each score interval we defined (or of length 2), we randomly sample 200 triples from all triples N times and label them to estimate the precision for each time: $P_i = C_i/A$, where C_i is the number of correct extractions that we label at i-th time and A is the total number of extractions. Finally, we calculate the average of them to get the overall precision: $P = \sum_{i=1}^{N} P_i/N$ (N denotes sampling n-th times).

Yield: Recall is highly important to evaluate the performance of extractor. However, evaluating recall is difficult at a large scale, since it is hard to identify all existed correct triples in a large corpus. Thus, most procedures in the literature either completely ignoring recall or using some kind of proxy, such as yield. Yield is the total number of extractions labeled as correct. And for each score interval, yield can be estimated by multiplying the total number of extractions with estimated precision. $yield = P * C$, where C is the total number of extractions.

4.3 Experimental Results

We use our model GAM to achieve more than 70,000 triples from 1,218 Baidu Baike documents and get their importance scores. Our results confirm that with the continuous increment of triples score, triples are more important to the document and more accurate. We evaluated the triples with scores above different thresholds and obtained the following results. The results show that there are 7726 triples with a threshold of more than 4 points, most of which are the location information of scenic spots, which are of high importance to the domain, and the accuracy rate can reach more than 70%. In addition, the accuracy rate of the threshold above 3 points can reach 66.7%. Table 2 shows the sample of triples we acquire from different documents. The scores are calculated by the GAM model and we can find the triples are important to the documents.

Table 2. The samples of triples in different documents

Score	Triples	Documents
10.050	颐和园/坐落/北京西郊	颐和园
10.050	Summer Palace/located in/Beijing western suburbs	Summer Palace
9.079	荷花/成为/著名景观	什刹海
9.079	lotus/becomes/famous scenery	Shichahai
7.275	恭王府/占地/61120 平方米	恭王府
7.275	Prince kung's Mansion/covers/61120 square meters	Prince kung's Mansion
5.010	东南角楼/建于/明朝	明城墙遗址公园
5.010	Southeast turret/built in/Ming	Wall Ruins Park
4.487	嘉靖/重建/中极殿	故宫博物院
4.487	Jiajing/rebuilt/Zhongji Hall	Palace Museum

Table 3 shows the number and the precision of the extracted triples for each score interval of length 2. As we can see, although most (nearly 69%) extractions fall into the lowest score interval $(0, -1)$, the extractions falling in a higher score interval will get higher precision. Thus, our method can get high-precision triples which can better meet the requirements of most downstream tasks.

Table 3. The precision and yield of extractions in different score interval.

Score interval	(14,12]	(12,10]	(10,8]	(8,6]	(6,4]	(4,2]	(2,0]	(0,−1)
No. of triples	1244	799	565	1765	3714	5453	8296	48272
Precision	0.95	0.89	0.87	0.85	0.70	0.53	0.35	0.21
Yield	1182	712	492	1501	2599	2890	2904	10137

Further, we plot the precision-yield curve for extractions with scores greater than 4 in Fig. 3. We can see that when the threshold is greater than 4, the accuracy of our triples reaches more than 70%, achieving excellent results.

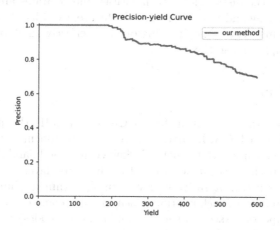

Fig. 3. The precision-yield curve for extractions with scores greater than 4.

Through the above analysis, we ascertain that setting 4 as the threshold can effectively extract high-precision (77%) extractions. Then, we extract triples from 1218 documents using our method with 3 and 4 as the threshold and compare with baseline. As we can see from Table 4, when the threshold is 4, the precision of and the yield of our extractions are 77% and 5949 respectively, while these of extractions from DSNFs [8] only are 58% and 5459, and from UnCORE [11] are 41.2% and 841. In summary, experimental results on travel corpus demonstrate the superiority of our method over the Chinese open IE models, both in terms of precision and yield.

Table 4. Extractions from 1218 documents.

Method	No. of triples	Precision	Yield
Our method (threshold = 4)	7726	**77%**	5949
Our method (threshold = 3)	**10154**	70%	**7107**
DSNFs [8]	9292	58%	5459
UnCORE [11]	2038	41.2%	841

4.4 Error Analysis

In the paper, we focus on extract the triples composed with two entities and a relation. However, the method of triples generation relies on our rules which have limitations. We find that most of the errors are from triples generation. They can be summarized as follow: 1) Entity coverage is incomplete, and some important entities are missing. 2) Entity recognition error. Some ambiguous words will make mistakes in word segmentation. 3) Relation words recognition error. In the part of word segmentation and POS tagging, *jiagu* will have recognition errors, such as recognizing some numbers or nouns as verbs. 4) Long-distance triples match incorrectly. When the distances between entities and relations are far away, the triples will be missing and mismatched. In the future, we will optimize our model such as introducing dependency parsing or other effective methods to improve the quality of triples generation.

5   Conclusion

In order to address the issues in the field of Chinese open IE, this paper proposes an unsupervised model: GAM. Firstly, the triples in the documents are obtained through predefined rules with the help of a knowledge graph, and then the triples are used as the basis to construct a graph with a type layer, entity layer, and relation layer. We use an iterative propagation algorithm to obtain the domain and range distribution of each relation which can also complete the schema, and we obtain the importance scores of entities and relations nodes of each document. Finally, we design a formula to combine the above scores to acquire the score of each triple, which measures the accuracy and importance of it. Our model can achieve high accuracy of triples, and the obtained triples have been verified more important to the documents, which can better enrich the knowledge graph.

Acknowledgements. This paper is supported by the Key Technology Develop and Research Project (SGZJDK00KFJS1900149) in STATE GRID Corporation of China.

References

1. Jurafsky, D., Martin, J.H.: Speech and Language Processing, 2Nd edn. Prentice-Hall Inc, Upper Saddle River, NJ, USA (2009)

2. Banko, M., Cafarella, M.J., Soderland, S., et al.: Open information extraction from the web. In: IJCAI 2007, Proceedings of the 20th International Joint Conference on Artificial Intelligence, Hyderabad, India, 6–12 January 2007. Morgan Kaufmann Publishers Inc. (2007)

3. Fader, A., Soderland, S., Etzioni, O.: Identifying relations for open information extraction. In: Proceedings of the conference on empirical methods in natural language processing, pp. 1535–1545. Association for Computational Linguistics (2011)

4. Schmitz, M., Bart, R., Soderland, S., et al.: Open language learning for information extraction. In: Proceedings of the 2012 Joint Conference on Empirical Methods in Natural Language Processing and Computational Natural Language Learning, pp. 523–534. Association for Computational Linguistics (2012)

5. Del Corro, L., Gemulla, R.: ClausIE: clause-based open information extraction (2013)

6. Tseng, Y.H., Lee, L.H., Lin, S.Y., et al.: Chinese open relation extraction for knowledge acquisition. In: Proceedings of the 14th Conference of the European Chapter of the Association for Computational Linguistics, volume 2: Short Papers, pp. 12–16 (2014)

7. Qiu, L., Zhang, Y.: Zore: a syntax-based system for Chinese open relation extraction. In: Proceedings of the. Conference on Empirical Methods in Natural Language Processing (EMNLP), vol. 2014, pp. 1870–1880 (2014)

8. Jia, S., E, S., Li, M., et al.: Chinese open relation extraction and knowledge base establishment. ACM Trans. Asian Low-Resource Lang. Inf. Process. (TALLIP), **17**(3), 1–22 (2018)

9. Jin, H., Li, C., Zhang, J., et al.: XLORE2: large-scale cross-lingual knowledge graph construction and application (2019)

10. Zhang, Z., Le, S., et al.: A joint model for entity set expansion and attribute extraction from web search queries (AAAI2016)

11. Qin, B., Liu, A., Liu, T.: Unsupervised Chinese open entity relation extraction. J. Comput. Res. Dev. **52**(5), 1029–1035 (2015)

12. Yao, L., Mao, C., Luo, Y.: Graph convolutional networks for text classification. In: Proceedings of the AAAI Conference on Artificial Intelligence, vol. 33, pp. 7370–7377 (2019)

IBRE: An Incremental Bootstrapping Approach for Chinese Appointment and Dismissal Relation Extraction

Lu Mao, Meiling Wang, Changliang Li[✉], Junfeng Fan, and Changyu Hou

AI Lab, KingSoft Corp, Beijing, China
{maolu,wangmeiling1,lichangliang,fanjunfeng,houchangyu}@kingsoft.com

Abstract. In the field of government affairs, Appointment and Dismissal Relation Extraction (ADRE) of officials from personnel news is crucial for updating government knowledge. However, ADRE faces great challenges, including extremely high accuracy demand, lack of data, tuple sparsity and conflict, and requiring incremental update. To address these challenges, we propose an Incremental Bootstrapping approach for Relation Extraction (IBRE) and apply it to real-time updating of personnel knowledge in government-affair Knowledge Graphs. IBRE starts with few seeds and trains with pattern generation, pattern evaluation, tuple prediction and seed augmentation in an iterative and incremental manner. First, we design a new strategy for defining seeds as document-tuple pairs to eliminate the effects of tuple sparsity and conflict. Then, a new definition of patterns with both word and part-of-speech is proposed to get high accuracy. Finally, we augment seeds with corrected tuples and apply incremental learning to continually improve performance with least training cost. We build a dataset called ADNP (Appointment and Dismissal News from People.cn) and compare our approach with baselines. Comparison results show that our approach performs the best. Moreover, experimental results demonstrate that incremental learning continuously improves the performance of IBRE.

Keywords: Relation extraction · Bootstrapping · Knowledge graphs · Incremental learning

1 Introduction

Relation Extraction (RE), which extracts semantic relationships from texts, is of great significance to various natural language processing applications such as knowledge base, question answering and summarization. Extracted relations usually occur between two entities of certain types and fall into categories [9]. Figure 1 displays examples of appointment and dismissal relations between person entities (PER) and position entities (POST). Successful RE requires detecting argument entities with their entity types such as PER and POST, and determining relation types such as appointment, dismissal and NA (i.e., nonexistence).

© Springer Nature Singapore Pte Ltd. 2021
H. Chen et al. (Eds.): CCKS 2020, CCIS 1356, pp. 132–144, 2021.
https://doi.org/10.1007/978-981-16-1964-9_11

C1: 任命姜万荣为住房和城乡建设部副部长。
C2: 免去张汉晖的外交部副部长职务。

E1: Appoint Jiang Wanrong as Vice Minister of Housing and Urban-Rural Development.
E2: Dismiss Zhang Hanhui from the post of Vice Minister of Foreign Affairs.

Appointment Tuple: (姜万荣, 住房和城乡建设部副部长)
Dismissal Tuple: (张汉晖, 外交部副部长)

Fig. 1. Examples of appointment and dismissal relation extraction.

Table 1. Occurrence statistics of tuples in 281 items of personnel news.

Type	Appointment tuples			Dismissal tuples		
Occurrences	$x = 1$	$1 < x <= 5$	$x > 5$	$x = 1$	$1 < x <= 5$	$x > 5$
Num	970	126	3	651	11	0
Percentage (%)	88.26	11.46	0.27	98.34	1.66	0

RE plays an essential role in constructing, enriching and updating Knowledge Graphs (KGs) [14]. In the field of government affairs, it is valuable to keep personnel knowledge of government officials updating in real time. For example, an official is appointed as or dismissed from a certain position according to official news, which is periodically published by official websites such as People.cn[1] and Xinhuanet.com[2]. Therefore, Appointment and Dismissal Relation Extraction (ADRE) from these personnel news to update personnel knowledge in government-affair KGs is a key task. However, there are four challenges in exploring a high-quality Chinese ADRE approach:

Extremely High Accuracy Demand. Government-affair area requires extremely high even 100% accuracy. As a result, human intervention is indispensable. Therefore, the higher accuracy ADRE algorithm can reach, the lower manual correction will cost.

Lack of Data. The official websites such as People.cn and Xinhuanet.com only display about 300 news items, as a result, the training data are limited. Therefore, it is unpractical to get large amount of unlabeled data, let alone sufficient labeled data, which is required by various supervised approaches.

Tuple Sparsity and Conflict. Bootstrapping methods are usually applied when lack of labeled data. As an outstanding bootstrapping method, Snowball usually uses multi-occurrence and consistent tuples as seeds to generate extraction patterns (see details in Fig. 2) [1]. However, as shown in Table 1, most tuples appear only once through all the personnel news documents. Besides, since the news updates continually, an appointment tuple in one document can be a dismissal tuple in another document, thus causing conflict.

---

[1] renshi.people.com.cn.
[2] www.xinhuanet.com/politics/rs.htm.

Incremental Update. As discussed before, the official websites periodically publish personnel news, thus training data can be gradually accumulated. Then, how to apply incremental data to improve the performance of ADRE approach is significant and challenging.

To address these challenges, we propose an Incremental Bootstrapping Relation Extraction approach (IBRE) and apply it to real-time updating of personnel knowledge in government-affair KGs. IBRE starts with a small amount of labeled data as initial seeds and it trains RE by pattern generation, pattern evaluation, tuple prediction and seed augmentation in an iterative and incremental manner. Firstly, we design a new strategy for defining seeds as document-tuple pairs, which eliminates the effects of tuple sparsity and conflict because seed tuples are only valid for the seed document where they are labeled. Then, in order to reach high accuracy, we define pairs of word patterns and part-of-speech (POS) patterns, which are used for word-level pattern matching and syntax-level pattern matching in pattern evaluation and tuple prediction. Finally, we augment seeds with corrected tuples and apply incremental learning to constantly improve RE performance with least training cost. We build a dataset called ADNP (Appointment and Dismissal News from People.cn) and compare IBRE with rule-based matching method and Snowball with respect to ADRE performance. Experimental results demonstrate that IBRE performs the best. We also conduct ablation study and experiments on incremental learning to further prove the effectiveness of our approach. Experimental results demonstrate that incremental learning continuously improves the performance of IBRE. Our main contributions include:

(1) We propose a new bootstrapping framework for scarce and incremental updated data, and the framework is applied to real-time updating of personnel knowledge in government-affair KGs, which is the first to our best knowledge.
(2) We propose a new seed definition of document-tuple pairs to eliminate the effects of tuple sparsity and tuple conflict in data.
(3) We adopt both word patterns and POS patterns for pattern matching in order to reach high accuracy.
(4) We apply incremental training to continually improve RE performance.

2   Related Work

Recent studies focus on extracting relations with neural networks by supervised approaches [10,11,16]. However, supervised approaches require large amount of training data, thus facing the usual difficulty of lacking sufficient labeled data. This limitation also makes supervised approaches hard to extend, because detecting new relation types requires new training data [7,8,15].

Rule-based methods, distant supervision and bootstrapping methods can be applied for RE when lack of labeled data. Rule-based matching methods require experts with professional skills to build a large-scale knowledge base and make rules manually, which is time-consuming, laborious and inefficient [4]. Distant supervision is an alternative approach and its main idea is to apply a large

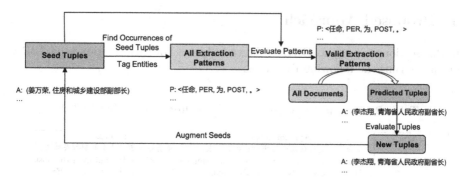

Fig. 2. Framework of Snowball.

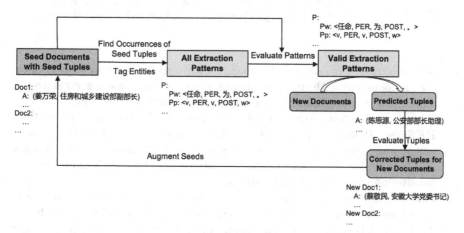

Fig. 3. Framework of IBRE.

knowledge base for automatically obtaining relation labels and then train with these assumed labeled data [6]. However, mislabeling problems should be solved properly [3,9,13].

Bootstrapping methods are usually applied when labeled data are scarce in practice, and they typically use a small dataset to learn how to extract relations in an iterative manner [2]. As a famous one of such bootstrapping methods, Snowball can achieve better and more stable results over multiple iterations [1]. As show in Fig. 2, starting with a few seed tuples, Snowball finds 5-tuple pattern of entity tags and left, middle, right contexts around the entities, and represents the contexts using word vectors. Then, after pattern evaluation and tuple evaluation, Snowball retrains reliable tuples to the next iteration. However, Snowball relies on multi-occurrence and consistent tuples through all documents, thus is not suitable for ADRE. In addition, Snowball applies similarity match on word vectors to evaluate patterns and tuples, which makes precision tend to be low.

3   Proposed Approach

In this section we present IBRE as shown in Fig. 3. For each relation type, IBRE primarily runs the following five steps:

Table 2. Examples of patterns generating from Snowball and IBRE.

Type	Appointment	Dismissal
Sentence	任命姜万荣为住房和城市建设部副部长。 Appoint Jiang Wangrong as Vice Minister of Housing and Urban-Rural Development.	免去张汉晖的外交部副部长职务。 Dismiss Zhang Hanhui from the post of Vice Minister of Foreign Affairs.
Relation Tuple	(姜万荣, 住房和城市建设部副部长)	(张汉晖, 外交部副部长)
Snowball Pattern	< 任命, PER, 为, POST, 。 >	< 免去, PER, 的, POST, 职务 >
IBRE Pattern	Word:< 任命, PER, 为, POST, 。> POS:<v, PER, v, POST, w>	Word:< 免去, PER, 的, POST, 职务 > POS:<v, PER, u, POST, n>

(1) Seed selection. Choose seed documents with seed tuples from labeled data, e.g., a document with an appointment tuple (姜万荣, 住房和城市建设部副部长). This strategy can eliminate effects of tuple sparsity and conflict.

(2) Pattern generation. Automatically generate word patterns (e.g., <任命, PER, 为, $POST$, 。>) and POS patterns (e.g., $<v, PER, v, POST, w>$, where v denotes verb and w denotes punctuation) from seed documents and seed tuples, which contributes to pattern matching with more comprehensive information and high precision.

(3) Pattern evaluation. Evaluate patterns by pattern matching with seed documents and keep only ones that are regarded as being sufficiently reliable.

(4) Tuple prediction. For new documents, predict tuples by pattern matching with the valid patterns, e.g., a predicted tuple (陈思源, 公安部部长助理).

(5) Seed augmentation. Evaluate the predicted tuples, and then add the corrected tuples and their corresponding documents into seeds, e.g., a new document with a corrected tuple (蔡敬民, 安徽大学党委书记), which gradually increases recall of RE.

3.1   Seed Selection

We first define a tuple for a given relation between entity type t_{e_1} and t_{e_2} as (e_1, e_2), where e_1 is the head entity with named entity tag t_{e_1}, and e_2 is the tail entity with named entity tag t_{e_2}. Further, we define an IBRE seed as a document-tuple pair in an innovative way. In detail, for each seed document d_S, all the labeled tuples of the relation type contained by d_S are defined as seed tuples $T_S = \{(e_1, e_2)\}$. Thus, IBRE does not require tuples to occur multiple times through all the documents. In this way, the conflict of seed tuples in different documents can also be ignored.

3.2   Pattern Generation

In this section, we present the process of generating all the patterns from seeds for a relation type. By applying Chinese NLP tools, IBRE marks texts with POS tags and entity tags.

For each seed tuple (e_1, e_2) of a given relation type, IBRE first finds sentences in the corresponding seed document where entity e_1 and entity e_2 occur close to each other. Then analyze the contexts of e_1 and e_2 to generate Snowball patterns like $<l, t_{e_1}, m, t_{e_2}, r>$, where l, m, r denotes left, middle, right contexts around the entities e_1 and e_2, and l and r are limited to x words to the left and the right of the entity pair. We extend the definition of Snowball pattern by adding POS pattern. In detail, we define an IBRE pattern as a pair of (p_w, p_p), where p_w is word pattern denoted as $<l_w, t_{e_1}, m_w, t_{e_2}, r_w>$ and p_p is POS pattern denoted as $<l_p, t_{e_1}, m_p, t_{e_2}, r_p>$. The whole process described above is defined as *CreateOccurrence*. As shown in Table 2, for a sentence '任命姜万荣为住房和城市建设部副部长。' with an appointment tuple (姜万荣, 住房和城市建设部副部长), the word pattern is extracted as $<$任命, $PER,$ 为, $POST,$ 。$>$, and the POS pattern is $<v, PER, v, POST, w>$.

3.3   Pattern Evaluation

We calculate confidence of IBRE patterns and discard the noisy patterns that have low confidence and tend to generate wrong tuples. For each seed document of a given relation type between entity type t_{e_1} and t_{e_2}, the evaluation process is described as follows:

(1) Firstly, identify sentences that include t_{e_1} and t_{e_2}, and scan each sentence by *CreateOccurrence* function. In detail, for each sentence, generate $s_p = <l'_p, t_{e_1}, m'_p, t_{e_2}, r'_p>$ and $s_w = <l'_w, t_{e_1}, m'_w, t_{e_2}, r'_w>$ as pattern generation in Sect. 3.2. If there exists a pair of (p_w, p_p) where p_w exactly matches s_w and p_p exactly matches s_p, then the tuple (e_1, e_2) is extracted.
(2) If an extracted tuple is included in seed tuples, then the tuple is considered as a positive match for the pattern, and otherwise as a negative match. The confidence of a pattern is defined as the percentage of the positive matches.
(3) Valid patterns are chosen according to pattern confidence threshold (τ_p). Note that, different from Snowball, IBRE does not discard patterns with low coverage because they will not bring additional noise to the results.

3.4   Tuple Prediction

For a new document d_N, IBRE scans d_N to predict tuples for a given relation between entity type t_{e_1} and t_{e_2} by matching with its valid patterns, and the basic algorithm is outlined in Algorithm 1. In detail, for each sentence containing t_{e_1} and t_{e_2}, $s_p = <l'_p, t_{e_1}, m'_p, t_{e_2}, r'_p>$ and $s_w = <l'_w, t_{e_1}, m'_w, t_{e_2}, r'_w>$ can be extracted by *CreateOccurrence*, and (e_1, e_2) is generated as a tuple of the relation if there is a valid pattern such that s_p matches p_p and s_w is similar enough to p_w. Note that we adopt two word similarity matching methods in experiments, see details in Sect. 4.2.

Algorithm 1. PredictTuples

Input: *valid_patterns, new document d_N*
Output: *predicted_tuples*
1: *predicted_tuples* = []
2: **for** *sentence* in d_N **do**
3:     $(e_1, e_2), s_p, s_w = CreateOccurrence(sentence)$
4:     $T_C = (e_1, e_2)$
5:     **for** (p_p, p_w) in valid_patterns **do**
6:         **if** $isMatch(s_p, p_p)$ && $isSimilar(s_w, p_w)$ **then**
7:             *predicted_tuples.append(T_c)*
8:         **end if**
9:     **end for**
10: **end for**

Algorithm 2. Algorithm for incremental learning.

Input: $d_{N1}, d_{N2}...d_{N\infty}, valid\_patterns$
Output: *valid_patterns*
1: **for** i in $1...\infty$ **do**
2:     *predicted_tuples = PredictTuples(valid_patterns)*
3:     $T_{COS} = EvaluateTuples(predicted\_tuples)$
4:     **if** $exists(T_{COS})$ **then**
5:         *new_patterns = GeneratePatterns(d_{Ni}, T_{COS})*
6:         *new_patterns_unseen = FilterPatterns(new_patterns, valid_patterns)*
7:         *new_patterns_valid = EvaluatePatterns(new_patterns_unseen)*
8:         *valid_patterns.extend(new_patterns_valid)*
9:     **end if**
10: **end for**

3.5 Seed Augmentation

For tuple prediction of a given relation, if mispredicted tuples or missing tuples are found by human evaluation, then the corrected tuples and their corresponding documents are added into seeds, hence IBRE patterns are constantly accumulated and IBRE operates in an incremental manner to improve performance. The incremental learning process is outlined in Algorithm 2:

(1) For each new document d_{Ni}, predict tuples with Algorithm 1 in Sect. 3.4, and the predicted tuples are evaluated manually.
(2) Tuples which are mispredicted or missed by IBRE are corrected to T_{COS}. Then T_{COS} and d_{Ni} are used to generate patterns as Sect. 3.2.
(3) If the generated patterns are unseen in existing valid patterns, then they are added into valid patterns after evaluation as Sect. 3.3.

Therefore, the performance of IBRE is improved with least cost of training new seeds and new documents, instead of training all the data from scratch.

Table 3. The distribution of documents cross tuples in ADNP dataset.

Type	Appointment tuples			Dismissal tuples		
Number	$x <= 1$	$1 < x <= 10$	$x > 10$	$x <= 1$	$1 < x <= 10$	$x > 10$
Train	81	97	22	134	60	6
Test	28	48	5	59	17	5
Total	109	145	27	193	77	11

Table 4. The distribution of tuples cross occurrences in ADNP dataset.

Type	Appointment tuples			Dismissal tuples		
Number	$x = 1$	$1 < x <= 5$	$x > 5$	$x = 1$	$1 < x <= 5$	$x > 5$
Train	755	96	2	492	5	0
Test	215	30	1	159	6	0
Total	970	126	3	651	11	0

4 Experiment

In this section, we first describe the dataset that we build for experiments in Sect. 4.1. Then, we conduct experimental evaluation to compare our approach with baselines in Sect. 4.2. We also conduct ablation study in Sect. 4.3. Finally, we evaluate the effects of incremental learning in Sect. 4.4.

4.1 Dataset

We build a new dataset called ADNP (Appointment and Dismissal News from People.cn). Daily updating Personnel Channel at People.cn displays about 300 latest news items of appointment and dismissal information. Our ADNP dataset contains 281 documents published between August and November, 2019, and we label these documents with the relation tuples of appointment and dismissal $(PER, POST)$. Since people always use historical experience to predict future, we arrange the documents in order of their publish time from oldest to latest, and specify the first 200 as the training set (historical data) and the rest as the test set (future data). The training set contains 1008 appointment tuples and 502 dismissal tuples, while the test set contains 295 appointment tuples and 172 dismissal tuples. In addition, the number of tuples contained by different documents also varies. As shown in Table 3, more than 50% documents contain 1 to 10 appointment tuples, while around 70% documents contain no more than 1 dismissal tuple. The occurrence frequency of appointment tuples and dismissal tuples are shown in Table 4. The results show that about 90% appointment tuples and 99% dismissal tuples occur only once through all documents.

Table 5. ADRE performance of IBRE and baselines in ADNP dataset.

Type	Appointment			Dismissal		
Approach	P(%)	R(%)	F1(%)	P(%)	R(%)	F1(%)
Rule-based	55.63	60.34	57.89	62.50	43.60	51.37
Snowball	27.86	53.06	36.53	14.06	41.28	20.97
IBRE I	87.70	74.92	80.80	86.78	61.05	71.67
IBRE II	**89.96**	**75.93**	**82.35**	**86.99**	**62.21**	**72.54**

4.2　Approach Comparison

Baselines

(1) Rule-based matching method. We implement a basic rule-based matching method, where $(PER, POST)$ tuples co-occurring with keywords such as '任命' (appoint), '任' (be appointed), '当选' (be elected) are regarded as tuples of appointment relation, and $(PER, POST)$ tuples co-occurring with keywords such as '免去' (dismiss), '不再担任'(no longer serve as), '此前担任' (previously serve as) are regarded as tuples of dismissal relation.

(2) Snowball. We re-implement Snowball and apply it to our data. As shown in Fig. 2, Snowball starts with a few seed tuples of $(PER, POST)$ to learn extracting appointment and dismissal relations and then use the extracted tuples for training in an iterative manner.

Experiment Settings. For Snowball, we use the same operational parameters as [1] except for the minimum tuple confidence, and we set the minimum tuple confidence as 0.9 in order to get higher precision.

For IBRE, we set pattern confidence threshold (τ_p) as 0.5 in order to get higher recall. In addition, We apply two word similarity matching methods, where the first applies word vectors with similarity greater than 0.6 (I), and the second applies word exact matching ignoring all function words and punctuation marks (II).

For both Snowball and IBRE, left and right contexts are set as one word, and initial seeds are generated from first 5 training documents with 46 appointment tuples and 16 dismissal tuples. Note that, Snowball seeds are only tuples, while IBRE seeds contain both tuples and documents. For all the approaches, Bi-GRU-CRF network [5] is applied to get POS tags and entity tags, and a pre-trained Chinese Word2vec [12] is applied in Snowball and IBRE when necessary.

Experimental Results. We use precision (P), recall (R) and F1-score (F1) of predicted tuples in test set to evaluate ADRE performance. Table 5 displays the evaluation results of ADRE in ADNP dataset, and the experimental results demonstrate that for both appointment and dismissal relations IBRE II performs the best among all the approaches:

Table 6. ADRE results of ablation study.

Type	Appointment			Dismissal		
Approach	P(%)	R(%)	F1(%)	P(%)	R(%)	F1(%)
IBRE	89.96	75.93	**82.35**	86.99	62.21	**72.54**
IBRE w/o pattern evaluation	76.45	75.93	76.19	70.86	62.21	66.25
IBRE w/o POS pattern	**92.69**	68.81	78.99	**87.93**	59.30	70.83
IBRE w/o word pattern	82.31	**77.29**	79.72	38.99	**62.79**	48.11

(1) IBRE I and IBRE II outperform rule-based matching method, because rule-based method only works on rules seen before and can not automatically learn new rules as data grow.
(2) IBRE I and IBRE II outperform Snowball, because pattern matching of Snowball ignores syntactic structure which leads to low precision, and pattern clustering of Snowball results in less patterns which leads to low recall.
(3) IBRE II performs better than IBRE I, which indicates that exact match of notional words such as nouns and verbs is useful and match of function words and punctuation marks is unnecessary.
(4) Among all approaches, the performance of appointment RE is always better than that of dismissal RE, probably due to obviously less dismissal tuples and patterns than appointment through ADNP dataset.

4.3   Ablation Study

Pattern evaluation and pattern matching are the key parts to guarantee the performance of IBRE. In this section, we conduct ablation study to evaluate whether and how each part contributes to IBRE.

(1) IBRE w/o pattern evaluation. Use all generated patterns instead of valid patterns to predict tuples.
(2) IBRE w/o POS pattern. Use only word patterns to predict tuples, exactly match all the words regardless of whether they are function words, punctuation marks or not.
(3) IBRE w/o word pattern. Use only POS patterns to predict tuples.

Table 6 lists ADRE results of ablation study. We find that:

(1) Pattern evaluation is indispensable because inaccurate patterns bring noise to predicted tuples, leading to low precision. Also, noisy patterns do not improve the recall.
(2) Only word-pattern matching reaches the highest precision but the recall is extremely low because of its bad generalization.
(3) Only POS-pattern matching results in the highest recall, which is slightly higher than full IBRE. However, the precision is quite low especially for dismissal relations.

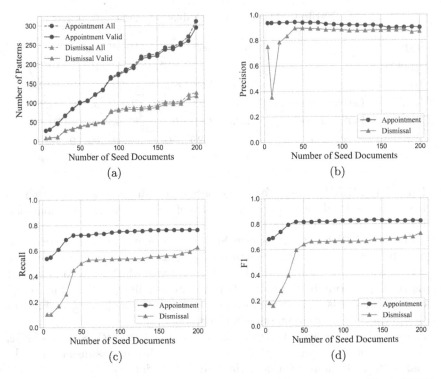

Fig. 4. Pattern number (a) and Precision (b), Recall (c) and F1-score (d) of ADRE as seed documents increase.

(4) IBRE works as a trade-off, which sacrifices a small part of precision to get quite higher recall. Thus it can result in the highest F1, especially in the situations of small dataset.

4.4   Effects of Incremental Learning

In order to explore the effectiveness of incremental learning on the performance of IBRE, we record the evaluation results of IBRE in test set when seeds are accumulated. In detail, we first arrange the documents of the training set in order of publish time from oldest to latest. Then we set the first 5 documents with 46 appointment tuples and 16 dismissal tuples as initial seeds. Finally, we run IBRE in order on the rest of documents, thus seeds will be accumulated over time.

.   Figure 4 displays the trends of pattern number and P, R, F1 of ADRE when seed documents increase:

(1) As shown in Fig. 4(a), when seed documents increase to 200, valid patterns for appointment increase to 294 (94.8% of all appointment patterns), and valid patterns for dismissal increase to 116 (92.8% of all dismissal patterns).

(2) Figure 4(b) reveals that the precision of appointment RE remains a relatively high precision with the growth of seed documents; for dismissal RE, the precision is fluctuate at the beginning mainly because the number of seed pairs for dismissal is too small.

(3) Figure 4(c) shows that there is a gradual increase in the recall of both appointment and dismissal RE.

(4) Figure 4(d) shows that F1 of appointment RE remains relatively high (about 80%) after 30 seed documents, while the dismissal RE reaches a relatively high F1 (about 66%) after 60 seed documents and still keep growing.

5   Conclusion

In this paper, we present a new bootstrapping approach IBRE to address the great challenges faced in ADRE. We first design a novel strategy for defining seeds to eliminate the effects of tuple sparsity and conflict. Then, we define pairs of word and POS patterns to get high accuracy. Finally, we augment seeds with corrected tuples and apply incremental learning to continually improve the performance with least training cost. We build ADNP dataset to compare IBRE with baselines, and experimental results show that IBRE performs the best. IBRE has been applied to real-time updating of personnel knowledge in government-affair KGs. In the future, we will explore IBRE on more relation extraction tasks in other fields.

References

1. Agichtein, E., Gravano, L.: Snowball: extracting relations from large plain-text collections. Proc. ACM DL **2000**, 85–94 (2000)
2. Brin, S.: Extracting patterns and relations from the world wide web. In: Atzeni, P., Mendelzon, A., Mecca, G. (eds.) WebDB 1998. LNCS, vol. 1590, pp. 172–183. Springer, Heidelberg (1999). https://doi.org/10.1007/10704656_11
3. Hoffmann, R., Zhang, C., Ling, X., Zettlemoyer, L., Weld, D.S.: Knowledge-based weak supervision for information extraction of overlapping relations. Proc. ACL-HLT **2011**, 541–550 (2011)
4. Jian, X., Zhixiong, Z., Zhenxin, W.: Review on techniques of entity relation extraction. Data Anal. Knowl. Disc. **24**(8), 18–23 (2008)
5. Jiao, Z., Sun, S., Sun, K.: Chinese lexical analysis with deep Bi-GRU-CRF network. arXiv pp. arXiv-1807 (2018)
6. Mintz, M., Bills, S., Snow, R., Jurafsky, D.: Distant supervision for relation extraction without labeled data. Proc. ACL-IJCNLP **2009**, 1003–1011 (2009)
7. Mooney, R.J., Bunescu, R.C.: Subsequence kernels for relation extraction. NIPS **2006**, 171–178 (2006)
8. Nguyen, T.H., Grishman, R.: Relation extraction: perspective from convolutional neural networks. Proc. NAACL-HLT **2015**, 39–48 (2015)
9. Pawar, S., Palshikar, G.K., Bhattacharyya, P.: Relation extraction: a survey. arXiv preprint arXiv:1712.05191 (2017)

10. dos Santos, C., Xiang, B., Zhou, B.: Classifying relations by ranking with convolutional neural networks. In: Proceedings of ACL-IJCNLP 2015, vol. 1, Long Papers, pp. 626–634 (2015)
11. Socher, R., Huval, B., Manning, C.D., Ng, A.Y.: Semantic compositionality through recursive matrix-vector spaces. Proc. EMNLP-CoNLL **2012**, 1201–1211 (2012)
12. Song, Y., Shi, S., Li, J., Zhang, H.: Directional skip-gram: explicitly distinguishing left and right context for word embeddings. In: Proceedings of NAACL HLT 2018, vol. 2 (Short Papers), pp. 175–180 (2018)
13. Yao, L., Haghighi, A., Riedel, S., McCallum, A.: Structured relation discovery using generative models. In: proceedings of EMNLP 2011, pp. 1456–1466 (2011)
14. Yoo, S., Jeong, O.: Automating the expansion of a knowledge graph. Expert Syst. Appl. **141**, 112965 (2020)
15. Zelenko, D., Aone, C., Richardella, A.: Kernel methods for relation extraction. J. Mach. Learn. Res. **3**, 1083–1106 (2003)
16. Zeng, D., Liu, K., Lai, S., Zhou, G., Zhao, J.: Relation classification via convolutional deep neural network. In: Proceedings of COLING 2014: Technical Papers, pp. 2335–2344 (2014)

Knowledge Extraction: Event

Hybrid Syntactic Graph Convolutional Networks for Chinese Event Detection

Xiaobo Ma[1,2], Yongbin Liu[1,2(✉)], and Chunping Ouyang[1,2]

[1] University of South China, Hunan, China
[2] Hunan Medical Big Data International Sci.&Tech. Innovation Cooperation Base, Hunan, China

Abstract. Event Detection (ED) is a task that aims to recognize triggers and identify the event type in sentences. Syntactic information plays a crucial role in event detection model accurately to recognize the triggers and event type. The previous works commonly use word embedding to obtain context representation that cannot fully exploit syntactic information. In this paper, we propose a novel model HSGCN (Hybrid Syntactic Graph Convolutional Networks) for Chinese event detection, which utilizes graph convolutional networks to generate sentence-level feature and exploit a task-specific hybrid representation considering both character-level feature and word-level feature. Our experiments demonstrate that HSGCN model can capture rich syntactic to improve identifying the triggers and event type. Compared with existing models, HSGCN can achieve efficient and accurate results on ACE 2005 and KBPEval2017 datasets. In trigger identification and type classification tasks, HSGCN respectively achieved 70.2% and 65.7% F-score with average 1.2% and 0.9% absolute improvement compare to state-of-the-art method.

Keywords: Event detection · Graph convolutional networks · Syntactic information · Hybrid representation

1 Introduction

The content of events that defined by ACE is consisted of event triggers and event arguments. Event Detection (ED) is a key step in event extraction, which aims to recognize triggers and identify the event type. For example, in the sentence "Clashes between the police and protesters killed at least five people", an event detection system will detect a death event triggered by "killed". Event trigger is a single token that represent the occurrence of an event. A token is not equivalent to a word, for example, the phrase "step down" is concatenated into a token "step-down" as the trigger of a "personnel" event. This design has been used in numerous studied on ED.

Chinese Event Detection is more complex than English Event Detection, because triggers are not always exactly match with a word. In many cases, a trigger can be the part of a word or cross multiple words. What's more, some

H. Chen et al. (Eds.): CCKS 2020, CCIS 1356, pp. 147–159, 2021.
https://doi.org/10.1007/978-981-16-1964-9_12

related words that we call cue words in the sentence can provide available information to assist the trigger classification. These cue words are the crucial factor to identify event type. However, sometimes an event may has more than one cue word, and different cue word is located in the different parts of sentence. Therefore, traditional word embedding is difficult to consider these cue words because they are scattered and far away from the triggers. To solve this problem, we use dependency tree to link cue words to the trigger words, which can connect a dependency word to a head word and provide their dependency relation.

Some previous works like [21,29] had employed syntactic dependency information to generate syntactic representation for event factuality prediction. Usually, these models generated a syntactic representation by one-hot encoding and concatenated syntactic representation to context representation in the final stage. Although integrating syntactic information into representations had been proved to be effective in ED task, using the simple combination would cause noise data appearing in syntactic and context representations, even lost some useful information. In order to address these problems, we introduce graph convolutional networks [13,17,20] to represent syntactic information flexibly, without requiring special encoding for dependency information as in previous works.

Compare with English, Chinese has a particularity that a word can be composed of multiple characters and these characters have their own meanings. For example, in Fig. 1, the word "遭袭" (be attacked) composed of the character "遭" (encounter) and the character "袭" (attack). In this sentence, the word "遭袭" (be attacked) which is a whole trigger would be divided into two characters in dependency parser. Hence combination of word embedding and character embedding could keep the semantic of Chinese vocabulary effectively.

联合国/ 南苏丹/ 特派团/ 在/ 阿科博/ 的/ 营地/ 遭/ 袭/ 。

UN mission camp in South Sudan in Akobo attacked.

Fig. 1. An example of attack event which the trigger word is divided into two characters after tokening.

In view of the above discussion, a model called HSGCN (Hybrid Syntactic Graph Convolutional Networks) is proposed for Chinese event detection task. We conduct experiments on ACE 2005 and KBPEval2017 event extraction datasets. The experiment results demonstrate HSGCN outperforms previous models. The contributions of this paper include: (i) an important feature for triggers identification, syntactic dependency, is introduced to our model, and exploiting graph convolutional networks which is well-suited for handling dependency tree to represent syntactic dependency, (ii) considering the semantic characteristics of Chinese vocabulary, combining word-level feature and character-level feature could keep the original semantic of Chinese word to improve the triggers identification, and (iii) provides state-of-the-art performance on ACE 2005 and KBPE-val2017 datasets in Chinese event detection.

2 Related Work

For event detection task, traditional classification models are mainly divided into two categories: feature engineering based model and machine learning based model. [4] presented a classification-based model towards single-slot and multi-slot information extraction which got a better performance than pattern-learning approaches. [1] employed maximum entropy to complete triggers detection and designed TiMBL classifier to distinguish event types. In follow-up studies, some researchers captured document-level features to improve ED model. [10] first exploited cross-document feature to break down the document boundaries. [9] used information about different types of events to make predictions, and designed a document-level prediction model to achieve effectively event extraction in document-level. Feature-based methods can realize a performance gain by enrich different forms of features. However, feature-based methods have a problem of heavily dependent on man-made, which is time consuming and error prone. To solve these problems, [25] learned features automatically by using machine learning model to combine linguistic and structural information. [12] used conditional random fields with carefully designed features for the task of identifying definition sentences.

As machine learning technology continues to improve, recently, researchers pay more attention on deep learning methods to accomplish event detection task. [11,19,24] introduced the non-consecutive convolution to traditional CNN models and achieved a great performance for event detection. [26] proposed a dynamic multi-pool convolutional neural network to encode sentence semantic clues on event extraction task. [22] enhance a recurrent neural network with dependency bridges to capture syntactic information for event extraction. [15] proposed three hybrid representation methods to combine character-level and word-level feature. [28] proposed a joint model to extract event mentions and entities by capture structural information. These models brought some valuable solutions to ED task. However, most of them were designed for English event detection task, which cannot be used directly in Chinese event detection task. Furthermore, the existing models take no account of complicated compositional structure and polysemy in Chinese, which resulted in performance degradation in Chinese event detection task. Inspired by the previous work [5,6,16,26], we design a hybrid syntactic graph convolutional networks (HSGCN) which represent syntactic dependency in sentence level based on GCN to improve the triggers identification. In additional, our method keeps as much as possible the original semantic of Chinese vocabulary by combining word-level feature and character-level feature, thus achieves better performance in Chinese event detection task.

3   HSGCN

Event detection generally can be regard as a multiclass classification problem. The target of event detection is recognizing trigger candidate and identifying the event type of trigger candidate. In Chinese event detection, a word can to be a trigger candidate or form a part of trigger candidate. Therefore, we use characters as basic units in our model to match the requirement of Chinese processing. It means that the classifier decide each character in the sentence whether forms a part of triggers. If so, the classifier would identify the position of the character in the triggers.

Our framework which is depicted in Fig. 2 consists of the following four components: (a) sentence encoding that represents the sentence in both word-level and character-level; (b) sentence-level feature extraction based on syntactic graph convolutional networks and lexical feature extraction; (c) hybrid representation learning by combining word embedding and character embedding; (d) trigger generator and event type classifier.

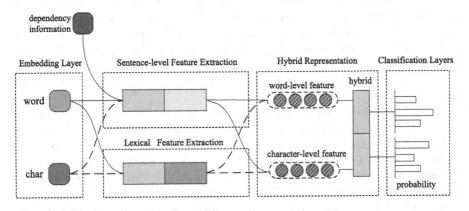

Fig. 2. The overall architecture of our model. The details of this architecture are depicted in Fig. 3, Fig. 4 and Fig. 6.

3.1   Sentence Encoding

Normally, the first step in event detection is to convert each word in the sentences into an embedding vector. In our work, we generate the final representation combining character-level representation and word-level representation. Therefore, the model encode sentences both in character-level and word-level.

We use the Skip-gram model [18] to obtain the character embedding and word embedding. Skip-gram model has been widely applied to many NLP tasks and achieve great performance. In particular, the sentence $S = (x_1, x_2, ..., x_n)$ would be fed in to Skip-gram model and trains the embedding of words or characters $x_1, x_2, ..., x_n$ by maximizing the average log probability,

$$\frac{1}{n}\sum_{t=1}^{n}\sum_{-w\le i\le w}\log p(x_{t+i}\mid w_t) \tag{1}$$

where w is the size of training window.

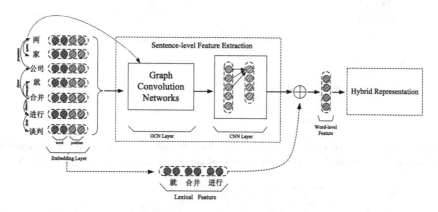

Fig. 3. The architecture that incorporating syntactic information in word-level feature In this example, the input is a sentence: 两家公司就合并进行谈判 (The two companies are negotiating a merger) and current word is "合并" (merger). ⊕ indicates a concatenation operation.

It should be noticed that the both character-level and word-level representation in our model is a lexical feature representation like [2] which can capture more rich semantics information than traditional word feature. Specifically, in Fig. 3, the current word is "合并" (merger) and its context tokens are "就" (as for) and "进行" (proceed). After generating three words embedding separately, our model concatenated the three words embedding into a whole word-level lexical feature to represent the word "合并". Similar to the case of word-level representation, in Fig. 4, the model also concatenated the character embedding of current character "并" and its context tokens into a whole character-level lexical feature.

3.2 Incorporating Syntactic Information to Sentence-Level Features

According to previous word [26], sentence-level clues can contribute significantly to event extraction. Concretely, in Fig. 5, a syntactic arc (along the *conj-arc* from conquered to attempted, along the *xcomp-arc* from attempted to invade) connect a trigger conquered to another trigger invade. From the trigger *conquered*, we obtain a clue to identify the event type about the trigger *invade*. Because of the clue word is far away from the candidate word, traditional method can't encode this dispersive context information. So the method considering sentence-level feature can effectively solve the above problem by learning the compositional semantic features of sentences. Taking full advantage of sentence-level feature,

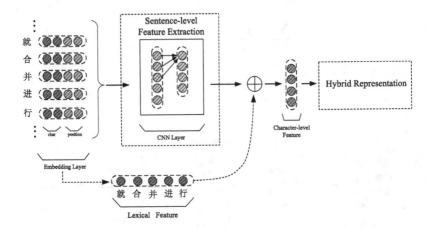

Fig. 4. The architecture to obtain character-level feature. $\oplus$ indicates a concatenation operation. The detailed architecture of hybrid representation is described in Fig. 6.

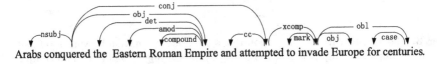

Fig. 5. The result of syntactic parser. The sentence contains two attack events respectively triggered by "conquered" and "invade".

we propose a novel feature extraction architecture to incorporate dependency information by using graph convolutional networks.

Let an directed graph $G = (V, E)$ be the dependency parsing tree for sentence $S = (x_1, x_2, ..., x_n)$, where V is the set of nodes ($|V| = n$) and E is the sets of edges. Each v_i in V indicates a token x_i. $(v_i, v_j, l_{ij}) \in E$ indicates the edge from node v_i to v_j with label l_{ij}. Due to the information of directed graph does not necessarily travel in its direction [13], we also add inverse edge (v_j, v_i, l_{ji}^{-1}) and all self-loops, i.e., (v_i, v_i). For the $k - th$ graph convolution network layer, model would calculate the vector $h_{v_j}^{(k+1)}$ of node v_j by:

$$h_{v_j}^{(k+1)} = f(\sum_{v_i \in N(v_j)} (W_{l_{ij}}^k h_{v_i}^k + b_{l_{ij}}^k)) \tag{2}$$

where $W_{l_{ij}}^k \in R^{d \times d}$ the weight matrix and $b_{lij}^k \in R^k$ is the bias for label. $N(v_j)$ indicates the set of neighbors of node v_j including v_j itself. In this work, we use Stanford CoreNLP toolkit [2] to generate the arcs in dependency parsing tree for sentences. It should be noticed that the first layer of graph convolutional networks is initialized by the output of embedding layer. As some of edges might be erroneous or irrelevant for the downstream task [23], we employ gates [17] on the edges to calculate their individual importance. For each edge (v_i, v_j), we calculate a score $g_{l_{ij}}^k$ indicating the importance for event detection by:

$$g_{l_{ij}}^k = \sigma(V_{l_{ij}}^k h_{v_i}^k + B_{ij}^k) \qquad (3)$$

where $V_{l_{ij}}$ and B_{ij}^k are the weight matrix and bias of the gate, respectively, σ is the sigmoid function. Based the gating mechanism, the updated GCN propagation rulecan be formulated as

$$h_{v_j}^{(k+1)} = f(\sum_{v_i \in N(v_j)} g_{l_{ij}}^k \times (W_{l_{ij}}^k h_{v_i}^k + b_{l_{ij}}^k)) \qquad (4)$$

In the proposed model, the GCNs are mainly be used to capture dependency information, while they can't capture sufficient information flow. In event detection, sequential context can be leveraged to expand information flow. Therefore, we feed output of GCN layer to a CNN layer and get the final sentence level feature.

3.3   Hybrid Representation Learning

As we shown in Fig. 3, the lexical feature and sentence-level feature make up the word-level feature. According to previous works [14,26], the word feature can provide more accurate semantic information, while the character feature can reveal the inner compositional structure of event triggers. Therefore, we use the similar architecture that described in Fig. 4 to obtain character-level feature. Then we incorporating word-level and character-level feature by a task-specific hybrid representation [15]. It should be noticed that only the word-level feature contain the dependency information in our case. In order to avoid excessively increasing the complexity of the model, we remove the GCN layer and the representation of dependency parser tree when we learn character-level feature.

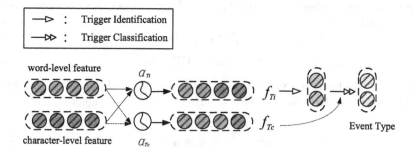

Fig. 6. Framework of event detection by hybrid representation learning. The word-level feature and character-level feature are respective the output of architecture in Fig. 3 and Fig. 4.

Chinese event detection can be divided into two processes: trigger identification and trigger classification. These two processes require different feature in a particular model. For example, trigger identification may rely on more structural information from character-level feature, while trigger classification need more semantic information from word-level feature. Consequently, two gates are trained to model the information flow for trigger identification and trigger classification. Figure 6 describes the framework of task-specific hybrid representation. We use f'_{word} and f'_{char} to separately represent word-level and character-level feature. The gates are calculated from a nonlinear transformation as:

$$\alpha_{Ti} = \sigma(W_{Ti}f'_{char} + U_{Ti}f'_{word} + b_{Ti}) \tag{5}$$

$$\alpha_{Tc} = \sigma(W_{Tc}f'_{char} + U_{Tc}f'_{word} + b_{Tc}) \tag{6}$$

where σ is the sigmoid function, $W \in R^{d' \times d'}$ and $U \in R^{d' \times d'}$ are weight matrixes, and b is the bias term.

Then base these gates, two different vector would be separately calculated by:

$$f_{Ti} = \alpha_{Ti} \odot f'_{char} + (1 - \alpha_{Ti}) \odot f'_{word} \tag{7}$$

$$f_{Tc} = \alpha_{Tc} \odot f'_{char} + (1 - \alpha_{Tc}) \odot f'_{word} \tag{8}$$

where f_{Ti} and f_{Tc} are the hybrid features respectively for trigger identification and trigger classification; $\odot$ is the elementwise product operation.

3.4 Training and Classification

Following the above works, we consider event detection as a multi-class classification problem and use stochastic gradient descent for training. To train the trigger generator, we dispose the training datasets by using negative sampling. Concretely, we divide all characters in sentences into positive and negative samples according to whether the characters are included in triggers. Given the current token w_i, we feed its feature F_i into a fully-connected network to predict the trigger label as:

$$T_i = f(W_T F_i + b_T) \tag{9}$$

$$y_{t_i} = softmax(WT_i + b) \tag{10}$$

where f is a non-linear activation and y_{t_i} is the output of the $i-th$ trigger label.

Then we can define the objective function to train the trigger classifier as:

$$J(\theta) = \sum_{i=1}^{n} log(p(y_{t_i}|\theta)) \tag{11}$$

where n is the number of tokens and θ is the network parameter. We employ stochastic gradient descent to maximize the $J(\theta)$.

4   Experiments

4.1   Dataset, Evaluation Metric and Hyperparameter Setting

We evaluate our model on ACE 2005 dataset and KBPEval2017 dataset for event detection. To compare with previous work ([3,7] and [27]), we adopt the same data split to the previous work. According to this data split, there are 569 documents in train set, 64 documents in development set and 64 documents in test set. KBPEval2017 is another common dataset that is widely used for event extraction. In KBPEval2017 dataset, there are 693 documents that derived from newswire and discussion forum.

The experiments are designed to evaluate the performance of our model in trigger identification and trigger type classification. We calculate recall (R), precision (P) and F-measure (F) that same as Feng et al., 2016 to evaluate our model.

In the process of preprocessing, we use the Stanford CoreNLP toolkit to conduct tokening and generate dependency parsing trees. For negative sampling, the ratio of positive to negative instances is set to 1:20. In embedding layer, we respectively limit to 220 and 130 tokens in character-level representation and word-level representation. The dimensions of word embedding and character embedding are both 100, while the dimension of position embedding is set to 10. We use a three-layer GCN and a one-layer CNN with 400 filters. We also set batch size to 32 and dropout rate to 0.5.

4.2   Comparison with Baselines

For evaluating our method, we compare our method with the following baselines:

C-BiLSTM: A deep learning model for Chinese event extraction based on bidirectional LSTM and CNN [27]. C-BiLSTM views the event detection as a sequence labeling problem instead of relying on complicated natural language processing tools.

FBRNN: A forward-backward recurrent neural networks for event detection [8]. FBRNN is one of the first attempts to solve the problem of multi-word events in event detection.

HNN: A language-independent neural networks proposed by [7] to capture both sequence and structure information. HNN is a hybrid neural network which incorporates Bi-LSTM and convolutional neural networks for event detection.

NPN: A recent state-of-the-art deep learning model for Chinese event detection which respectively use three kinds of hybrid representation learning methods (concat, general and task-specific) to generate representation for event detection [15].

In these models, C-BiLSTM is a character-based model, while FBRNN and HNN are word-based model. Our model and NPN are based on both word and character. Table 1 presents the performance of the models on the test sets. As we can see from the table, character-based models like C-BiLSTM or word-based models like FBRNN and HNN have a relatively low F1 scores. It means that whether single word-based or single character-based models are not provide sufficient feature information for event detection. NPN performs better than HSGCN in recall value and HNN have a higher precision than NPN and HSGCN. However, HSGCN get competitive results on precision and recall that result in a highest F1 score among all models. Specifically, our model improve the F1 scores by 1.2% and 0.9% in trigger identification and trigger classification. What's more, the experiment results of HNN, NPN and HSGCN show that our model do not need sacrifice the recall to improve precision

Table 1. Experiment results on ACE 2005 dataset. (*) indicates the result taken from the original paper.

Model	Trigger identification (%)			Trigger classification (%)		
	P	R	F1	P	R	F1
C-BiLSTM	65.6	66.7	66.1	60.0	60.9	60.4
FBRNN	64.1	63.7	63.9	59.9	59.6	59.7
HNN	**74.2**	63.1	68.2	**77.1**	53.1	63.0
NPN (Task-specific)*	64.8	**73.8**	69.0	60.9	**69.3**	64.8
HSGCN	71.5	68.9	**70.2**	66.9	64.6	**65.7**

4.3   Performance on Specific Event Subtype

To further analyze the performance and effectiveness of HSGCN, we evaluate our model in different specific event subtypes. In this subsection, we evaluate HSGCN on KBPEval2017 dataset and test the model on five most common event subtype (Attack, Broadcast, Transport Ownership, Die and Transfer Money) of KBPEval2017 dataset. For comparison purposes, we select NPN (Task-specific) as baseline in this subsection. Table 2 show the specific result of HSGCN and NPN (Task-specific) in trigger identification and trigger classification.

As we can observe from the result, our model achieve better performance for almost every event subtype, either in trigger identification or trigger classification. The experimental results show the effectiveness of our model. Regarding the results of baseline on KBPEval2017 dataset, HSGCN is further improved, achieve 0.85% and 1.14% improvements of F1 score in trigger identification and trigger classification. In the same setting for datasets and performance measures, HSGCN gets better precision and F1 score, in spite of the model has a relatively lower recall. The same trends are observed when test HSGCN in specific event subtypes. As the result of Tables, we can deem both semantic and syntactic information are necessary in event detection. We believe that HSGCN has superior performance to capture semantic and syntactic information. Due to this improvement, HSGCN achieve a better performance in the experiments.

Table 2. The result of HSGCN and baseline in trigger identification and trigger classification. (ΔF1) is equal to the F1 score of HSGCN minus the F1 score of baseline.

Event type	Baseline (%)			HSGCN (%)			ΔF1
	P	R	F1	P	R	F1	
Trigger identification							
Attack	61.93	55.81	58.71	68.12	52.44	59.26	+0.55
Broadcast	65.29	48.20	55.46	68.21	48.81	56.90	+1.44
Transport ownership	66.79	45.22	53.93	70.10	46.25	55.73	+1.80
Die	64.17	55.45	59.49	66.93	54.49	60.07	+0.58
Transfer money	69.55	50.20	58.31	71.63	51.12	59.66	+1.35
All event type	63.72	53.56	58.20	67.91	52.24	59.05	+0.85
Trigger classification							
Attack	58.10	47.29	52.14	62.06	46.91	53.43	+1.29
Broadcast	60.07	42.55	49.81	63.57	43.65	51.76	+1.95
Transport ownership	55.75	48.76	52.02	59.85	47.78	53.14	+1.12
Die	62.83	49.73	55.52	67.68	47.53	55.84	+0.32
Transfer money	59.90	49.12	53.98	62.74	46.94	53.70	−0.28
All event type	56.97	47.68	51.91	63.65	45.47	53.05	+1.14

5 Conclusions

In this paper, we propose a hybrid syntactic graph convolutional networks (HSGCN) for Chinese event detection that exploit the syntactic information to effectively model the sentence-level feature. Our model also consider the different effects of words and characters in trigger identification and trigger classification. Compared with traditional event detection methods, our approach efficiently capture syntactic dependency and generate a sentence-level feature based on GCN, thus can take advantage of event information that scattered in the sentence. Furthermore, in order to keep the original semantic of Chinese vocabulary, word embedding and character embedding are concatenated, which improve the accuracy of triggers identification. Experimental results show that our method achieve a competitive performance on ACE 2005 and KBPEval2017 benchmark datasets.

Acknowledgements. This research was funded by the National Natural Science Foundation of China, grant number 61402220, the Philosophy and Social Science Foundation of Hunan Province, grant number 16YBA323, the Scientic Research Fund of Hunan Provincial Education Department for excellent talents, grant number 18B279, the key program of Scientic Research Fund of Hunan Provincial Education Department, grant number 19A439, the Project supported by the Natural Science Foundation of Hunan Province, China, grant number 2020JJ4525.

References

1. Ahn, D.: The stages of event extraction. In: Proceedings of the Workshop on Annotating and Reasoning about Time and Events, pp. 1–8 (2006)
2. Chen, D., Manning, C.D.: A fast and accurate dependency parser using neural networks. In: Proceedings of the 2014 Conference on Empirical Methods in Natural Language Processing (EMNLP), pp. 740–750 (2014)
3. Chen, Z., Ji, H.: Language specific issue and feature exploration in Chinese event extraction. In: Proceedings of Human Language Technologies: The 2009 Annual Conference of the North American Chapter of the Association for Computational Linguistics, Companion Volume: Short Papers, pp. 209–212 (2009)
4. Chieu, H.L., Ng, H.T.: A maximum entropy approach to information extraction from semi-structured and free text. In: AAAI/IAAI 2002, pp. 786–791 (2002)
5. Chung, T., Liu, Y., Xu, B.: Monotonic alignments for summarization. Knowledge-Based Syst. **192**, 105363.1–105363.10 (2020)
6. Chung, T., Xu, B., Liu, Y., Ouyang, C., Li, S., Luo, L.: Empirical study on character level neural network classifier for Chinese text. Eng. Appl. Artif. Intell. **80**, 1–7 (2019)
7. Feng, X., Qin, B., Liu, T.: A language-independent neural network for event detection. Sci. China Inf. Sci. **61**(9), 092106 (2018)
8. Ghaeini, R., Fern, X.Z., Huang, L., Tadepalli, P.: Event nugget detection with forward-backward recurrent neural networks. arXiv preprint arXiv:1802.05672 (2018)
9. Hong, Y., Zhang, J., Ma, B., Yao, J., Zhou, G., Zhu, Q.: Using cross-entity inference to improve event extraction. In: Proceedings of the 49th Annual Meeting of the Association for Computational Linguistics: Human Language Technologies, vol. 1, pp. 1127–1136. Association for Computational Linguistics (2011)
10. Ji, H., Grishman, R.: Refining event extraction through cross-document inference. In: Proceedings of ACL-2008: HLT, pp. 254–262 (2008)
11. Jin, H., Li, C., Zhang, J., Hou, L., Li, J., Zhang, P.: XLORE2: large-scale cross-lingual knowledge graph construction and application. Data Intell. **1**(1), 77–98 (2019)
12. Jin, Y., Kan, M.Y., Ng, J.P., He, X.: Mining scientific terms and their definitions: a study of the ACL anthology. In: Proceedings of the 2013 Conference on Empirical Methods in Natural Language Processing, pp. 780–790 (2013)
13. Kipf, T.N., Welling, M.: Semi-supervised classification with graph convolutional networks. arXiv preprint arXiv:1609.02907 (2016)
14. Li, P., Zhou, G., Zhu, Q., Hou, L.: Employing compositional semantics and discourse consistency in Chinese event extraction. In: Proceedings of the 2012 Joint Conference on Empirical Methods in Natural Language Processing and Computational Natural Language Learning, pp. 1006–1016 (2012)
15. Lin, H., Lu, Y., Han, X., Sun, L.: Nugget proposal networks for Chinese event detection. arXiv preprint arXiv:1805.00249 (2018)
16. Liu, Y., Ouyang, C., Li, J.: Ensemble method to joint inference for knowledge extraction. Expert Syst. Appl. **83**, 114–121 (2017)
17. Marcheggiani, D., Titov, I.: Encoding sentences with graph convolutional networks for semantic role labeling. arXiv preprint arXiv:1703.04826 (2017)
18. Mikolov, T., Chen, K., Corrado, G., Dean, J.: Efficient estimation of word representations in vector space. arXiv preprint arXiv:1301.3781 (2013)

19. Nguyen, T.H., Grishman, R.: Modeling skip-grams for event detection with convolutional neural networks. In: Proceedings of the 2016 Conference on Empirical Methods in Natural Language Processing, pp. 886–891 (2016)

20. Nguyen, T.H., Grishman, R.: Graph convolutional networks with argument-aware pooling for event detection. In: Thirty-second AAAI conference on artificial intelligence (2018)

21. Rudinger, R., White, A.S., Van Durme, B.: Neural models of factuality. arXiv preprint arXiv:1804.02472 (2018)

22. Sha, L., Qian, F., Chang, B., Sui, Z.: Jointly extracting event triggers and arguments by dependency-bridge RNN and tensor-based argument interaction. In: Thirty-Second AAAI Conference on Artificial Intelligence (2018)

23. Vashishth, S., Bhandari, M., Yadav, P., Rai, P., Bhattacharyya, C., Talukdar, P.: Incorporating syntactic and semantic information in word embeddings using graph convolutional networks. arXiv preprint arXiv:1809.04283 (2018)

24. Wan, H., Zhang, Y., Zhang, J., Tang, J.: AMiner: search and mining of academic social networks. Data Intell. 1(1), 58–76 (2019)

25. Westerhout, E.: Definition extraction using linguistic and structural features. In: Proceedings of the 1st Workshop on Definition Extraction, pp. 61–67. Association for Computational Linguistics (2009)

26. Yubo, C., Liheng, X., Kang, L., Daojian, Z., Jun, Z., et al.: Event extraction via dynamic multi-pooling convolutional neural networks (2015)

27. Zeng, Y., Yang, H., Feng, Y., Wang, Z., Zhao, D.: A convolution BiLSTM neural network model for Chinese event extraction. In: Lin, C.-Y., Xue, N., Zhao, D., Huang, X., Feng, Y. (eds.) ICCPOL/NLPCC-2016. LNCS (LNAI), vol. 10102, pp. 275–287. Springer, Cham (2016). https://doi.org/10.1007/978-3-319-50496-4_23

28. Zhang, J., Qin, Y., Zhang, Y., Liu, M., Ji, D.: Extracting entities and events as a single task using a transition-based neural model. In: Proceedings of the 28th International Joint Conference on Artificial Intelligence, pp. 5422–5428. AAAI Press (2019)

29. Zhang, T., Ji, H., Sil, A.: Joint entity and event extraction with generative adversarial imitation learning. Data Intell. 1(2), 99–120 (2019)

Meta Learning for Event Argument Extraction via Domain-Specific Information Enhanced

Hang Yang[1,2(✉)], Yubo Chen[1], Kang Liu[1,2], and Jun Zhao[1,2]

[1] Institute of Automation, Chinese Academy of Sciences, Beijing, China
[2] University of Chinese Academy of Sciences, Beijing, China
{hang.yang,yubo.chen,kliu,jzhao}@nlpr.ia.ac.cn

Abstract. As a sub-task of event extraction, event argument extraction (EAE) aims at identifying arguments and classifying the roles they play in an event. Existing supervised learning based methods for EAE require large-scale labeled data which prevents the adaptation of the model to new event types. In this paper, we propose a meta-learning based EAE method, which aims at learning a good initial model that can adapt quickly to new event types with limited samples for training. For taking advantage of the consistency of event structures under the same domain, we introduce dynamic memory networks (DMN) to learn the domain-specific information. We conduct experiments under meta-learning setting to explore the scalability of our methods on EAE. The experimental results show that our method can learn general and transferable information that can be applied to the EAE of new event types which only contain few samples.

Keywords: Event argument extraction · Meta-learning · Domain-specific information

1 Introduction

Event extraction (EE) aims at extracting event information from unstructured texts which can be divided into two subtasks: event detection (ED) and event argument extraction (EAE). ED aims to detect trigger words from individual sentences and determine the event type of them. EAE aims to identify the event arguments and classify the roles they play in an event. For instance, given a sentence *"all 225 people aboard were killed."*, ED aims to identify *"killed"* as a trigger and determine the event type as *"Attack"*. EAE aims to identify that *"225 people"* is an event argument with role type *"Victim"* in an *Attack* event. In this article, we focus on EAE.

Recently, the EAE models [2,13,20] based on deep-learning have achieved acceptable results in the predefined event domain. However, these supervised methods rely heavily on large-scale labeled data which need costly collecting

© Springer Nature Singapore Pte Ltd. 2021
H. Chen et al. (Eds.): CCKS 2020, CCIS 1356, pp. 160–172, 2021.
https://doi.org/10.1007/978-981-16-1964-9_13

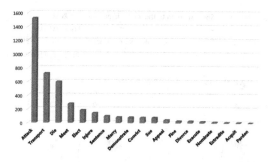

Fig. 1. Statistics of instances for different event types on ACE 2005 English dataset.

and annotating. Such expensive annotation has hindered the application of EAE on new event types especially the low-frequency events with low event sources. Moreover, as shown in Fig. 1, the ACE 2005 corpus [4] which is widely-used for exploring EE follows a long-tailed distribution where some event types contain a few instances. So we call for a transferable EAE method to learn quickly and effectively on new event types with limited samples.

Humans beings are born with the ability to learn quickly whether in recognizing objects after observation of a few samples or learning new skills quickly after just minutes of experiments [5]. Meta-learning (or learn to learn) [5,16,17] is based on this inspiration and it can quickly learn a new task supported by a few instances. For exploring the transferable of EAE, we propose a meta-learning based method to learn general and transferable information for EAE that can be applied to new event types with limited samples.

Besides, the existing EAE methods ignore the relationship among different event types under the same event domain. Specifically, on the benchmark dataset ACE 2005 for EE, there are 8 event types and 33 event subtypes, and the subtype events that belong to the same event type (i.e., event domain) tend to have similar event structures [8]. For example, the event subtype *Marry* and *Divorce* that belongs to the same event type *Life* contain nearly identical event structures (both of them have shared event arguments such as *Time-within, Person* etc.) as shown in Fig. 2. We hope to model the interaction among different event types under the same event domain to enhance the learning for EAE especially on the event types with low-resource.

In this paper, to explore the meta-learning for EAE, we consider different event types as different tasks, which is fundamentally different from traditional learning which aims at finding the optimal parameters for all event types. An intuition of the difference between these two learning approaches is shown in Fig. 2. We aim to learn an initial EAE model based on existing event types (high-resource tasks) that could be quickly and effectively adapted to new event types (low-resource tasks) given limited samples. Specifically, we consider the 8 event types as 8 event domains, and 33 subtypes as 33 tasks on ACE 2005 for the research of the meta-learning on EAE. And model-agnostic meta-learning

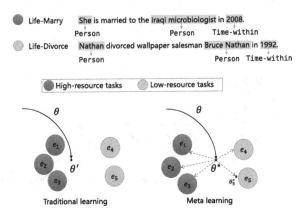

Fig. 2. Differences between traditional learning and meta-learning. Traditional learning aims to find the optimal parameters for all event types and meta-learning aims to find the initial parameters which can faster adapt to new tasks. Specifically, We aim to learn an initialization model though the training on high-resource tasks (*"Marry"*) which can adapt to the low-resource tasks (*"Divorce"*).

(MAML) [5], a general algorithm for meta-learning, is used to learn the transferable of EAE. But there is a problem that MAML only learns the common knowledge shared within the tasks [6], and it can't taking advantage of the relationship among them especially the commonness of events in the same domain. To better learn the structural similarity of tasks under the same event domain, we use a dynamic memory network (DMN) to encode the domain-specific information which benefits the meta-learning for EAE. The main contributions of this work are as follows:

– We introduce meta-learning on the task of EAE for exploring the transferable of it.
– We combine the MAML algorithm to learn a good initialization EAE model that can be quickly adapted to low-resource event types supported by limited training data. And we devise the DMN to encode domain-specific information to enhance the meta-learning for EAE.
– We conduct experiments for exploring meta-learning for EAE and results demonstrate the effectiveness of our method.

2 Problem Formulation

2.1 Task Definition

For EAE, the input instance c is an event mention containing a trigger word with its event type k and a set of entities (entity seen) or not (entity unseen). The output is the structured event information containing the arguments with their role types in the event mention. In the formulation of meta-learning for EAE, we take each event subtype as an individual task. Following standard meta-learning setting [5] which consists of three phases, **meta-training, meta-test** and **meta-dev** phases. We denote the K different meta-training tasks as: $\mathcal{T}_{meta-training} = \{\mathcal{T}_1, \mathcal{T}_2, \cdots, \mathcal{T}_K\}$ where each task $\mathcal{T}_k$ has its own training and validation set $\{D_k^{train}, D_k^{valid}\}$.

$$D_k^{train} = \left\{ \left(c_n^{(k)}, r_n^{(k)}, k \right), n = 1 \ldots N_k \right\} \tag{1}$$

During the training process on the meta-training phrase containing K difference tasks, we get an initilize model $\mathcal{M}_{train}$:

$$\mathcal{M}_{train} : C \times K \rightarrow O \tag{2}$$

where C is the input and O is the output.

Similarly, we denote the meta-testing tasks as: $\mathcal{T}_{meta-test} = \{\mathcal{T}_1', \mathcal{T}_2', \cdots, \mathcal{T}_T'\}$ where each task $\mathcal{T}_t'$ has its own training and testing set $\{D_t^{train}\prime, D_t^{test}\prime\}$.

$$D_t^{train}\prime = \left\{ \left(c_n^{(t)}, r_n^{(t)}, t \right), n = 1 \ldots N_t \right\} \tag{3}$$

where N_t is the number of training samples which means N_t-shot in the meta-test phase and $N_t << N_k$. For the adaptation, we fine-tune the model $\mathcal{M}_{train}$ with target task training data $D_t^{train}\prime$ and obtain a new task-specific model $\mathcal{M}_{test}(t)$.

$$\mathcal{M}_{test}(t) : C_{test} \times T \rightarrow O \tag{4}$$

The goal of meta-learning for EAE is to learn an initilize model $\mathcal{M}_{train}$ that can quickly adapt on the new target tasks T in $\mathcal{T}_{meta-test}$ with limited data.

3   Method

This section describes our approach to extract event arguments under the meta-learning setting. Most existing EE methods [2,13,20] set the entity as gold and treat the EAE as a classifier for entities. But in reality, the entity in a new instance is unknown, and the use of the NER tool will also bring errors. For exploring the meta-learning for EAE reliably, we take different methods under the setting of entity unseen or entity seen. In this section, we will first introduce the basic EAE methods for two situations (entity unseen and entity seen). Secondly, we will introduce the DMN which is used to establish relationships between tasks that are in the same event domain. Finally, we introduce how to combine the MAML algorithm and our method for the meta-learning of EAE.

3.1   EAE Model with Entity Seen

For entity seen, being consistent with most current methods [2,13,18], each entity in the event mention is a candidate argument. Following [2], we take the lexical-level and sentence-level features to represent a candidate argument for argument role classifier as shown in Fig. 3(a).

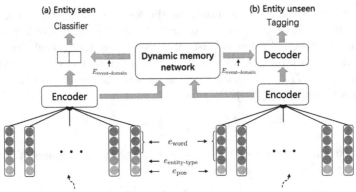

Life-Divorce: *Nathan divorced wallpaper salesman Bruce Nathan in 1992.*

Fig. 3. There are two methods to handle different situations in event argument extraction. (a) For entity seen, we use DMCNN based model for candidate argument classification. (b) For entity unseen, we use Bi-LSTM based model for instance tagging. For both methods, we introduce the dynamic memory networks (DMN) to encode domain-specific information.

Instance Encoder. We denote an event mention C_k from task $\mathcal{T}_k$ as an instance and it is formalized as $E_k = \{w_1, \ldots w_t, \ldots, w_e, \ldots, w_n\}$, where w_t denote the trigger word and w_e denote the candidate argument (the gold entity in the event mention). We select dynamic multi-pooling convolutional neural networks (DMCNN) [2] to learn the representation of sentences.

$$h = \text{Encoder}\{w_1, \ldots w_t, \ldots, w_e, \ldots, w_n\} \tag{5}$$

where $w = [e_{word}; e_{entity\text{-}type}; e_{pos}]$, e_{word} is the word embedding of w, $e_{entity\text{-}type}$ represents the entity type embedding of w (None if w is not an entity) and e_{pos} denotes the position embedding which represents the distance between word w and trigger w_t.

Augment Role Classifier. We concatenate the embedding h of instance sentence and the entity embedding w_e as the input of the argument role classifier.

$$p(r|w_e) = softmax(W_c \cdot [h; w_e] + b_c) \tag{6}$$

where $W_c \in R^{r \times d_h}$ and $b_c \in R^{r \times d_h}$ are learned parameters of the classification layer and r is the number of event role categories.
The objective function is defined as followes:

$$L(\theta) = -\sum_l logp(r_l|x_l) \tag{7}$$

where l is the number of instances.

3.2   EAE Model with Entity Unseen

For entity unseen, as shown in Fig. 3(b), we formulate EAE as a sequence labeling task where each token in a sequence is tagged as the beginning (B), inside (I) or outside (O) of an argument.

Instance Encoder. For each instance $E_k = \{w_1, \ldots w_t, \ldots, w_n\}$, where w_t denote the trigger word and others denote the words in the instance. We use Bi-directional Long Short-Term Memory (Bi-LSTM) to encoder sentence into hidden embedding,

$$\{h_1, h_2, \ldots, h_n\} = \text{Encoder}\,\{w_1, \ldots w_t, \ldots, w_n\} \tag{8}$$

where $w = [e_{\text{word}}; e_{\text{pos}}]$.

Decoder and Argument Role Tagging. We use Bi-LSTM to decode and get the output.

$$\{d_1, d_2, \ldots, d_n\} = \text{Decoder}\{h_1, h_2, \ldots, h_n\} \tag{9}$$

For each word in sentence, predicted tag as follows:

$$p(r_i|d_i) = softmax(W_t \cdot d_i + b_t) \tag{10}$$

where $W_t \in R^{r \times d_h}$ and $b_t \in R^{r \times d_h}$ are learned parameters of the classification layer and r is the number of tag categories.

The objective function is defined as followes:

$$L(\theta) = -\sum_l logp(r_l|x_l) \tag{11}$$

where l is the number of instances.

3.3   Dynamic Memory Network

One problem of MAML is that it only learns the common knowledge shared within the tasks and ignore the close relationship between tasks. To better take advantage of similar tasks, we introduce a DMN to memory the domain-specific information. DMN, a neural network architecture that processes input sequences and questions, forms episodic memories and generates relevant answers for question answering problems [10]. In this paper, we simplify the DMN and use it to encode domain-specific information. Our DMN consists of two modules: a memory module and an output module. For the above two methods, whether the entity is seen or unseen, we take the hidden embedding of the instance as the input of DMN.

Episodic Memory Module. For each event domain, we initialize a memory vector and then update it. We denote the instance encoder as the input, the memory module chooses which parts of the input to focus on through attention mechanism and then updates itself. After each iteration, the memory module retrieves event information and produces a new domain-specific "Memory". Formally, the memory module consists of three components: the attention gate, the attentional GRU, and the memory update gate.

Attention Gate. Given the inputs h and initial memory m, the attention gate determines how much the memory should pay attention to.

$$a = [h \circ m, |h - m|] \tag{12}$$

where $\circ$, $-$ are element-wise product, subtraction respectively. Then the gate score function G:

$$g = \sigma \left(W^{(2)} \cdot \tanh \left(W^{(1)} \cdot a \right) \right) \tag{13}$$

where σ is the sigmoid non-linear function, $W^{(1)}$ and $W^{(2)}$ are parameters of the perceptron.
We employ a modified GRU get compute the new memory m_t.

$$m_t = g \cdot GRU(h; m_{t-1}) + (1 - g) \cdot m_{t-1} \tag{14}$$

Output Module. Depending on the updated domain-dependance memory and the sentence representation from the encoder, we can get an output vector.

$$E_{\text{event-domain}} = W^{(3)}[m_t; h] \tag{15}$$

where ; is concatenation operator, $W^{(3)} \in R^{d_h \times d_h}$. For the entity seen model, we concatenate the $E_{\text{event-domain}}$ and the entity feature as the input of the argument role classifier. For the entity unseen model, we denote the $E_{\text{event-domain}}$ as the initial representation of the decoder for tagging.

3.4   Meta-Learning for EAE with Domain-Specific Information

MAML aims at learning an initial model that can adapt to a new target task through a few gradient updates. We propose to use DMN to get the domain-specific information so that the model can leverage the relationship between tasks under the same event domain.

First, we randomly initialize the parameters θ of our EAE model. Then, we sample instance c_{ij} from task T_i and compute the domain-specific memory m_i. In the normal update step, we use gradient descent to update θ_i for each task with iteration k.

$$\theta_i' = \theta - \alpha \nabla_\theta \mathcal{L}_{T_i} \left(f_\theta(c_{ij}, m_i) \right) \tag{16}$$

In the meta-update step, we sum up the updated loss value from all tasks and update the initial θ to minimize the meta-object loss function.

$$\theta \leftarrow \theta - \beta \nabla_\theta \sum_{T_i \sim p(T)} \mathcal{L}_{T_i}(f_{\theta_i'}(c_{ij}', m_i)) \tag{17}$$

Finally, we can get a transferable θ of EAE model and memory m for each event domain.

In the meta-test phase, we fine-tune θ through the training on the few-shot samples for a new task. Then the updated model makes a prediction on testing samples.

4   Experiments

To facilitate the evaluation of the EAE models under the standard meta-learning setting, we construct a dataset tailored particularly for it based on existing corpus ACE 2005. And we compare it with several baselines to verify the effectiveness of our proposed method. To measure argument identification performance under the same conditions for compared systems, we set the entity seen which is a common setting in EE [2,13]. In another method, to make the EAE more consistent with the reality where the entity is unknown, we set the entities as unseen.

4.1   Datasets and Evaluation Metrics

We construct the meta-learning setting based datasets for EAE from widely used ACE 2005. ACE2005 contains 599 documents, which are annotated with 8 event types, 33 event subtypes. For exploring the meta-learning for EAE, we consider each event subtype as a task and define the event mention which contains event trigger and event arguments as an instance. To utilize the domain-specific information, we consider different event types as different domains and get 8 event domains: *Business, Conflict, Contact, Justice, Life, Movement, Personnel, Transaction.* For each event domain, we select a task which only contains a few instances as $\mathcal{T}_{meta-test}$ or $\mathcal{T}_{meta-dev}$ and others as $\mathcal{T}_{meta-training}$. Through the way, we get 4/33, 4/33 and 25/33 tasks for meta-test/meta-dev/meta-train on ACE 2005. In the meta-test phases, we take 5 and 10 instances as training samples for the 5-shot and 10-shot learning setting and take 10 instances for testing. We follow the standard evaluation procedures: an event argument is correct if its span and role match a reference argument [9]. For evaluation, we report the micro-F1 score for each event type on the meta-testing phase and the macro-F1 score averaged across event types.

4.2   Baselines

To evaluate the effectiveness of our method, we compare it with the following baselines.

DMCNN: [2] propose a dynamic multi-pooling CNN to extract sentence-level features automatically.

HMEAE: [18] propose a hierarchical modular event argument extraction model to provide effective inductive bias from the concept hierarchy of event argument roles.

Transfer Learning: It aims to learn an extract event argument model based on the meta-training tasks, and tune parameters on the meta-testing tasks.

4.3   Implementation Details

The hyper-parameters are set as follows. For all experiments, we use the pre-trained GloVe word embedding with a dimension of 100. During the training, we set the batch size to 32 and the dropout rate to 0.5, and stochastic gradient descent (SGD) optimizer is used with 50 iterations. In our method where entity unseen, we apply the Bi-LSTM with a hidden size of 200 to construct the encoder and decoder and the initial memory embedding is set to 400 for DMN. When entity seen, we set the window size as 3, the number of feature map as 300 for DMCNN and the initial memory embedding is set to 300 for DMN. For MAML, we set $\alpha = 0.01$ and $\beta = 0.001$ with 5 adaption step applied to compute the updated parameters.

Table 1. The experimental results on 5-shot and 10-shot setting with micro-F1 score (%) on the meta-test tasks with the setting of **entity unseen**.

Entity unseen	5-shot				10-shot			
Method	End-Org	Demonstrate	Divorce	Transport	End-Org	Demonstrate	Divorce	Transport
LSTM	15.37	29.43	28.50	5.58	24.90	28.10	29.32	9.74
LSTM + Transfer	20.65	31.67	26.22	7.86	25.41	31.25	36.87	10.92
LSTM + MAML	23.45	**38.76**	34.78	**11.25**	24.78	41.35	41.33	**13.20**
Ours	**27.32**	38.23	**37.42**	–	**27.81**	**45.69**	**42.40**	–

Table 2. The experimental results on 5-shot and 10-shot setting with micro-F1 score (%) on the meta-test tasks with the setting of **entity seen**.

Entity seen	5-shot				10-shot			
Method	End-Org	Demonstrate	Divorce	Transport	End-Org	Demonstrate	Divorce	Transport
DMCNN	32.47	37.21	36.89	11.63	42.18	35.48	52.64	15.39
HMEAE	35.22	39.38	37.20	12.88	39.27	38.40	55.31	17.61
DMCNN + Transfer	35.40	44.52	42.60	11.33	40.67	36.82	53.87	16.21
DMCNN + MAML	42.67	56.32	50.27	**18.32**	45.18	62.59	55.69	**19.35**
Ours	**46.50**	**58.40**	**59.61**	–	**53.24**	**63.21**	**57.94**	–

4.4   Results

We conduct our experiments under a 5-shot and 10-shot learning setting where there are 5 and 10 training samples for each task in $\mathcal{T}_{meta-test}$ and $\mathcal{T}_{meta-dev}$. For transfer learning, we train our model on $\mathcal{T}_{meta-test}$ and adjust the parameters on $\mathcal{T}_{meta-dev}$, then fine-tune our model and get the final test results on $\mathcal{T}_{meta-test}$. For MAML, we find the best initialization model through $\mathcal{T}_{meta-dev}$ and then fine-tune the model on $\mathcal{T}_{meta-test}$ to get the test results. Our experiment results are shown in Table 1 and Table 2, we have the following observations:

(1) Whether the results comes from sequence labeling based methods as shown in Table 1 or entity classification based methods as shown in Table 2, our method performs the best of all. First, MAML-based methods achieve improvements (about 5.5% in F1 on 5-shot and 4.5% in F1 on 10-shot respectively) compared with transfer learning, which based on the almost same network framework except for the gradient update rules. But one problem of MAML is that it only learns the common knowledge shared within the tasks without making use of the relationship between tasks especially the tasks under the same event domain. We devise a DMN to encode domain-specific information for leveraging the relationship among events under the same domain. The results prove that the effectiveness of the DMN which can learn the domain-specific information to improve the meta-learning of EAE.

(2) Whether it is a 5-shot or a 10-shot setting, our method can achieve the best. Our method aim at learning an initial model through training on $T_{meta-train}$, which achieves the best performance on new tasks $T_{meta-test}$ through training on a few samples (5-shot or 10-shot). The results prove that our methods learn a transferable EAE model that can quickly and effectively adapt to target event types with low-resource. It also proves that our methods can alleviate the long-tail problem in EAE.

(3) For validating our method on a cross-domain, we test our method on the *"Movement"* event domain which only contains one event type *"Transport"*. So that there is no domain-specific information encoder for this event type. And the experiments also show that our MAML-based method can perform better on cross-domain.

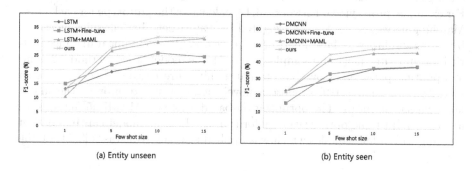

(a) Entity unseen   (b) Entity seen

Fig. 4. The macro-F1 score of each method under different few-shot size (1, 5, 10, 15) setting.

4.5 Few-Shot Size

To validate the impact of the few-shot size on the performance of our method, we evaluate it by various few-shot sizes: 1, 5, 10, 15. From the result as shown in Fig. 4, the F1-score evaluation for each method with different few-shot sizes, we have the following observations: (1) with the increase of training instances, the performance of these methods achieves better. (2) We can see that our methods (whether entity seen or unseen) performs the best of all where the few-shot size reaches 5 and more. (3) The performances of MAML-based methods remain nearly the same after the few-shot size reaches 10, it means that our method can be better transferable to new event types with limited training instances. (4) The DMN is beneficial for the meta-learning of EAE especially even the few-shot size is small.

5   Related Work

Event Argument Extraction. Most of the previous EAE methods based on supervised learning that relied on large-scale training data. Such works can achieve high quality for pre-defined types, but can not adapt to new event types without annotation. To improve the scalability of EAE, [8] proposed a transferable model, which transfers the knowledge from the existing types to the extraction of unseen types. Besides, to address the problem of insufficient training data, [1] proposed to automatically label training data via external knowledge. [20] proposed a generation method to promote EE based on pre-trained language models.

Meta-learning in NLP. Meta-learning aims at adapting to new tasks quickly with little data through the meta-training on a set of tasks. Several meta-learning models have been proposed recently which can be divided into three types: (1) The Metric-based method, which aims at learning the distance distribution among classes, is usually used for few-shot learning in classification tasks [7,21]. For example, prototypical networks [6,17] calculates the prototype for each class. Matching networks [22] learn a weighted K-nearest neighbor classifier measured by cosine similarity. Relation networks [16] learn a deep non-linear distance metric for comparing query and sample items. (2) The model-based method [12] aims at updating parameters by the internal architecture or controlled by another meta-learning model. [19] propose a novel meta-learner to improve the multi-label classification. (3) The optimization-based approach MAML [5] intends for a way to adjust the optimization algorithm so that the model can be good at learning with a few samples which was applied to many NLP tasks [3,11,14,15].

6   Conclusion

We introduce the meta-learning formalization for exploring the transferable of EAE where limited samples are available on new event types. Specifically, we propose a MAML-based method to learn an initial EAE model that can quickly adapt to a new type and devise a DMN to encode domain-specific information for improvement. Experiments on ACE 2005 under meta-learning setting validate the scalability and effectiveness of our method.

Acknowledgements. This work is supported by the National Key R&D Program of China (2020AAA0105200), the National Natural Science Foundation of China (No. 61533018, No. 61976211, No. 61806201) and the Key Research Program of the Chinese Academy of Sciences (Grant NO. ZDBS-SSW-JSC006). This work is also supported by Beijing Academy of Artificial Intelligence (BAAI2019QN0301) and independent research project of National Laboratory of Pattern Recognition.

References

1. Chen, Y., Liu, S., Zhang, X., Liu, K., Zhao, J.: Automatically labeled data generation for large scale event extraction. In: Proceedings of the 55th Annual Meeting of the Association for Computational Linguistics, pp. 409–419 (2017)
2. Chen, Y., Xu, L., Liu, K., Zeng, D., Zhao, J.: Event extraction via dynamic multipooling convolutional neural networks. In: Proceedings of the 53rd Annual Meeting of the Association for Computational Linguistics, pp. 167–176 (2015)
3. Deng, S., Zhang, N., Kang, J., Zhang, Y., Zhang, W., Chen, H.: Meta-learning with dynamic-memory-based prototypical network for few-shot event detection. arXiv preprint arXiv:1910.11621 (2019)
4. Doddington, G.R., Mitchell, A., Przybocki, M.A., Ramshaw, L.A., Strassel, S.M., Weischedel, R.M.: The automatic content extraction (ACE) program-tasks, data, and evaluation. In: LREC, vol. 2, p. 1. Lisbon (2004)
5. Finn, C., Abbeel, P., Levine, S.: Model-agnostic meta-learning for fast adaptation of deep networks. In: Proceedings of the 34th International Conference on Machine Learning, vol. 70, pp. 1126–1135. JMLR. org (2017)
6. Gao, T., Han, X., Liu, Z., Sun, M.: Hybrid attention-based prototypical networks for noisy few-shot relation classification. In: Proceedings of the Thirty-Second AAAI Conference on Artificial Intelligence, (AAAI-2019), New York, USA (2019)
7. Geng, R., Li, B., Li, Y., Zhu, X., Jian, P., Sun, J.: Induction networks for few-shot text classification. In: Proceedings of the 2019 Conference on Empirical Methods in Natural Language Processing, November 2019
8. Huang, L., Ji, H., Cho, K., Dagan, I., Riedel, S., Voss, C.: Zero-shot transfer learning for event extraction. In: Proceedings of the 56th Annual Meeting of the Association for Computational Linguistics, July 2018
9. Ji, H., Grishman, R.: Refining event extraction through cross-document inference. In: 46th Annual Meeting of the Association for Computational Linguistics: Human Language Technologies, ACL-2008: HLT, pp. 254–262 (2008)
10. Kumar, A., et al.: Ask me anything: dynamic memory networks for natural language processing. In: International Conference on Machine Learning, pp. 1378–1387 (2016)
11. Lv, X., Gu, Y., Han, X., Hou, L., Li, J., Liu, Z.: Adapting meta knowledge graph information for multi-hop reasoning over few-shot relations. In: Proceedings of the 2019 Conference on Empirical Methods in Natural Language Processing, November 2019
12. Munkhdalai, T., Yu, H.: Meta networks. In: Proceedings of the 34th International Conference on Machine Learning, vol. 70, pp. 2554–2563. JMLR. org (2017)
13. Nguyen, T.H., Cho, K., Grishman, R.: Joint event extraction via recurrent neural networks. In: Proceedings of the 2016 Conference of the North American Chapter of the Association for Computational Linguistics, pp. 300–309 (2016)

14. Obamuyide, A., Vlachos, A.: Meta-learning improves lifelong relation extraction. In: Proceedings of the 4th Workshop on Representation Learning for NLP, August 2019
15. Qian, K., Yu, Z.: Domain adaptive dialog generation via meta learning. In: Proceedings of the 57th Annual Meeting of the Association for Computational Linguistics, July 2019
16. Santoro, A., Bartunov, S., Botvinick, M., Wierstra, D., Lillicrap, T.: Meta-learning with memory-augmented neural networks. In: International Conference on Machine Learning, pp. 1842–1850 (2016)
17. Snell, J., Swersky, K., Zemel, R.: Prototypical networks for few-shot learning. In: Advances in Neural Information Processing Systems, pp. 4077–4087 (2017)
18. Wang, X., et al.: HMEAE: hierarchical modular event argument extraction. In: Proceedings of the 2019 Conference on Empirical Methods in Natural Language Processing, pp. 5781–5787 (2019)
19. Wu, J., Xiong, W., Wang, W.Y.: Learning to learn and predict: a meta-learning approach for multi-label classification. In: Proceedings of the 2019 Conference on Empirical Methods in Natural Language Processing, pp. 4345–4355 (2019)
20. Yang, S., Feng, D., Qiao, L., Kan, Z., Li, D.: Exploring pre-trained language models for event extraction and generation. In: Proceedings of the 57th Annual Meeting of the Association for Computational Linguistics, pp. 5284–5294 (2019)
21. Ye, Z.X., Ling, Z.H.: Multi-level matching and aggregation network for few-shot relation classification. In: Proceedings of the 57th Annual Meeting of the Association for Computational Linguistics, July 2019
22. Yu, M., et al.: Diverse few-shot text classification with multiple metrics. In: Proceedings of the 2018 Conference of the North American Chapter of the Association for Computational Linguistics, June 2018

A Survey on Event Relation Identification

Ya Liu[1(✉)], Jiachen Tian[1], Lan Zhang[2], Yibo Feng[3], and Hong Fang[2]

[1] College of Intelligence and Computing, Tianjin University, Tianjin, China
[2] Shanghai Polytechnic University, Shanghai, China
[3] School of Mathematics and Statistics, Kashi University, Kashi 844000, China

Abstract. Event relation identification aims to identify relations between events in texts, including causal relation, temporal relation, sub-class relation and so on. Most of the research focuses on temporal relation and causal relation. Extracting events and the relation between events is an essential step to build an event-centric knowledge graph, which plays an important role in story ending prediction and decision-making. The form of causal and temporal relation in natural language text is diverse, sparse and complex which brings challenges to relation identification. In recent years, the integration of deep learning and knowledge has promoted the relation identification progress. This paper describes in detail the characteristics of causal and temporal relations in natural language texts and their connections. What is more, this paper surveys existing approaches based on pattern matching, machine learning and deep learning. Besides, this paper analyzes corpus and points out the future development direction and contributes ideas to further improve relation identification between events. To our knowledge, this is the first paper to survey the event relation identification.

Keywords: Event causality identification · Temporal relation identification · Joint relation extraction

1 Introduce

Event relation identification (ERI) aims to identify relations between events in texts, including causal relation, temporal relation, sub-class relation and so on. Extracting events and the relation between events is an essential step to build an event-centric knowledge graph, which plays an important role in story ending prediction and decision-making.

Most of the research focuses on causal relation and temporal relation. The causality in natural language text generally exists between events or between entities. According to the mapping relationship between events, Causality can be divided into one-to-one, one-to-many, many-to-one, and many-to-many. There are two types of causality, explicit and implicit causality. The former means that causality is guided by some explicit causal connectives, such as "cause", "lead to" and "because of". The latter means that there are no explicit causal connectives in the sentence. Causality can be expressed in one sentence or multiple adjacent

© Springer Nature Singapore Pte Ltd. 2021
H. Chen et al. (Eds.): CCKS 2020, CCIS 1356, pp. 173–184, 2021.
https://doi.org/10.1007/978-981-16-1964-9_14

sentences. Causality can be described in precise semantics or hidden in logic, which means the causality could be inferred. For example, "A causes B" and "B causes C," then "A causes C" is the causality hidden in the logical chain. The various forms of causality make the identification of causality challenging.

In natural language, temporal information can be used to express the relation between events, and reasoning about temporal relations can help figure out the temporality and duration of events, and summarize the timeline of a series of events. Generally speaking, temporal relations are categorized into three types [1]: the E-E TLINKs (those between a pair of events), the T-T TLINKs (those between a pair of times), and the E-T TLINKs (those between an event and a time). This paper focuses on the first type, E-E TLINKs. The temporal relation of events represents how different events expressed in an article's context are related to each other in chronological order. Common trigger words for temporal relations are:before, after, overlapping, and some hidden temporal relations.

In general, most of the original methods are based on pattern matching, and the machine can only identify the relation with explicit trigger words. With the development of machine learning, especially deep learning, and the increase of annotation data, the machine can identify implicit relations, and ERI has also expanded from the sentence level to the document level.

This paper introduces the classic data sets in this field in the second part and respectively introduce different methods of identifying the causal relation and temporal relation in the third part and fourth part. In the fifth part, it introduces the method of joint extraction of the two relations. In the sixth part, it studies the performance of different methods and analyzes the reasons. In the last part, it summarizes the paper and put forward some directions.

2   Corpus

Several well-known corpora are often used in relation recognition. Table 1 and Table 2 respectively, show the statistical analyses of causal and temporal relation datasets.

Event-Causality. [2] annotated a corpus called Event-Causality, which consists of 25 documents, 1,134 events, and 414 of 887 event pairs are causally related. Many researchers have extended this dataset for different tasks [3,4].

Causal-TimeBank. [5] annotated event-causal relations in the TempEval-3 corpus and released a corpus called CausalTimeBank, which contains 184 documents, 6,813 events, and 318 of 7,608 event pairs are causally related. On the other hand, it is also one dataset with temporal and causal annotations to identify temporal and causal relations [3,4]. However, Causal-TB is sparse, in which causal relations were annotated based on only explicit causal connectives (e.g., "A happened because of B").

Japanese Dataset. [6] mainly applied the excitation filter [7], where they only kept the event causality candidates that contained excitation templates in each phrase to all event causality candidates from the web, acquiring 2,451,254 event causality candidates.

EventStoryLine. [8] had built a corpus called EventStoryLine, which contains 258 documents and more than 5,000 causal relations. The EventStoryLine corpus is the largest dataset for causal relation identification until now with comprehensive event-causal relations annotated, both intra-sentence and cross-sentence, which presents unique challenges for causal relation identification. This paper showed that only 117 annotated causal relations in this dataset are indicated by explicit causal cue phrases while the others are implicit. This dataset is widely used in causality extraction [9,10].

TempEval-3 and Related Corpus. The TempEval3 (TE3) workshop [11] provided datasets for TimeBank [12], AQUAINT [13], Sil-VER and Platinum, TB and AQ for training and TE3-PT for testing. [14] gave a detailed introduction to TempEval-3.

TBAQ and Korean TimeBank Corpus. To better evaluate the model's performance, [15] used English and Korean datasets. It used the TBAQ corpus in English-base models, and used the Korean TimeBank corpus in Korean-base models.

TimeBank Dense. [16] proposed a new annotation process, TimeBank Dense used 14 relations based on the Allen interval relations and the authers annotated 36 random documents each with at least two annotators. The corpus set on six relations: before, after, includes, is included, simultaneous, and vague.

Table 1. Corpora of causality.

Name	Doc	Event	Causal relation
Event-Causality	25	1134	414
CausalTimeBank	184	6813	318
Japanese dataset	–	–	2,451,254 (event causality candidates)
EventStoryLine	258	258	5000

3   Event Causality Identification

In NLP, there are two main aspects of causality, including causality extraction [2,6,7,19,20], which focuses on extracting event-causal pairs from the web or document collections, and causality identification, which focuses on judging whether a given event pair is causal.

Table 2. Corpora of temporality.

Name	Doc	Event	T-Link	Words
TimeBank	*183*	6714	5121	61418
AQUAINT	*73*	4431	5797	33973
Korean TimeBank	*725*	10519	4629	50674
TB-Dense	*36*	1600	5700	–

Semi-supervised and Unsupervised-Based Methods. The characteristic of this type of method is to construct rules based on the common pattern of causal candidate pairs. [21] proposed a causality extraction method based on part-of-speech, syntactic analysis and causal relation templates. The paper used the semi-supervised idea first to extract some sentence templates expressing causality from sentences on Wikipedia that clearly contained causality, and then used these templates to extract causal candidate pairs in other sentences, and then iterate continuously. [22] described the textual causality connectors. [20] selected the most frequently used connectors, "because", "because of", "lead to", "after" respectively, and constructed corresponding rules to extract causal candidate pairs. [2,6] created corresponding templates from semantic relations directly or indirectly related to causality, and extracted causal candidate pairs.

Obviously, this method is better for the identification of causality with explicit connectives, and it is poor for causal relations that do not have obvious causal triggers and that require reasoning and background knowledge. Furthermore, this method is not widely used.

Machine Learning-Based Methods. The characteristic of this type of method is to find suitable features to build a classifier. The features for the event causality classifier are various. Causation and prevention are explicit relations directly related to causality, and material is an implicit relation indirectly related to causality. Context features mean other text fragments of the sentences where the event causality candidate appears. Association features could measure the association strength between cause and effect, based on the statistical analysis characterized by point-wise mutual information (PMI) or the count of events co-occurrence on large corpora. Base features refer to basic features such as part of speech and syntactic features.

[23] used a Bayesian classifier based on the part-of-speech, semantics and dependency features to filter incorrect candidate pairs, which can only extract from sentences with the explicit connective. [6] improved the performance by constructing a classifier based on SVM, leveraging different combinations of features mentioned above, especially the semantic relationship that did not seem directly relevant to causality. [24] trained two separate logistic classifiers, one for causality detection within sentences, and one for causality detection between sentences.

Feature-based machine learning methods can expand features from different aspects and increase the diversity of features. Besides, the use of carefully selected features can improve the accuracy of recognition in the case of insufficient corpus. However, these methods also have some disadvantages. They have a certain dependence on features by learning the features of the annotated corpus, and the performance is often related to the size of the corpus. Generally, the larger the corpus, the better the recognition effect.

Deep Learning-Based Methods. Deep learning is beneficial in solving long-distance dependence of text, semantic understanding, and learning common language representations. The integration of deep learning and knowledge promotes the improvement of NLP downstream tasks, including event causality identification. [9] thought that causality existed in a wide range of forms and lacked clear connectives. And the integration of more background knowledge could promote the identification of causality. The paper introduced multi-column convolutional neural networks (MCNNs)[25] to combine context features and background knowledge from different sources, including short binary patterns, answers from a why-QA system, and one or two consecutive sentences with clue terms. The experimental results showed that the neural network-based method outperformed the method based on SVMs. Moreover, the recent work [10] employs BERT architecture, pre-trained from scratch based on causality-rich texts, which can learn context-dependent representation for the task and achieves superior performance. Furthermore, the paper also mentioned that when the pre-training corpus is insufficient, background knowledge can effectively promote the performance of BERT. [3] believed that existing methods cannot use background knowledge and are usually not well generalized to new previously invisible data. Therefore, the paper introduced knowledge for reasoning to enrich the representation of events and leveraged a mechanism of event mention masking generalization to mine context-specific patterns that are irrelevant to the event, greatly solving the above two problems.

4   Event Temporal Relation Identification

The work related to relation extraction can be roughly divided into three categories: pattern matching, machine learning (including unsupervised, supervised, etc.), deep learning.

Pattern Matching-Based Methods. Previous methods of temporal relation extraction are mostly based on features. [27] used MaxEnt classifier to classify temporal relations in the corpus of hand-tagged features, including tense, aspect, and so on. Later [28] proposed a two-stage classifier. Firstly, it learned incomplete event attributes. The second step was to combine other language features to classify. In a dependency tree, because the full paths were too sparse, [29] relied on the 3-gram path between two event mentions as features. If the entire path is turned into a sequence, more evidence about the temporal relations will be generated.

Machine Learning-Based Methods. However, the feature-based method is not suitable for two events that are far apart or separated by other events. This method cannot be coded with discrete features. Many scholars use machine learning classifiers [30, 31] for temporal relation extraction. These classifiers are trained based on predefined, limited, and fixed relation type patterns. However, these classifiers' disadvantage is that when there are different sentences or different texts in the temporal relation, it cannot identify all event relation types. [32] proposed an unsupervised method that used the Open IE system to extract the temporal relation. In this paper, it constructed a network structure of events by this system. However, the shortage that can only identify similar event nodes in the graph restricted the system's performance. [33] developed a structured learning approach to identify temporal relations in natural language text, and it had the advantage of capturing the global nature of the problem. Besides, the global nature made better use of unmarked data, which further improved the proposed method.

Deep Learning-Based Methods. With the excellent performance of deep learning in various fields, it began to be applied to event temporal relation recognition. From the perspective of natural language processing, there have been many studies based on deep learning models in recent years. Between events in the same sentence, [34] extracted a series of certain words that represented the temporal relation between two events, then built bidirectional recurrent neural network (LSTM) models which took context word sequence, part of speech tag sequence, and dependency sequence as input. [15] extend the BERT model, which divided the input sentence into single words for labeling, and converted it into an embedding vector, thereby extracted the temporal relation from the input sentence. However, the above model cannot solve the problem of extracting relations across multiple sentences. In addition, the temporal relation of cross-language events is difficult to obtain because the sentence contains hidden semantics, and different language texts are difficult to represent in the same feature space. A Bi-LSTM cross-attention (BLCATT) model [35] was proposed to solve the temporal relation extraction of cross-language events. The model explored the logical hierarchy of temporality, it used word vectors to represent text and map two languages to the same space and then obtained event semantic information, including temporal series features through attention mechanism and inter-event rule features. This is a document-level implementation of the advancement of event temporal relation extraction.

5   Joint Relation Extraction

The temporal relation and causal relation are two important types of events relations, and there is a certain connection between them. For example, a cause event usually occurs before an effect event. How to connect them and use the characteristics of the two relations are the focuses of joint reasoning.

[37] introduced a new semantic annotation framework: CatreRs. Taking into account the causality of the temporal factor of the FFECT event "after" the CAUSE event, the paper proposed the temporal relation: Before, Overlaps and During, and judged the causality of events in the text based on these three relations. Although this paper proposed an annotation method of causal relation based on temporality, this framework was more suitable for script learning and story comprehension tasks than a comprehensive analysis of all temporality and causality in the text.

In [31], TimeML was optimized through SVM and feature engineering for the temporal relation. In the causality extraction, the paper proposed similar annotation criteria to TimeML. According to manual annotation, <CLINK> represented the cause and result of the event, and the <C-SINGLE> tag represented the text element of the causal relation. According to the above method, the result of extracting causal relation from TempEval-3 corpus showed that the causal relation extraction involving the temporal factor was improved in both F1 and Recall.

[4] established a joint reasoning framework based on Constraint Condition Model (CCM). The feature of this paper was that the joint problem was expressed as an integer linear programming (ILP) problem, which enforced the inherent constraints of temporality and causality. The causal relation was further integrated to complete the TCR joint reasoning framework on the basis of the temporal module with constraints. The advantage of this work was the development of a jointly annotated data set, which solved a key obstacle to the lack of common annotated data sets for temporality and causality.

6   The State of the Art

Table 3, Table 4, Table 5, show the performance of each model on different datasets. For sentence-level causality identification, [3] is the state of the art on EventStoryLine datasets and EventCausality datasets. In the context of low-resource environments, the knowledge-enhanced event causality identification proposed effectively utilized the extra knowledge in the knowledge graph, and the semantic representation based on BERT. [10] is the most advanced method on Japanese datasets.

Besides, existing methods mainly identify sentence-level causality, but they lack the document-level causality recognition. One reason is that different mentions of the same event are scattered in multiple sentences, and the effect of event extraction seriously affects the extraction of causality between events. The other reason is that event causality is sparse among all possible event pairs in the document. In addition, the explanation of causality is not clear. [24] added global constraints to the document-level causality identification by adopting various measures such as extracting relationships around main events, sentence location, and event co-referencing. The combination of global structure and fine-grained structure, segmented processing, event co-referencing, event fusion, and topic-based all help enrich the document-level causal structure.

Table 3. Empirical results of different models on EventStoryLine. Pre., Rec. and F1 indicate precision (%), recall (%) and F1-score (%) respectively; Bold denotes best results;

METHODS	PRE.	REC.	F1.
OP [8]	22.5	98.6	36.6
LSTM [36]	34.0	41.5	37.4
Seq [34]	32.7	44.9	37.8
LR+ [24]	37.0	45.2	40.7
LIP [24]	38.8	52.4	44.6
MFULL [3]	**41.9**	**62.5**	**50.1**

Table 4. Empirical results of different models on EventCausality datasets. Pre., Rec. and F1 indicate precision (%), recall (%) and F1-score (%) respectively; Bold denotes best results;

METHODS	PRE.	REC.	F1.
PMI [2]	26.6	20.8	23.3
ECD [2]	40.9	23.5	29.9
CEA [2]	**62.2**	28.0	38.6
MFULL [3]	34.1	**68.2**	**45.4**

Table 5. Empirical results of different models on Japanese datasets. Pre., Rec., F1. and Avg.P indicate precision (%), recall (%), F1-score (%) and average precision (%) respectively; Bold denotes best results;

METHODS	PRE.	REC.	F1.	Avg.P
SVM [6]	–	–	–	45.96
MCNN [9]	**61.1**	40.2	48.5	55.13
BERT [10]	52.0	**64.0**	**57.4**	**57.4**

For the extraction of temporal relation at the sentence level, [15] designed the temporal information extraction model, it could predict not only backward relations in temporality (i.e., forward and backward), but also temporal information of specific events (i.e., including). Due to the limited data set, and in different languages the representation of temporal relations is different. When some trigger words can not be found and hidden temporal relations between events cannot be identified, If we want to extract the temporal relation between multiple events, it is also worth exploring how to grasp the whole and improve the accuracy.

7   Application of Relation Extraction

Event relation extraction is currently used to construct Event Logic Graph (ELG). [19] formally put forward the concept of event logic graph (ELG) for the first time, in which the nodes represented the events and the edges represented the relations between events, and described the process of constructing ELG. Event relation extraction could be used to construct ELG. On September 10, 2018, HIT-SCIR officially released the "Financial Event Logic Graph v1.0", a large-scale Chinese event logical knowledge about the financial field. Subsequently, the "Financial Event Logic Graph v2.0" was launched, which extracted events from more sources, significantly increasing the scale of data, the number of specific events, and the number of specific causal relations.

[39] proposed a novel context-aware variational autoencoder (CWAVE) effectively learning event background information to guide the If-Then reasoning. The training process, similar to BERT, consisted of two periods. In the pre-train period, CWAVE was trained on an auxiliary dataset, consisting of three narrative story corpora and containing rich event background knowledge, to learn the event background information. In the finetune period, CWVAE was trained on the task-specific dataset to adapt the event background information to each specific aspect of If-Then inferential target. Similar ideas can also be applied to other text reasoning techniques, e.g., tractive reasoning based on knowledge graphs, which means given an observed event, it is required to select the possible cause of the observed event.

8   Concluding Remarks

Despite recent advances in event relation identification, state-of-the-art identification results are still far from satisfactory. Most of the work is limited to sentence-level relation identification and regards relation identification between events as a classification task. Due to the complexity of the form of event relations, document-level event relation extraction is still a difficult task. In a low-resource environment, there are not enough tool resources to learn effective features to detect events, parameters and corresponding relations between events, and there are not enough annotation samples to train the event and relation extraction. We can consider the following points. (1) Joint extraction. We can consider the joint extraction and optimization of entities, events and event relations. Joint extraction can allow each task to correct each other's errors and make full use of the information and dependencies between entities and events. (2) Integrating knowledge graph. Add external knowledge to enrich event expressions, which can be used to solve relation extraction that requires reasoning. The connections between entities not only explicitly exist in the knowledge graph, but also implicitly exist in the pre-training language model. (3) Combination of global structure and local structure. We can learn from the idea of generating abstracts to extract the relation based on topic distribution and segmentation processing. Articles are composed of paragraphs, and different paragraphs focus on different topics and events. Modeling paragraphs to form the global structure of the article is beneficial to document-level event extraction.

References

1. Ning, Q., Feng, Z., Roth, D.: A structured learning approach to temporal relation extraction. In: Proceedings of the 2017 Conference on Empirical Methods in Natural Language Processing, Copenhagen, Denmark, pp. 1027–1037 (2017)
2. Do, Q., Chan, Y.S., Roth, D.: Minimally supervised event causality identification. In: Proceedings of the 2011 Conference on Empirical Methods in Natural Language Processing, pp. 294–303 (2011)
3. Liu, J., Chen, Y., Zhao, J.: Knowledge enhanced event causality identification with mention masking generalizations
4. Ning, Q., Feng, Z., Wu, H., Roth, D.: Joint reasoning for temporal and causal relations. arXiv preprint arXiv:1906.04941 (2019)
5. Mirza, P., Sprugnoli, R., Tonelli, S., Speranza, M.: Annotating causality in the TempEval-3 Corpus, pp. 10–19. Gothenburg, Sweden (2014)
6. Hashimoto, C., et al.: Toward future scenario generation: extracting event causality exploiting semantic relation, context, and association features. In: Proceedings of the 52nd Annual Meeting of the Association for Computational Linguistics, pp. 987–997 (2014)
7. Hashimoto, C., Torisawa, K., De Saeger, S., Oh, J.H.: Excitatory or inhibitory: a new semantic orientation extracts contradiction and causality from the web. In: Proceedings of the 2012 Joint Conference on Empirical Methods in Natural Language Processing and Computational Natural Language Learning, pp. 619–630 (2012)
8. Caselli, T., Vossen, P.T.J.M.: The event StoryLine corpus: a new benchmark for causal and temporal relation extraction. In: Proceedings of the Workshop Events and Stories in the News, Vancouver, Canada, pp. 77–86 (2017)
9. Kruengkrai, C., Torisawa, K., Hashimoto, C., Kloetzer, J., Oh, J.H., Tanaka, M.: Improving event causality recognition with multiple background knowledge sources using multi-column convolutional neural networks. In: Thirty-First AAAI Conference on Artificial Intelligence (2017)
10. Kadowaki, K., Iida, R., Torisawa, K., Oh, J.H., Kloetzer, J.: Event causality recognition exploiting multiple annotators' judgments and background knowledge. In: Proceedings of the 2019 Conference on Empirical Methods in Natural Language Processing and the 9th International Joint Conference on Natural Language Processing (EMNLP-IJCNLP), pp. 5820–5826 (2019)
11. UzZaman, N., Llorens, H., Allen, J., Derczynski, L., erhagen, M.V., Pustejovsky, J.: SemEval-2013 Task 1: TEMPEV AL-3: evaluating time expressions, events, and temporal relations. In: Second Joint Conference on Lexical and Computational Semantics, vol. 2, pp. 1–9 (2013)
12. Pustejovsky, J., et al.: The timebank corpus. In: Corpus linguistics, p. 20, March 2013

13. Graff, D.: The AQUAINT corpus of English news text. Linguistic Data Consortium, Philadelphia (2002)
14. UzZaman, N., Llorens, H., Allen, J., Derczynski, L., Verhagen, M., Pustejovsky, J.: Tempeval-3: evaluating events, time expressions, and temporal relations (2012)
15. Lim, C.G., Choi, H.J.: Temporal relationship extraction for natural language texts by using deep bidirectional language model. In: 2020 IEEE International Conference on Big Data and Smart Computing (BigComp), pp. 555–557. IEEE, February 2020
16. Cassidy, T., McDowell, B., Chambers, N., Bethard, S.: An annotation framework for dense event ordering. Carnegie-Mellon Univ Pittsburgh PA (2014)
17. Prasad, R., Miltsakaki, E., Dinesh, N., Lee, A., Joshi, A., Webber, B.L.: The Penn discourse treebank 1.0 annotation manual (2006)
18. Palmer, M., Gildea, D., Kingsbury, P.: The proposition bank: an annotated corpus of semantic roles. Comput. Linguist. **31**(1), 71–106 (2005)
19. Ding, X., Li, Z., Liu, T., Liao, K.: ELG: an event logic graph. arXiv:1907.08015 (2019)
20. Zhao, S., et al.: Constructing and embedding abstract event causality networks from text snippets. In: Proceedings of the Tenth ACM International Conference on Web Search and Data Mining, pp. 335–344 (2017)
21. Ittoo, A., Bouma, G.: Extracting Explicit and Implicit Causal Relations from Sparse, Domain-Specific Texts. In: Muñoz, R., Montoyo, A., Métais, E. (eds.) NLDB 2011. LNCS, vol. 6716, pp. 52–63. Springer, Heidelberg (2011). https://doi.org/10.1007/978-3-642-22327-3_6
22. Wolff, P., Song, G.: Models of causation and the semantics of causal verbs. Cogn. Psychol. **47**(3), 276–332 (2003)
23. Sorgente, A., Vettigli, G., Mele, F.: Automatic extraction of cause-effect relations in natural language text, pp. 37–48 (2013)
24. Gao, L., Choubey, P. K., Huang, R.: Modeling document-level causal structures for event causal relation identification. In: Proceedings of the 2019 Conference of the North American Chapter of the Association for Computational Linguistics: Human Language Technologies, pp. 1808–1817 (2019)
25. Ciregan, D., Meier, U., Schmidhuber, J.: Multi-column deep neural networks for image classification. In: 2012 IEEE Conference on Computer Vision and Pattern Recognition, pp. 3642–3649. IEEE (2012)
26. Li, W., He, L., Zhuge, H.: Abstractive news summarization based on event semantic link network. Association for Computational Linguistics (2016)
27. Mani, I., Verhagen, M., Wellner, B., Lee, C., Pustejovsky, J.: Machine learning of temporal relations. In: Proceedings of the 21st International Conference on Computational Linguistics and 44th Annual Meeting of the Association for Computational Linguistics, pp. 753–760, July 2006
28. Chambers, N., Wang, S., Jurafsky, D.: Classifying temporal relations between events. In: Proceedings of the 45th Annual Meeting of the Association for Computational Linguistics Companion Volume Proceedings of the Demo and Poster Sessions, pp. 173–176, June 2007
29. Laokulrat, N., Miwa, M., Tsuruoka, Y., Chikayama, T.: UTTime: temporal relation classification using deep syntactic features. In: Second Joint Conference on Lexical and Computational Semantics (* SEM), Proceedings of the Seventh International Workshop on Semantic Evaluation (SemEval 2013), pp. 88–92, June 2013
30. Laokulrat, N., Miwa, M., Tsuruoka, Y.: Stacking approach to temporal relation classification with temporal inference. Inf. Media Technol. **11**, 53–78 (2016)

31. Mirza, P., Tonelli, S.: CATENA: causal and temporal relation extraction from natural language texts. In: Proceedings of COLING 2016, the 26th International Conference on Computational Linguistics: Technical Papers, pp. 64–75, December 2016

32. Vo, D.-T., Bagheri, E.: Extracting Temporal Event Relations Based on Event Networks. In: Azzopardi, L., Stein, B., Fuhr, N., Mayr, P., Hauff, C., Hiemstra, D. (eds.) ECIR 2019. LNCS, vol. 11437, pp. 844–851. Springer, Cham (2019). https://doi.org/10.1007/978-3-030-15712-8_61

33. Ning, Q., Feng, Z., Roth, D.: A structured learning approach to temporal relation extraction. arXiv preprint arXiv: 1906.04943 (2019)

34. Choubey, P.K., Huang, R.: A sequential model for classifying temporal relations between intra-sentence events. arXiv preprint arXiv: 1707.07343 (2017)

35. Wang, J., Guo, J., Yu, Z., Gao, S., Huang, Y.: Temporal Relationship Recognition of Chinese and Vietnamese Bilingual News Events Based on BLCATT. In: Sun, Y., Lu, T., Yu, Z., Fan, H., Gao, L. (eds.) ChineseCSCW 2019. CCIS, vol. 1042, pp. 497–509. Springer, Singapore (2019). https://doi.org/10.1007/978-981-15-1377-0_39

36. Cheng, F., Miyao, Y.: Classifying temporal relations by bidirectional LSTM over dependency paths. In: Proceedings of the 55th Annual Meeting of the Association for Computational Linguistics, pp. 1–6 (2017)

37. Mostafazadeh, N., Grealish, A., Chambers, N., Allen, J., Vanderwende, L.: CaTeRS: causal and temporal relation scheme for semantic annotation of event structures. In: Proceedings of the Fourth Workshop on Events, pp. 51–61, June 2016

38. Pustejovsky, J., et al.: TimeML: robust specification of event and temporal expressions in text. New Dir. Question Answering 3, 28–34 (2003)

39. Graff, D.: The AQUAINT corpus of English news text: [content copyright] Portions 1998–2000 New York Times Inc., 1998–2000 Associated Press Inc., 1996–2000 Xinhua News Service. Linguistic Data Consortium (2002)

Employing Multi-cues to Identify Event Factuality

Liumin Zhang, Peifeng Li, Zhong Qian[✉], Xiaoxu Zhu, and Qiaoming Zhu

School of Computer Sciences and Technology, Soochow University, Jiangsu, China
20185227086@stu.suda.edu.cn, {pfli,qianzhong,xiaoxu,
qmzhu}@suda.edu.cn

Abstract. Event factuality represents the factual nature of events in texts, and describes whether an event is a fact, a possibility, or an impossible situation. Previous work usually used the embedding of event sentence to represent the event factuality, ignoring the other helpful evidence, such as negative words, speculative words and time words. To address the above issue, this paper introduces various kinds of effective cues, i.e., time cue, negative cue and speculative, to a BERT-based convolutional neural network to identify Chinese sentence-level event factuality. Experimental results on the Chinese DLEF corpus showed that our model outperforms the baseline BERT on macro and micro F1 by 3.64% and 3.77%, respectively. Moreover, the training time of our model is just one-fifth of the benchmark.

Keywords: Event factuality identification · BERT · Time cue

1 Introduction

Event factuality expresses the commitment of relevant sources towards the factual nature of events, conveying whether an event is characterized as a fact, a possibility, or an impossible situation, and is useful for deep NLP applications, such as text analysis [1], rumor detection [2], and so on. Event factuality identification (EFI) can be divided into sentence-level and document-level task. To be specific, sentence-level EFI is to identify the factuality of the events according to the information within the current sentence, while document-level EFI is to identify events according to the information among the current document. This paper focuses on Chinese sentence-level EFI. In principle, event factuality is related to various factors, including time, speculative and negative cues. Take the following sentences for examples:

(S1) *McCulley, a famous economist, doubts that the tax rate will **increase**soon.*

(S2) *The report added that, because they did not understand Gondola's **driving skills**, they can't change directions between the moored boats.*

(S3) *It is reported that another Israeli driver was **killed**in a shooting in the Gaza Strip on the 13th.*

In S1, the factuality of the event "*increase*" is possible to happen (PS+) according to the predicate "*doubts*", a speculative cue, and "*will*", an adverb word, representing

© Springer Nature Singapore Pte Ltd. 2021
H. Chen et al. (Eds.): CCKS 2020, CCIS 1356, pp. 185–195, 2021.
https://doi.org/10.1007/978-981-16-1964-9_15

an expectation that will happen. In S2, because the event *"driving skill"* is governed by the negative cue *"not"*, its event factuality is certainly not to happen (CT−). In S3, Since *"killed"* is modified by the time *"13th"*, which represents the past, as we know, it is relatively easy to identify the fact because this event occurred in the past. Since it is a past time, we can determine whether it is a fact according to the grammar habit and exclude the other event factuality such as possible to happen (PS+). So the event factuality of *"killed"* is certain to happen (CT+), according to the semantic characteristics of the sentence and the time. In a word, we can find that negative cues, speculative cues, time words and adverb cues play an important role in identifying the factuality of events.

Previous methods employed rules [3, 4], traditional machine learning models [5, 6], and deep learning models [7, 8], which relied on annotated information, and were also time-consuming. Although some researches [7, 8] considered negative or speculative cues, they still ignored the time and adverb words and did not consider modeling the above cues together.

Recent work on nature language process always used deep learning methods. However, some studies have found that existing deep learning models might not really understand the natural language texts [9] and vulnerably suffer from adversarial attacks [10]. Through their observation, the existing deep learning models on event factuality identification paid great attention to non-significant words and ignored important ones, such as negative, time cues and so on.

BERT, a pre-trained deep bidirectional transformer model, achieves state-of-the-art performance across various tasks. Tenney et al. [11] and Liu et al. [12] showed that BERT embeddings encode information about parts of speech, syntactic chunks and roles. However, the knowledge of syntax in BERT is partial, since probing classifiers could not recover the labels of distant parent nodes in syntactic trees.

To address the above issues, this paper proposes an event factuality identification model to identify event factuality from raw texts. It first extracts basic factors related with factuality automatically, and then utilizes a BERT-based Convolutional Neural Network (Bert_CNN) with the integration of speculative cues, negative cues, time words and adverbs, which can make up for the inability of BERT to learn long-distance syntactic knowledge and pay more attention to the important words in the sentence. The contributions of this paper can be summarized as the following aspects:

1. We introduce the combination of negative, speculative, time and adverb cues arriving event to event factuality identification, which can effectively filter the noise comparing with the event sentence.
2. Our model can identify event factuality from raw texts, which does not rely on annotated information, and is beneficial for application in industry.
3. The training time of our model is just one-fifth of the Bert benchmark.

The remaining contents of this paper are organized as follows: Sect. 2 briefly introduces the related work in this field. Section 3 describes the model architecture and implementation details. Section 4 reports the experimental settings and results. Section 5 summarizes the work and prospects for the future work.

2   Related Work

This section mainly introduces the related work on sentence-level events factuality identification from two aspects: the event factuality corpora and methods related to event factuality identification.

2.1   Corpora

For the English sentence-level event factual corpus, Saurí et al. [13] constructed an event factuality-oriented factbank corpus, based on the time-labeled corpus timebank. The event factuality value in factbank consists of two parts: modality and polarity, which mainly includes CT+, CT−, PR+, PS+, Uu, etc., the factuality of events in the factbank corpus is determined according to different sources. Lee et al. [6] annotated the UW factual corpus on the TempEval-3 corpus, using [−3, 3] to represent the situation from when the event is determined not to happen to the event definitely happens.

In terms of Chinese sentence-level event factual corpus, Cao et al. [4] annotated the event factuality values and their related attributes based on the ACE2005 Chinese event corpus. Qian et al. [14] built a bilingual document-level event factuality corpus DLEF from *China Daily* and *Sina* bilingual news.

2.2   Sentence-Level Event Factuality Identification

Sentence-level event factuality identification is divided into three categories: Rule-based methods, statistical methods and neural networks.

In the rule-based method and statistical methods, Saurí and Saurí et al. [15] proposed an algorithm to calculate the event factuality. Cao et al. [4] constructed a corpus about event factuality and extracted three-dimensional models using rules to infer the value of Chinese event factuality. Saurí et al. [15] and Werner [16] have adopted a method based on support vector machine SVM and feature engineering.

In the neural network-based method, Qian et al. [8] proposed a method based on generative adversarial networks on the FactBank corpus, which divides sentence-level event factual identification into two steps. The first step was to extract features from raw corpus, and then proposed generative recognition model of assisted classification against network events; He et al. [7] used the semantic features of sentences on the corpus labeled by Cao et al. [4], combined with the convolutional neural network to infer event factuality; Qian et al. [14] used BiLSTM and Attention to obtain sentence features, and then used the adversarial neural network to identify the event factuality of the document and sentence. Veyseh et al. [17] used GCN, whose the weight matrix A was the linear combination of the sematic structure and the syntactic structure with the trade-off λ, to identify the event factuality.

Compared with the previous work, this paper is the first one to identify Chinese event factuality by BERT on raw texts, and the first one to introduce the integration of different cues to this task. A s we know, adverbs are special for Chinese, which can enhance sentence semantics and facilitate the capture of Bert semantics.

3   Model of Event Factuality Identification

Our model is mainly divided into two parts: one part uses BERT to extract the semantic features of the sentence, and the input is the sentence itself; the other part is to obtain the syntactic features through the combination of BERT and CNN, and the input of the syntactic feature part is composed of time words, adverbs, negative and speculative cues, all of two parts use max-pooling to obtain the last global feature.

3.1   Input Feature

Let $(x1, x2, ..., xn)$ be a sentence, where n is the number of words/tokens and xi is the i-th token in the sentence. Other cues' representations are shown below:

(1).   The paths from the cues to events in the sentence are recorded as TIME_AD: firstly, using the pyltp[1] to perform part-of-speech, and then obtaining time words if 'nt' or 'd', which represents time words or adverbs., in the result of part-of-speech, such as "*13th*" in S3, and then form the path of cue words to the event which obtained through the dependency parsing tree, such as "killed in 13th".

(2).   The paths from the cues to events in the sentence are recorded as NEG and POS: the negative and speculative cues used in this paper are obtained by matching the cue words in the CNeUn corpus, and then the dependency parsing tree is constructed by pyltp tool. We obtain the syntax paths from cues to events according to the dependency parsing tree. For example, from the sentence S1, we can obtain the dependency syntactic structure like Fig. 1 by using pyltp and then locate the event "*increase*" and the speculative cue "*doubts*" in the Fig. 1. Hence we can obtain the syntactic sentence "*doubts will increase*".

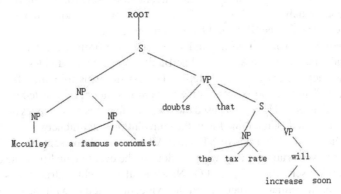

Fig. 1. Dependency syntactic structure diagram.

The above two types of syntactic paths are concatenated together, and then BERT [18] and CNN are used for word segmentation and encoding to obtain the semantic and adjacent syntactic features of the sentence.

---

[1] https://pyltp.readthedocs.io/zh_CN/latest/.

3.2 BERT-Based Neural Network for EFI

To obtain a universal representation for both the sentence semantics and cues, we utilize BERT, a pre-trained deep bidirectional transformer model that achieves state-of-the-art performance across various tasks, as the encoder.

BERT is a multi-layer transformer pre-trained on next sentence prediction and masked word prediction using extremely large datasets. Specifically, we generate the input sequence by concatenating a [cls] token, the tokenized sentence and a [sep] token. For each token in the sequence, its input representation is the elementwise addition of WordPiece embeddings, positional embeddings, and segment embeddings. BERT fine-tuning is straightforward. In the baseline Bert_Sen, we use the last hidden representation of [cls] as the basis of classification.

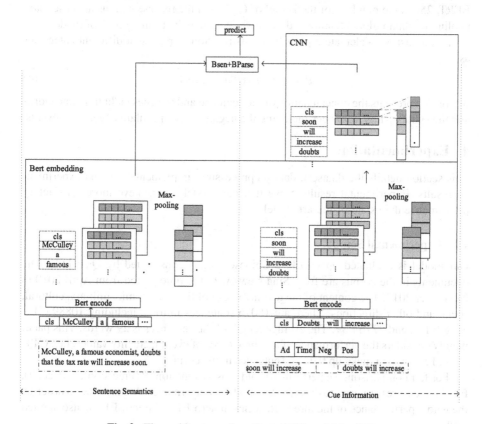

Fig. 2. The architecture of our Bert_CNN model for EFI

This paper uses BERT to obtain sequences representation BSen and BParse (where BSen represents the BERT encoding information of the sentence, BParse represents the vector code after the BERT encoding the syntactic path from the negative, speculative, the time and adverb cues to event). We design Bert_CNN for event factuality identification shown in Fig. 2. The loss function in the model is a cross-entropy function. As the output of layer BERT is expressed as follow:

$$BSen = BERT(cls, x1, x2, ..., xn, sep)$$ (1)

$$BParse = BERT(cls, TIME\_AD \oplus NEG \oplus POS, sep)$$ (2)

where x1, x2, ... represents character in the event sentence. We use the contextualized token representations from the last layer of BERT by max-pooling to obtain global feature of cues. And as the output of layer CNN [19] is expressed as follow:

$$Rp = BParse * W1 + b1$$ (3)

where $W1 \in R^{(768*256)}$, $b1 \in R^{(1*256)}$, where 768 represents 768 hidden units of BERT, 256 is the numbers of the kernel of CNN and the representations also use max-pooling to obtain global feature and meanwhile reduce the training time of model.

At the last, we calculate a probability distribution to represent different categories as:

$$P = softmax(BSen \oplus Rp)$$ (4)

where $\oplus$ represents the concatenation of the sentence and the cues including time words and the syntax paths from cues including the negative and speculative words, to events.

4 Experimentation

This section details the datasets, data preprocessing, implementation, and experimental results. Experimental results prove that our model can achieve more satisfactory performance than state-of-the-art model.

4.1 Experimental Setup

Our model is evaluated on DLEF, a Chinese corpus constructed by Qian [14]. The documents in the corpus are from Sina News with the time period from April 2010 to November 2017. The content of the text involves political, economic, military, cultural, social and other directions. There are 18076 sentences in total, including 10889 CT+, 3718 CT−, and 2776 PS+. The average length of the sentences is 29 words. This paper mainly considers the performance of the three types of event factuality values, i.e., CT+, CT−, PS+, which make up 97% of sentences in the corpus.

For fair comparison, we perform 10-fold cross-validation. P (Precision), R (Recall), F1 (F1-measure) are used to describe the model's performance. Meanwhile, to describe the whole performance of the three categories, macro-F1 and micro-F1 are also applied [20].

The length of sentences and syntax sentences we constructed is set to 70, the window size of the convolutional neural network is 1, which is in order to be focus on the character level, the number of convolutional kernels is 256, and the features of the convolutional neural network are extracted by max-pooling. We use the pre-trained uncased BERT-base model for embedding. The number of transformer blocks is 12, the hidden layer size is 768, the number of self-attention heads is 12, and the total number of parameters for the pre-trained model is 110M.

4.2 Experimental Results

To evaluate our model on the task of event factuality identification, we use the following two models for comparison.

1) **Qian** [14]: It used BiLSTM and the attention mechanism to extract sentence features and the adversarial training through the whole process to add more semantic information.
2) **BERT**: It used Bert to encode sentence and did not use any additional information.

The input of the above two models is all the sentence itself, the input of our model BERT_CNN is the sentence itself and the integration of the path from the different cues to the events. Table 1 shows the performance comparison and it shows that our BERT_CNN outperforms the other two baselines both on micro-F1 and Macro-F1. This result verified the effectiveness of our model BERT_CNN on the task of event factuality identification.

Table 1. Performance comparison between our BERT_CNN and the Baselines

	CT+			CT−			PS+			Overall	
	P	R	F1	P	R	F1	P	R	F1	Micro-F1	Macro-F1
Qian	86.57	88.95	87.73	73.89	74.92	74.20	72.78	65.65	68.88	81.98	76.94
Bert_Sen	87.58	88.96	88.27	75.07	**77.51**	76.27	78.22	63.08	69.84	83.30	78.36
Bert_CNN	**89.23**	**92.86**	**91.01**	**84.43**	76.42	**80.23**	**79.12**	**77.42**	**78.26**	**87.08**	**83.24**

From Table 1, we can find that:

- The performance of CT+ is better than that of CT− and PS+ on all the models. According to the analysis of the corpus, the reasons are as follows: firstly, there are 18076 sentences in the corpus, of which 10889 are CT+, accounting for about 60%, which are relatively large in number and easy to predict; secondly, because the corpus is news, the words are generally characteristic, for example, in S2, there is a definitely negation for the event *"driving"*, the adjective *"not"* changes the factuality value of the whole sentence into negation, which has strong characteristics.
- Compared with Qian, our BERT_CNN improves the macro F1 and micro F1 by 5.61 and 6.30, respectively. Compared with Bert, BERT_CNN improves the macro F1 and micro F1 by an average of 3.78 and 4.88, respectively by combining syntactic information, time words and adverb cues. The main reason is that our BERT_CNN introduces the three cues to further discover the modality and polarity in event sentences.
- Our model BERT_CNN improves the F1 scores on three categories and the improvement on the category PS+ is the largest one, especially its Recall with an increase of 14.34. This result shows our model can prove the effective syntactic feature and cue features. For example, in S1, for the event *"increase"*, the syntactic path constructed is *"doubt will increase"*, shortens the distance between the cue *"doubt"* and event *"increase"*. Hence, our model can effectively filter the noise in sentences.

Besides, due to the using of max-pooling and CNN to extract global sentence information, the training time of our BERT_CNN is 207 s, while the training time of the BERT benchmark model is 1122 s, which is about 5 times of our model.

4.3 Analysis

Table 2 shows the detail performance comparison on different input features, where Sen, Pos, Neg and Time_AD refer to the event sentence, the path from speculative cue to event, the path from negative cue to event, and the path from time and adverb cue to event, respectively.

Table 2. Performance comparison on different simplified models

	CT+			CT−			PS+			Micro-F1	Macro-F1
	P	R	F1	P	R	F1	P	R	F1		
Bert_CNN	89.23	92.86	91.01	84.43	76.42	80.23	79.12	77.42	78.26	87.08	83.24
Bert_Sen	87.58	88.96	88.27	75.07	77.51	76.27	78.22	63.08	69.84	83.30	78.36
Sen_Pos	88.48	89.88	89.17	74.86	72.63	73.73	79.25	75.27	77.21	84.38	80.05
Sen_Time_AD	88.55	88.55	88.55	73.40	74.80	74.09	78.37	79.21	78.79	84.18	80.48
Sen_Neg	88.08	91.45	89.70	85.80	76.69	80.86	74.82	73.48	74.14	85.71	81.64
Sen_Neg_Pos	88.42	92.53	90.43	86.25	74.80	80.12	75.45	74.91	75.18	86.23	82.04

From Table 2, we can find out that:

- If we only use the sentence to represent event (Bert_Sen), without using any cue information, both Micro-F1 and Macro-F1 drop rapidly. This result indicate the effectiveness of the negative, speculative, and time cue to identify event factuality. Besides, the increases of CT− and PS+ are much larger than that of CT+, and this indicates that the above three cues are helpful to identify the CT− and PS+ events.
- Compared with Sen, Sen_Time_AD (adding the path of time and adverb cue to the event) can improve the Macro-F1 and Micro-F1 by 2.12 and 0.88, respectively, especially for the PS+ events with the improvement of 8.95. The reason is that the time and adverb cues can help the model to correct those pseudo samples in PS+ (actually those samples are CT+ events) and classify them as CT+. Therefore, Sen_Time_AD can improve the Recall of CT+ and the precision of PS+. For example, in the case of event *"wanted"* in S3, because the time *"13th"* was separately identified as the feature of the time word, the tense was identified as the past, which increases the possibility of the event being identified correctly.
- Compared with Sen, Sen_Neg (adding the path of negative cue to the event) and Sen_Pos (adding the path of speculative cue to the event) are both effective for event factuality identification. Especially, Sen_Pos improves the Micro-F1 and Macro-F1 by 1.08 and 1.69, respectively, while Sen_Neg improves the micro-F1 and Macro-F1 by 2.41 and 3.28, respectively. Hence, Sen_Neg is more effective than Sen_Pos in identifying event factuality. Besides, we also can find out that Sen_Neg is better to recognize those CT− events with a significant improvement of 4.59 in F1 score,

while Sen_Pos is better to recognize those PS+ events with an improvement of 7.37 in F1 score. This is consistent to the nature of those two type cues. If we combine the negative cue and speculative cue together to identify event factuality, the performance can be further improved. This indicates these two type cues can cooperate each other. For example in S4, there are three different cue words, i.e., *"never"* (negative cue), *"consider"* (speculative cue) and *"denied"* (negative cue). The event *"resigning"* in the sentence has actually been negated by the speculative cues *"consider"* if we only use the speculative cue. Thus, the factuality will be misidentified as PS+. Therefore, when the above two type cues are introduced in our model, it can identify this event factuality correctly. Besides, we find out that the priority of the negative cues is higher than that of the speculative cues when identifying sentence-level event factuality.

(S4) *The Secretary of state of the United States said he had <u>never considered</u> resigningand <u>denied</u>that he had called trump an idiot.*

- When our model contains all four cues (BERT_CNN), it achieves the highest performance. This indicates the above four cues are not only effective but also complement each other, because they prefer to recognize different type event factuality, as mentioned above.

Besides, our improvements come from the representation of cues. Tenney et al. [11] and Liu et al. [12] found that Bert embedding can learn short-distance syntactic and semantic role information. Hence, our model can learn a part of the syntactic knowledge in a short distance, but it cannot learn the syntactic knowledge in a long distance. For example in S1, for the event *"increase"*, the speculative cue *"doubts"* changes the facticity of the whole sentence into PS+. However, this cue is a little far away from the event *"increase"*, and BERT cannot learn them well. On the contrary, our path from the cue to the event is *"doubts will increase"* and it greatly shortens the distance between speculative cue and event word. Our model uses CNN to learn short distance syntactic information, which can filter noises in the sentence. Hence, our model BERT_CNN can improve the performance of event factuality.

5 Conclusion

This paper proposes an event factuality identification model to identify event factuality from raw texts. It first extracts basic factors related with factuality, and then utilizes a BERT-based Convolutional Neural Network (Bert_CNN) with the integration of speculative cues, negative cues, time words and adverb cues which can make up for the inability of BERT to learn long-distance syntactic knowledge and pay more attention to the important words in the sentence. In the future work, we will focus on how to increase the information in the adjacent sentence to enrich the semantic features of the event sentence and how to use the syntactic tree structure to learn the information.

Acknowledgments. The authors would like to thank the three anonymous reviewers for their comments on this paper. This research was supported by the National Natural Science Foundation of China (No. 61836007, 61772354 and 61773276), and the Priority Academic Program Development of Jiangsu Higher Education Institutions (PAPD).

References

1. Chu, X., Zhu, Q., Zhou, G.: A study on the relationship between the primary and secondary text in natural language processing. Comput. Sci. **40**(4), 842–860 (2017)
2. Jin Z., Cao J., Jiang Y., et al. News credibility evaluation on microblog with a hierarchical propagation model. In: Proceedings of the 2014 IEEE International Conference on Data Mining, Shenzhen, pp. 230–239 (2014)
3. Roser, S., James, P.: Are you sure that this happened? Assessing the factuality degree of events in text. Comput. Linguist. **38**(2), 1–39 (2012)
4. Cao, Y., Zhu, Q., Li, P.: Construction method of Chinese event factuality information corpus. Chin. Inf. J. **27**(6), 38–44 (2012)
5. Marie, C., Christopher, D., Christopher, P.: Did it happen? The pragmatic complexity of veridicality assessment. Comput. Linguist. **38**(2), 301–333 (2012)
6. Kenton, L., Yoav, A., Yejin, C., Luke, Z.: Event detection and factuality assessment with non-expert supervision. In: Proceedings of the 2015 Conference on Empirical Methods in Natural Language Processing, pp. 1643–1648. The Association for Computational Linguistics, Lisbon (2015)
7. He, T., Li, P., Zhu, Q.: A Chinese event factuality recognition method. Comput. Sci. **44**(005), 241–244, 256 (2017)
8. Qian, Z., Li, P., Zhang, Y., et al.: Event factuality identification via generative adversarial networks with auxiliary classification. In: Proceedings of 27th International Joint Conference on Artificial Intelligence. International Joint Conferences on Artificial Intelligence, Stockholm, pp. 4293–4300 (2018)
9. Mudrakarta, P.K., Taly, A., Sundararajan, M., et al.: Did the model understand the question? In: Proceedings of the 56th Annual Meeting of the Association for Computational Linguistics, pp. 1896–1906. Association for Computational Linguistics, Melbourne (2018)
10. Jia, R., Liang, P.: Adversarial examples for evaluating reading comprehension systems. In: Proceedings of the 2015 Conference on Empirical Methods in Natural Language Processing, pp. 2021–2031. Association for Computational Linguistics, Copenhagen (2017)
11. Tenney, I., Xia, P., Chen, B., et al.: What do you learn from context? Probing for sentence structure in contextualized word representations. In: Proceedings of 7th International Conference on Learning Representations. OpenReview, New Orleans (2019)
12. Liu, Y., Ott, M., Goyal, N., et al.: RoBERTa: a robustly optimized BERT pretraining approach. CoRR, 1907.11692(2019)
13. Saurí, R., Pustejovsky, J.: FactBank: a corpus annotated with event factuality. Lang. Resour. Eval. **43**(3), 227–268 (2019)
14. Qian, Z., Li, P., Zhu, Q., et al.: Document-level event factuality identification via adversarial neural network. In: Proceedings of the 2019 Conference of the North American Chapter of the Association for Computational Linguistics: Human Language Technologies, pp. 2799–2809. Association for Computational Linguistics, Minneapolis (2019)
15. Saurí, R.: A factuality profiler for eventualities in text. Ph.D. thesis, Brandeis University, Waltham, MA, USA (2008)

16. Werner, G.J., Prabhakaran, V., Diab, M., et al.: Committed belief tagging on the FactBank and LU corpora: a comparative study. In: Proceedings of the 2nd Workshop on Extra-Propositional Aspects of Meaning in Computational Semantics, pp. 32–40. Association for Computational Linguistics, Denver (2015)

17. Veyseh, A.P.B., Nguyen, T.H., Dou, D.: Graph based neural networks for event factuality prediction using syntactic and semantic structures. In: Proceedings of the 57th Annual Meeting of the Association for Computational Linguistics, pp. 4393–4399. Association for Computational Linguistics, Florence (2019)

18. Devlin, J., Chang, M.W., Lee, K., et al.: BERT: pre-training of deep bidirectional transformers for language understanding. In: Proceedings of the 2019 Conference of the North American Chapter of the Association for Computational Linguistics: Human Language Technologies, pp. 4171–4186. Association for Computational Linguistics. Minneapolis (2018)

19. Kim, Y.: Convolutional neural networks for sentence classification. In: Proceedings of the 2014 Conference on Empirical Methods in Natural Language Processing, pp. 1746–1751. Association for Computational Linguistics, Doha (2014)

20. Manning, C.D., Schütze, H.: Foundations of Statistical Natural Language Processing. The MIT Press, Cambridge (1999)

End-to-End Event Factuality Identification via Hybrid Neural Networks

Jinjuan Cao[1], Zhong Qian[1,2], Peifeng Li[1,2], Xiaoxu Zhu[1,2], and Qiaoming Zhu[1,2(✉)]

[1] School of Computer Science and Technology, Soochow University, Suzhou, China
{qianzhong,pfli,xiaoxzhu,qmzhu}@suda.edu.cn
[2] AI Research Institute, Soochow University, Suzhou, China

Abstract. Event factuality identification (EFI) is a task to judge the factuality of events in texts, and is also the basic task of many related applications in the field of Natural Language Processing (NLP), such as information extraction and rumor detection. Previous research on EFI relied on annotated information, which cannot be applied to real world applications directly, and some studies only considered the default source AUTHOR. To address the above issues, this paper launches an end-to-end EFI model considering different event-related sources, which constructs the candidate event sets from raw texts to capture various kinds of event-related information, and then proposes a hybrid neural network model on GCN and BiLSTM to learn semantic and syntactic features, respectively. The experimental results on FactBank show that our proposed approach outperforms the baselines.

Keywords: Event factuality identification · End-to-End model · Hybrid neural networks · FactBank

1 Introduction

Event factuality refers to the factuality of the event in the text for the event-related source. It is of great significance to judge the factuality of events for many NLP applications, since those events are not always credible, especially for events in news texts.

There are many factors that can affect the event factuality value, such as speculative and negative cues, some syntactic structures (e.g., conditional clauses), and sources of event. The following three examples show the impact of different information on event factuality, where the events are shown in **bold**.

E1:*Northbridge is a cool, calculating and clever criminal who could* **strike** *again.*
E2:*If it* **rains** *tomorrow, I may* **stay** *at home.*
E3:*He* **indicated** *that some assets might be* **sold** *off to* **service** *the debt.*

The factuality value of the event *strike* in E1 is PS+ (possible) under the influence of the speculative cue *could*. In E2, since event *rains* is in a conditional adverbial clause, its factuality value is Uu (underspecified), while the factuality value of event *stay* is PS+ under the influence of the speculative cue *may*. The event *indicated* in E3 is a Source

H. Chen et al. (Eds.): CCKS 2020, CCIS 1356, pp. 196–208, 2021.
https://doi.org/10.1007/978-981-16-1964-9_16

Introducing Predicate (SIP) which introducing the source *He*. The event *sold* and *service* have the factuality value PS+ due to the influence of the speculative cue *might* for the source *He*. For the default source AUTHOR who does not express views and opinions, the factuality values of event *sold* and *service* are Uu.

Previous research relied heavily on the annotated information, which cannot be applied to real world applications directly. This paper focuses on the end-to-end event factuality identification (EFI), which is more valuable and challenging. Currently, the main issue on EFI is that a large number of pseudo samples will be produced without the annotated information and can result the low performance. Moreover, the factuality of an event varies among different sources (cf. E3). To address these two issues above, we design a hybrid EFI model, which first uses a CNN-based model to filter pseudo samples, and then integrates GCN and BiLSTM to identify event factuality. Experimental results on FactBank show that our proposed approach outperforms the baselines. The main contributions of this paper are as follows.

- We propose an end-to-end model for event factuality identification, which identifies event factuality from raw texts without using annotated information.
- We introduce the event-related information, sentence and dependency paths to represent event factuality from different aspects and propose a hybrid model based on GCN and BiLSTM to learn semantic and syntactic features better.

2 Related Work

At present, there are relatively few event factuality corpora. Saurí et al. [1] constructed an English corpus FactBank based on TimeBank whose factuality values are composed of modality and polarity. Specifically, modality includes four types, i.e., certain (CT), probable (PR), possible (PS), and underspecified (U), while polarity includes three types, i.e., positive (+), negative (−), and underspecified (u). Therefore, the event factuality values can be CT+ , CT−, PR+ , PR−, PS+ , PS−, Uu, etc. Lee et al. [2] constructed a more fine-grained corpus UW to map the value of event factuality to [−3, 3]. Based on UW, the classification task can be transformed into a regression task to quantify the probability of a given event [3–5]. Minard et al. [6] proposed a multilingual event and time corpus MEANTIME that can be used for EFI. Cao et al. [7] annotated the event factuality values and their related attributes based on the ACE2005 Chinese event corpus. Qian et al. [8] built a bilingual document-level event factuality corpus based on China Daily and Sina bilingual news.

Early research on EFI mainly used rule-based or machine learning-based methods. The former formulated a series of rules to identify the event factuality, which is more targeted, but relied heavily on the annotated information. For example, Saurí et al. [9] directly used annotated information (e.g., upper predicates, negative words, modal words.) to build a top-down traversal dependency syntactic tree and applied various rules to calculate event factuality. Lotan et al. [10] also proposed a rule-based model on dependent syntax tree to calculate event factuality. The latter relied heavily on annotated corpora and manually selected features. For example, de Marneffe et al. [11] proposed a maximum entropy classifier using various kinds of word-level and sentence-level features, such as the upper predicate types, event subjects, modal words, negative words.

Saurí et al. [12] proposed a SVM-based classifier. Qian et al. [13] proposed a method combining rules and machine learning.

In recent years, neural network models have been applied to event factuality identification. For example, He et al. [14] introduced a CNN-based model to recognize Chinese event factuality. Qian et al. [15] proposed a hybrid neural network model which combines BiLSTM and CNN for event factuality identification. Qian et al. [16] proposed a neural model of event factuality identification via generative adversarial networks with auxiliary classification. Veyseh et al. [3] proposed a graph based neural network for event factuality prediction using syntactic and semantic structures.

Compared with previous research depended on annotated information, this paper conducts end-to-end event factuality research. Unlike Qian et al. [16] that used the pipeline model to get event-related features, this paper utilizes an end-to-end model constructing candidate set with various event-related features for EFI.

3   Event Factuality Identification

Previous studies on EFI mainly used feature-engineering methods, which needs various manually developed features, e.g., Qian et al. [16] proposed a neural network model and used the pipeline model that leads to the issue of error propagation. To alleviate this problem, we propose an end-to-end neural network model for EFI. The main challenge is that this model may produce a lot of pseudo samples. To solve this issue, we first use a CNN-based model to filter pseudo samples, and then propose a hybrid model based on GCN [3, 17] and BiLSTM to learn syntactic and semantic features. The framework of our model is shown as Fig. 1, which includes the three components: 1) candidate set selection, 2) CNN-based sample filter, and 3) event factuality identification with a hybrid neural network with GCN and BiLSTM.

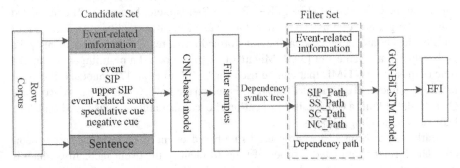

Fig. 1. The Framework of our end-to-end event factuality identification model

3.1   Candidate Set Selection

Since our model is an end-to-end method that does not rely any annotated information, the first step is to construct samples consisting of events and their related information from raw texts. As shown in E3, the event has different factuality value under the different event-related sources, and event factuality was influenced by SIP. As the analysis in Sect. 1, event factuality can be influenced by speculative and negative cues. Hence, we extract event, SIP, upper SIP, event-related source, speculative and negative cue as event-related information by the part-of-speech of words and the sentence structure. In particular, a sample is identified as positive one only when all its items are correctly identified. The event-related information used in this paper are as follows:

Event: In FactBank corpus, events are represented as event triggers (e.g., *"strike"* in E1), which are words that have a certain degree of factuality and can determine whether they have occurred. To obtain events, we simply select all of the nouns, verbs and adjectives as candidate words since most event corpora only annotated the above three types of words as triggers.

Source Introducing Predicate (SIP): SIP is a special event that can introduce a new event-related source, and can also affect the factuality of the event. For example, in E4, *speculates* introduces the event-related source *detective*. Under the influence of *speculates*, the source *detective* evaluates the factuality of the event *lives* as PS+ . We simply list all of the non-proprietary nouns, past participles, verbs with past tense, and adjectives as candidate words. If the event is SIP, it is marked as 1, otherwise it is marked as 0.

E4: *The detective **speculates** that he **lives** in 201.*

Upper SIP: For the current event, upper SIP introduces the source of the current event. As an example in E4, the upper SIP of the event *lives* is *speculates* which introduces the source *detective* for this event. We simply select all of the non-proprietary nouns, past participles, verbs with past tense, adjectives, and the leading clauses containing the current event as candidates.

Event-related Source: The event-related source is the evaluator of the event. Different event sources may have different views on the same event, which means the factuality of the event varies among different sources. The text author AUTHOR is the default source for all events, and other event-related sources for events are introduced by SIPs. In this paper, the source is obtained through the dependency syntax tree, which is usually the subject of the upper SIP. In particular, if the non-personal structure or passive voice appears in the upper SIP, the generalized source GEN is added as a candidate word, and if the sentence where the event appears as a direct quotation in quotation marks, the virtual source DUMMY is added as a candidate word.

Speculative Cue: Speculative cue is a word with uncertain meanings (e.g., *may*, *possible* and *probable*). We get the vocabulary of speculative cues from BioScope, then match speculative cues in the sentences as candidate words.

Negative Cue: Negative cue is a word with negative meanings (e.g., *not, deny*, and *fail)*. We get the vocabulary of negative cues from BioScope, then match negative cues in the sentences as candidate words.

3.2   CNN-Based Sample Filter

Following the Subsect. 3.1, we create a large-size candidate set with a large amount of negative samples. Hence, we propose a CNN-based filter to remove those pseudo samples from the candidate set, due to its fast training speed, its ability to effectively learn the characteristics of sentences, and the difficulty of overfitting.

We enumerate all possible lists combining events, SIP, upper SIP, event-related sources, speculative cue words, negative cue words, and event sentence as model input, and each one can be expressed as $(d_1, d_2, \ldots, d_6, x_1, x_2, \ldots, x_n)$, where d_1–d_6 refer to six type of event-related information mentioned above, $(x_1, x_2, \ldots, x_n)$ represents the event sentence where x_i is a word in the sentence. Then we map it into the low-dimensional space through the pre-trained Glove word vector to obtain $(e_1, e_2, \ldots, e_{n+6})$. In this paper, we use the CNN model which including convolutional layer, pooling layer and fully connected layer to learn to obtain semantic feature representation of event-related information and event sentences. The feature obtained by the CNN model is represented as f, and finally through the softmax to calculate the probability of the event-related information is correct.

3.3 Event Factuality Identification on GCN-BiLSTM

In order to better capture the connection between words in the sentence, and learn rich features from dependency path, we use a hybrid model combining GCN and BiLSTM for predicting event factuality. As shown in Fig. 2, the prediction model mainly includes four components: input layer, embedding layer, feature layer and output layer.

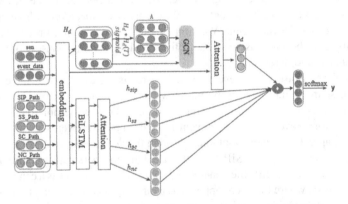

Fig. 2. Event factuality identification on GCN and BiLSTM

Input Layer: Due to the effectiveness of the dependency syntax tree in EFI task [3, 5], we employ the dependency syntax paths of event-related information as syntactic features, and use the following information as input: the event related information, the event sentence and dependency syntax paths. As shown in Fig. 3, the event related information used here includes the event, SIP, upper SIP, event-related source, speculative and negative cue, and their dependency syntax paths noted as SIP_Path, SS_Path, SC_Path, NC_Path, respectively.

Embedding Layer: We use the pretrained word vector to initialize the input, mapping each word into a low dimension vector through the Glove pre-trained word embeddings.

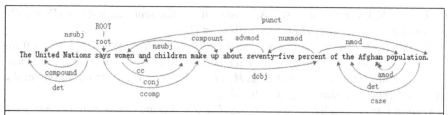

Event: *make*, the current event.

SIP: *1*, whether the current event introduces a predicate to the source.

Upper SIP: *says*, upper sip introduces the relevant source of the current event

Event related sources: *United*, event related source is introduced by the upper SIP.

Speculative cue: *0*, speculative cue that affect the factuality of the current event.

Negative cue: *0*, negative cue that affect the factuality of the current event.

Sen: *The United Nation says women and children make up about seventyfive percent of the Afghan population* the sentence where the current event is.

SIP_Path: *says ccomp make*, the dependency path of upper SIP to the current event.

SS_Path: *says nsubj United*, the dependency path of upper SIP to the event related sources.

SC_Path: *null*, the dependency path of speculative cue to the current event.

NC_Path: *null*, the dependency path of negative cue to the current event.

Fig. 3. Dependency of an example sentence and all features of an event in a sentence

Feature Layer: We use GCN-based model to obtain semantic features and utilize attention based BiLSTM model to obtain syntactic features from the syntactic paths. In order to learn rich features, this paper uses BiLSTM to learn features. BiLSTM obtains forward and backward text information $(h_1, h_2, \ldots, h_t)$ by stitching the features obtained by forward LSTM and backward LSTM. We express $(h_1, h_2, \ldots, h_t)$ as H, where t is the length of the text sequence.

To learn the semantic features, we input the word vector of event sentence and event related information into BiLSTM, then we get their rich representation H_s and H_e, respectively. To learn more semantic features, we first try to use the matrix A which is generated by LSTM and dependency tree [3]. According to the performance results of EFI, we finally use the vector representation H_s to calculate the relationship between h_i and h_j to obtain the weight matrix A, which is generated by LSTM as follows.

$$H_s = (h_1, h_2, \ldots, h_n) \tag{1}$$

$$h_i' = tanh(W_1 h_i) \tag{2}$$

$$a_{ij} = sigmoid\left(W_2\left[h_i', h_j'\right]\right) \tag{3}$$

$$A = (a_{ij})_{i,j=1..n} \tag{4}$$

Given the weight matrix A, we use a layer of GCN to enhance the representation of words and contextual representation in event sentences to obtain sentence features

$(h_1^g, h_2^g, \ldots, h_n^g)$, expressed as H^g. And we combine the sentence features H^g and H_e to obtain semantic features H_d through attention [18].

To learn the syntactic features, we use BiLSTM to learn the syntactic features from SIP_Path, SS_Path, SC_Path and NC_Path, and obtain $H_{sip}, H_{ss}, H_{uc}, H_{nc}$. To obtain more important information from $H_{sip}, H_{ss}, H_{uc}, H_{nc}$, we use attention to obtain the important syntactic features $h_{sip}, h_{ss}, h_{uc}, h_{nc}$.

Output Layer: Finally, the features $h_{sip}, h_{ss}, h_{uc}, h_{nc}$ are concatenated to obtain the feature representation f, and the feature f is calculated by the softmax classifier to obtain the output vector o:

$$f = h_d \oplus h_{sip} \oplus h_{ss} \oplus h_{nsc} \oplus h_{nc} \tag{5}$$

$$o = softmax(Wf + b) \tag{6}$$

The output vector o is used to calculate whether the event-related information is correct and at the same time to calculate the event factuality value. We utilize cross entropy to compute the loss:

$$L = -\sum_i y_i \log(p(y_i)) \tag{7}$$

y_i denotes the true labels and $p(y_i)$ represents the probability of annotated labels.

4   Experimentation

4.1   Experimental Settings

Our model evaluated on the English FactBank corpus and mainly considers the performance of CT+ , CT−, PR+ , PS+, Uu five factuality values, following Saurí et al. [9], de Marneffe et al. [12] and Qian et al. [16]. For event-related information, we construct the candidate samples of event-related information by the arrangement and combination of candidate words. Table 1 shows the statistics of the five factuality values and pseudo samples in the candidate set.

Table 1. Statistics of each category of candidate set

	CT+	Uu	CT−	PR+	PS+	Pseudo sample
All source	7165	4184	398	335	200	5291460
Default source (AUTHOR)	5037	3410	195	103	81	140062
Other source (no AUTHOR)	2128	774	203	232	119	5151398

We evaluate our model by 10-fold cross-validation on FactBank. As experimental settings, we set the learning rate as 0.001, the dropout as 0.5, the mini-batch size as 50. F1-measure is used to describe the performance, and micro-average and macro-average are used to describe the comprehensive performance of all the categories of factuality value. Word embeddings are initialized by Glove with 100 dimensions and the LSTM hidden layer has a dimension of 32.

4.2 Experimental Results

To evaluate our model on the task of event factuality identification, we use the following models for comparison. It is worthy to notice that we do not utilize pipeline model, because we identify event and source by LSTM model, then we can only get the recall 62.64% of samples, which means we cannot get better performance. We do not utilize the model of Veyseh et al. [3], because they only consider the factuality of the event under the default source.

BiLSTM + Att: It randomly extracts data with a positive and negative sample ratio of 1:20 from the candidate set as the training set, and use the attention based BiLSTM model for training. The model structure is the same as the model in this paper;

CNN + Hybrid_NN_1: First, it uses the CNN model to filter the candidate set, which is the same as the filter model in this paper. Then, it uses the Hybrid_NN_1 which proposed by Qian et al. [15]. We do not give the performance of Hybrid_NN_2 because it did not get good performance under the influence of the pseudo samples.

CNN + CNN: First, it uses the CNN model to filter the candidate set, then it uses the CNN model to make predictions which the input is same as the prediction model in this paper;

CNN + BiLSTM: First, it uses the CNN model to filter the candidate set. Then it uses the BiLSTM model to make predictions which the input is same as the prediction model in this paper;

CNN + BiLSTM + Att: Unlike CNN + BiLSTM, the attention mechanism is added;

CNN + GCN + BiLSTM: Our End-to-end hybrid model;

Table 2 shows the experimental results between different models and we analyze the experimental results from the following aspects.

Table 2. Experimental performance of various models

	CT+	Uu	CT−	PR+	PS+	Micro-F1	Macro-F1
BiLSTM + Att	44.41	41.48	20.79	21.66	18.86	40.27	29.44
CNN + Hybrid_NN_1	46.86	42.53	30.64	**28.26**	**27.38**	44.44	35.14
CNN + CNN	58.73	53.89	23.20	18.62	17.44	54.86	34.37
CNN + BiLSTM	61.14	55.82	29.18	22.43	18.09	57.23	37.33
CNN + BiLSTM + Att	62.11	56.70	31.49	23.59	18.43	58.13	38.47
CNN + GCN_BiLSTM (ours)	**63.50**	**59.09**	**35.53**	25.33	26.71	**59.97**	**42.03**

In comparison with BiLSTM + Att, the other models achieve the significant improvement from 4.17% to 19.70% in micro-F1. The low performance of BiLSTM + Att is due to the large-scale pseudo samples. This result shows that our CNN-based filtering model can effectively filter out the pseudo samples to improve the performance rapidly.

We can see that CNN + Hybrid_NN_1 does not get the good performance on Micro-F1 and Macro-F1, but it gets best performance on PR+ and PS+ , which means the model can learn more information from lower samples. Compared with CNN + CNN, CNN + BiLSTM improves the Micro-F1 and Macro-F1 scores by 2.37% and 2.96%, respectively. CNN + CNN learns n-gram fragments as features, and the sequential relationship between the features cannot be fully learned. On the contrary, CNN + BiLSTM considers the context information in word sequence. Besides, the performance gap between CNN + BiLSTM and CNN + BiLSTM + Att shows that the attention mechanism can effectively obtain important information from the features.

It can be seen from the table that our CNN + GCN_BiLSTM achieves the best performance both on Micro-F1 and Macro-F1, while the micro-F1 and macro-F1 scores increase by 1.84% and 3.56%, respectively. This indicates that the use of GCN-based network to obtain semantic features is more effective in obtaining word-to-word connections.

4.3 Analysis on Different Representations

To analyze the effectiveness of our proposed features in our model, we conduct the following simplified models for comparison:

BaseInput: The input is event related information and event sentence, and the model is BiLSTM + Attention.

Dep: The input uses dependency paths of SIP_Path, SS_Path, SC_Path and NC_Path the model is BiLSTM + Attention.

Basedata + Dep: The input is event related information and event sentence and the dependent paths of SIP_Path, SS_Path, SC_Path and NC_Path, the model is BiLSTM + Attention. Table 3 shows the experimental results of different inputs based on the BiLSTM + Attention model.

Table 3. The results based on the BiLSTM + Attention model (w/f: using the CNN-based filter; w/o f: not using the CNN-based filter), BaseInput means that the input is event related information and event sentence, and Dep means that the input is dependent path SIP_Path, SS_Path, SC_Psath, NC_Path

	CT+	Uu	CT−	PR+	PS+	Micro-F1	Macro-F1
BaseInput (w/o f)	42.20	39.43	8.80	7.49	4.79	38.09	20.54
Dep (w/o f)	35.17	29.64	22.93	16.50	21.83	32.02	25.22
BaseInput + Dep (w/o f)	44.41	41.48	20.79	21.66	18.86	40.27	29.44
BaseInput (w/f)	54.07	49.49	7.70	3.77	3.68	50.29	23.74
Dep (w/f)	45.97	38.37	32.08	22.41	**22.54**	42.50	32.27
BaseInput + Dep (w/f)	**62.11**	**56.70**	**31.49**	**23.59**	18.43	**58.13**	**38.47**

Table 2 and 3 show that all the models can obtain satisfactory results on CT+ and Uu due to their majority, but get lower results on CT−, PR+ and PS+ . On one hand, CT+ and Uu have a larger number of samples, while CT−, PR+ , and PS+ have fewer samples. On the other hand, the events annotated with CT−, PR+ , or PS+ label are indicated by negative and speculative cues. However, we do not use any annotated speculative and negative information. Our simple method of matching word with cue vocabulary may get the word which is not cue, like *appear* in E5, *appear* is matched as cue; Some cues do not affect the current event but other events within the sentence, as shown in E6, the cue *probable* acts on the event *occurred*, not on the event *said*, and when the event is *said*, the vocabulary will also match *probable*; Some events are not determined by cue, are semantically negative and speculative, as shown in E7, the event *buy* is influenced by the *too ... to* structure, and its semantic meaning is negative, so the event *buy* factuality value is CT−; These will increase the difficulty of EFI.

E5:*The two thieves will **appear** at court tomorrow morning.*
E6:*He **said** it was <u>probable</u> that a deadlock had **occurred** in the program.*
E7:*The book is <u>too</u> expensive for me to **buy**.*

Table 3 gives the results on BiLSTM + Attention model with different inputs. Comparing the models without the CNN-based filter (w/o f, i.e., the three upper columns in Table 3), the models using the CNN-based filter (w/f, i.e., the three below columns in Table 3) can improve the performance significantly. This result further ensures the effectiveness of our filter, which can filter a large number of pseudo samples and exclude a large amount of interference information.

Compared with BaseInput which only uses event sentences and event-related information, and Dep which only uses dependency path, BaseInput + Dep gets higher performance. Both the event sentences and event-related information can provide rich semantic information for this task, while the dependency path can provide the interaction between events and cues. Therefore, Dep can achieve better performance on CT−, PR+ and PS+ than the other values, while BaseInput performs to identify the types CT+ and Uu. Besides, Table 3 also shows that the combination of Dep and BaseInput (i.e., BaseInput + Dep) can further improve the performance and this indicates these two types are complementary.

4.4 Analysis on CNN-Based Filter

Table 4 shows the statistics of the five factuality values and pseudo samples using our CNN-based filter. Compared with Table 1, we can find that CNN-based filter can effectively filter 98% of pseudo examples and the number drops from 5291460 to 105815. However, it also will filter out 9.51% of positive samples at the same time. Especially, it filters about 40% of CT−, PR+ and PS+ , which results in the lower performance on CT−, PR+ and PS+ (F1: 18.43–31.49). The main reason for this result is that the number of CT−, PR+ and PS+ samples is too small, our model cannot learn enough features for them.

Table 4. Statistics of each category of filter set

	CT+	Uu	CT−	PR+	PS+	Pseudo sample
All source	6596	3951	258	189	112	105815
Default source (AUTHOR)	4989	3375	170	89	60	47709
Other source (no AUTHOR)	1607	576	88	100	52	58106

4.5 Error Analysis

The errors of our model are mainly divided into the following three categories: 1) errors in candidate set extraction; 2) errors in filter; 3) errors in event factuality identification.

Errors in candidate set extraction. These errors is due to the upstream errors in the stage of the event-related information extraction and account for 36.27%. Due to the large size of candidate set, there are a large amount of pseudo samples in the candidate set although a CNN-based filter is applied to reduce those pseudo samples. For example, almost 43.47% of the identified events are pseudo ones because most words are polysemes. Take E8 and E9 as examples, the word *watch* in E8 triggers an event, while this word in E9 refers to an item, not an event.

E8:*Police kept* **watch** *outside and about 25 undercover police were* **scattered** *throughout the crowd.*
E9: *This could cause significant* **harm** *to* watch *producers in the United States and the Virgin Islands.*

Errors in Filter. These errors come from incorrect event-related information and account for 51.46%. 52.26% of them are due to unidentified events, and 47.74% are due to the other five types of event-related information was identified incorrectly.

Errors in Event Factuality Identification. These errors come from incorrect identified event factuality and account for 12.27%. Among them, the classification errors in CT+ and Uu account for 98.82%. The reason is that the discrimination of CT+ and Uu mainly depends on semantic information and is hard to make a decision, while other values can be derived from explicit cues.

5 Conclusion

In this paper, we propose an end-to-end event factuality identification model, which first uses a CNN-based model to filter pseudo samples, and then integrates GCN and BiLSTM to identify event factuality. The experimental results on FactBank have verified the effectiveness of our proposed model. In our future work, one is to consider relevant rules to further remove those pseudo samples or use other effective methods to obtain event-related information, and the other is to consider further improving the model and learning richer features to increase the accuracy of event factuality recognition.

Acknowledgments. The authors would like to thank the three anonymous reviewers for their comments on this paper. This research was supported by the National Natural Science Foundation of China (No. 61772354, 61836007 and 61773276.), and the Priority Academic Program Development of Jiangsu Higher Education Institutions (PAPD).

References

1. Saurí, R., Pustejovsky, J.: Factbank: a corpus annotated with event factuality. Lang. Resour. Eval. **43**(3), 227–268 (2009)
2. Lee, K., Artzi, Y., Choi, Y., Zettlemoyer, L.: Event detection and factuality assessment with non-expert supervision. In: Proceedings of the 2015 Conference on Empirical Methods in Natural Language Processing, pp. 1643–1648 (2015)
3. Veyseh, A.P.B., Nguyen, T.H., Dou, D.: Graph based neural networks for event factuality prediction using syntactic and semantic structures. In: Proceedings of the 57th Annual Meeting of the Association for Computational Linguistics, pp. 4393–4399 (2019)
4. Stanovsky, G., Eckle-Kohler, J., Puzikov, Y., Dagan, I., Gurevych, I.: Integrating deep linguistic features in factuality prediction over unified datasets. In: Proceedings of the 55th Annual Meeting of the Association for Computational Linguistics, vol. 2, pp. 352–357 (2017)
5. Rudinger, R., White, A.S., Van Durme, B.: Neural models of factuality. In: Proceedings of the 2018 Conference of the North American Chapter of the Association for Computational Linguistics: Human Language Technologies, vol. 1, pp. 731–744 (2018)
6. Minard, A.L., et al.: Meantime, the newsreader multilingual event and time corpus. In: Proceedings of the Tenth International Conference on Language Resources and Evaluation (LREC 2016), pp. 4417–4422 (2016)
7. Cao, Y., Zhu, Q.M., Li, P.F.: The construction of chinese event factuality corpus. J. Chinese Inf. Process. **27**(6), 38–44 (2012)
8. Qian, Z., Li, P.F., Zhu, Q.M., Zhou, G.D.: Document-level event factuality identification via adversarial neural network. In: Proceedings of the 2019 Conference of the North American Chapter of the Association for Computational Linguistics: Human Language Technologies, vol. 1, pp. 2799–2809 (2019)
9. Saurí, R.: A factuality profiler for eventualities in text. PhD dissertation, Brandeis University (2008)
10. Lotan, A., Stern, A., Dagan, I.: Truthteller: annotating predicate truth. In: Proceedings of the 2013 Conference of the North American Chapter of the Association for Computational Linguistics: Human Language Technologies, pp. 752–757 (2013)
11. De Marneffe, M.C., Manning, C.D., Potts, C.: Did it happen? The pragmatic complexity of veridicality assessment. Comput. Linguist. **38**(2), 301–333 (2012)
12. Saurí, R., Pustejovsky, J.: Are you sure that this happened? assessing the factuality degree of events in text. Comput. Linguist. **38**(2), 261–299 (2012)
13. Qian, Z., Li, P.F., Zhu, Q.M.: A two-step approach for event factuality identification. In: Proceedings of the 2015 International Conference on Asian Language Processing (IALP), pp. 103–106 (2015)
14. He, T.X., Li, P.F., Zhu, Q.M.: Identifying chinese event factuality with convolutional neural networks. In: Proceedings of the Chinese Lexical Semantics - 18th Workshop (CLSW 2017), pp. 284–292 (2017)
15. Qian, Z., Li, P., Zhou, G., Zhu, Q.: Event factuality identification via hybrid neural networks. In: Cheng, L., Leung, A.C.S., Ozawa, S. (eds.) ICONIP 2018. LNCS, vol. 11305, pp. 335–347. Springer, Cham (2018). https://doi.org/10.1007/978-3-030-04221-9_30

16. Qian, Z., Li, P.F., Zhang, Y., Zhou, G.D., Zhu, Q.M.: Event factuality identification via generative adversarial networks with auxiliary classification. In: Proceedings of the Twenty-Seventh International Joint Conference on Artificial Intelligence, pp. 4293–4300 (2018)
17. Kipf, T.N., Welling, M.: Semi-supervised classification with graph convolutional networks. In: Proceedings of the 5th International Conference on Learning Representations (2017)
18. Vaswani, A., Shazeer, N., Parmar, N., et al.: Attention is all you need. In: Proceedings of the Advances in Neural Information Processing Systems 30: Annual Conference on Neural Information Processing Systems, pp. 5998–6008 (2017)

Knowledge Applications: Question Answering, Dialogue, Decision Support, and Recommendation

On Building a Knowledge Graph of Open Courses

Yuxuan Fan, Wenbo Hu, Wencheng Huang, and Zhichun Wang[✉]

School of Artificial Intelligence, Engineering Research Center of Intelligent
Technology and Educational Application, Ministry of Education,
Beijing Normal University, Beijing, China
{yxfan,hwb,wencheng}@mail.bnu.edu.cn, zcwang@bnu.edu.cn

Abstract. To help online learners to efficiently get open courses they
want to learn, we build a large knowledge graph of open courses. Our
knowledge graph contains structured information of a large number of
open courses, and also provides important links among concepts and
courses. Based on the extracted prerequisite relations in the knowledge
graph, learning paths of concepts and courses can be recommended to
learners. More specifically, we define an ontology for open courses and
extract structured information of courses guided by the ontology. Concept linking, prerequisite extraction, and course linking are performed,
which enables the knowledge graph to provide helpful assistance to learners.

Keywords: Open courses · Knowledge graph · Course linking ·
Prerequisite extraction

1 Introduction

Open courses on the web have become important learning materials for internet
users. Popular open courses including MOOCs (massive online open courses),
open courses of universities, and also courses on video sharing platforms. There
are a large number of open courses on the web, enabling online learners to
choose whatever they want to learn. However, a huge number of courses are not
well classified and organized. They are distributed on different platforms and
are varied in quality and target learners. Though lots of courses can be easily
accessed on the web, it is really difficult and time consuming for learners to find
the most suitable courses for themselves.

There have been some efforts to build knowledge graphs for MOOCs, including MOOCCube [17], MOOC-KG [5], etc. However, they just cover MOOCs
on the web, ignoring a large number of university open courses. And the existing work mainly focuses on representing course information in structured forms
(e.g. RDF). Deep analysis including concept linking, prerequisite extraction,
course linking are not well studied. What's more, how to use knowledge graph
to help learners in their learning activities also needs further exploration. To deal

© Springer Nature Singapore Pte Ltd. 2021
H. Chen et al. (Eds.): CCKS 2020, CCIS 1356, pp. 211–224, 2021.
https://doi.org/10.1007/978-981-16-1964-9_17

with the above challenges, we propose to build a new knowledge graph for open courses, aiming to help online learners to efficiently find suitable courses they need. In our work, we first define an ontology for the domain of open courses, and then extract structured information of courses from MOOC platforms and course websites of universities. Concept linking, prerequisite extraction, and course linking are performed based on the extracted course data. Using the built knowledge graph, we also build a web application that enables users to search information of concepts and courses, and recommends learning paths of concepts and courses to users.

The rest of this paper is organized as follows, Sect. 2 introduces the data collection process, Sect. 3 describes the key steps of building the knowledge graph, Sect. 4 presents the web application based on our knowledge graph, and Sect. 5 presents the detailed results of the knowledge graph, Sect. 6 reviews some related work, and Sect. 7 is the conclusion.

2   Data Collection

Collected data is divided into three parts, one is the course information collected from various open MOOC platforms, the other is the course information extracted from multiple universities, and the last part is the collection of concepts from Wikipedia.

2.1   MOOCs

We collect MOOC data from famous online learning platforms including Coursera[1], edX[2], XuetangX[3], and for each MOOC course, we extract the course name, duration, platform, school, description and category, etc. Totally we have 1988 MOOCs from 37 platforms.

In the process of MOOC courses collection, taking edX as an example, Fig. 1(a) shows the homepage of course "Introduction to C++". It mainly includes course name, teacher, introduction, subject, prerequisites, etc. This information is saved for next part of the work.

2.2   University Courses

University courses are open courses offered by universities on the Internet. Their schedule and related materials are given on their homepages. These materials provide students with structured information of one course and serve as a great guide for students' learning.

We altogether choose four universities as an example: Stanford, MIT, Caltech and the University of Washington. We collect the information about computer

---

[1] https://www.coursera.org.

[2] https://www.edx.org.

[3] https://xuetangx.com.

science related courses from 2012 to 2019, and obtain the course name, course number, school, teacher, description, URL from the homepage of each course, as is shown in Fig. 1(b). We then use semi-automated method including regular expression to extract prerequisite knowledge and main topics of each course from its description. For courses with rich learning materials, we also record the course slides, recommended books and related courses as additional learning guidance. By the above approach, we eventually obtained 560 computer science related courses from these universities.

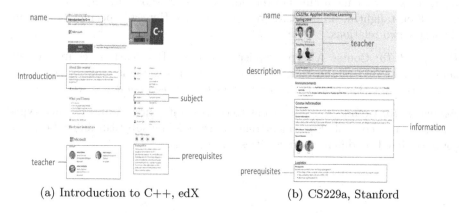

(a) Introduction to C++, edX   (b) CS229a, Stanford

Fig. 1. Course homepage examples

2.3   Concepts

Learning a course can be seen as a process of mastering a series of concepts, so concepts are also important in building an integrated knowledge graph of open courses. We use Wikipedia[4] as an additional concept data resource. We treat each Wikipedia page as a concept [7] and include it into our concept graph. We also record the category of each Wikipedia page and the links between pages to develop concept relations later. Through the above method, a total of 15397041 concepts are collected.

3   KG Building

We build our open course knowledge graph through ontology defining, entity linking, prerequisite extraction and course linking. In this section, we will introduce these steps in detail.

3.1   Ontology

The ontology of our knowledge graph has been manually created based on the most commonly used information in open online learning resources.

---

[4] https://www.wikipedia.org.

Classes. We define 4 classes and 2 subclasses as basis of our knowledge graph.

1) *Course:* relates to single course. It has two subclasses, one is *MOOC*, relates to courses on online learning platforms, and the other is *university course*, relates to open courses from universities. 2) *Concept:* concepts from Wikipedia which also appear as key topic in certain course. 3) *University:* university that offer certain course. 4) *Platform:* platform on which certain online course opens. Properties of each class are shown in Fig. 2.

Relations. We define 5 kinds of relations which are shown in Fig. 2: offeredBy, openedOn, taughtIn, prerequisiteTo, equivalentTo. Here we specifically explain the latter two relations.

PrerequisiteTo: Course A is prerequisite to course B if learning course A will help students understand contents in course B. The same is true for concepts. Prerequisite relation reveals the learning order between courses or concepts.

EquivalentTo: Course A is equivalent to course B if course A and B teach typically the same content. Equivalent relation reveals the substitutability between courses offered by different universities and platforms.

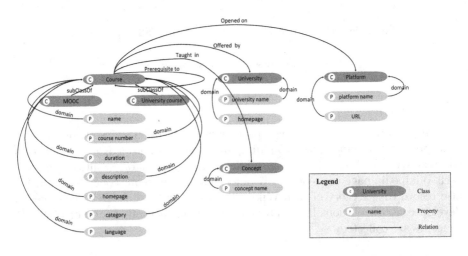

Fig. 2. Ontology of our knowledge graph for open courses

3.2 Concept Linking

Concept linking is to create connections between online course data and our Wikipedia concepts. We achieve so by extracting key concepts from homepages of MOOCs and university courses and linking them to our concept graph [8]. We first select course name, chapter headings and course description as the source text. Then we adopt DBpedia-Spotlight [4], a tool for annotating mentions of

Wikipedia concepts in the text, to identify key concepts and their Wikipedia links. Through these Wikipedia links, these key concepts can be easily linked to the concept graph we collected in Sect. 2.

Even though DBpedia-Spotlight annotates concepts with high accuracy, some of the concepts are irrelevant to the central topic of the course because DBpedia-Spotlight can not distinguish between concepts with different semantics. So after obtaining all the candidate concepts of one course, we use word embedding to capture the semantic relatedness between concepts and the course [14]. More specifically, we use fastText [2] to learn word representation and also obtain word vectors for out-of-vocabulary words. Then we calculate the cosine similarity between each concept and the course name. The higher the similarity score, the more likely that this concept is related to the central topic of the courses, so concepts with a similarity score higher than a threshold are selected and linked to our concept graph.

3.3 Prerequisite Extraction

Prerequisite extraction is to discover the prerequisite relationship between courses and concepts. It can be divided into two separate tasks: the first is to extract the prerequisite relationship between courses, and the second is to explore the prerequisite relationship between concepts.

Prerequisite Relationship of Courses. We record the prerequisite knowledge part of each course from the homepage, and use pattern matching method [6] to obtain the prerequisite information of the courses from plain text. Since most of the prerequisite knowledge presented on the course homepage is course number instead of course name, we also collect the course numbering rules of selected universities and map each course number to its course name. We then manually correct some inconsistent prerequisite relations.

Prerequisite Relationship of Concepts. Unlike prerequisite relation between courses, concept relations are not that explicit. Mainly following the work of Jiang et al. [9], we choose concepts appeared in the same course description as candidate concept pair, and for each concept pair, we calculate several features including the in degree and out degree of concepts in the concept graph, the number of times the concepts appear in MOOC video subtitles simultaneously and so on. For each potential concept pair, we use pre-trained logistic regression model from Jiang et al.' work to classify whether the concept pair has the prerequisite relation.

3.4 Course Linking

Courses offered by different universities may have the same name, or they may have different names but teach roughly the same content. The same university may offer the same course in different years. Course linking is to dig out courses which focus on the same topic, and link these courses with equivalent relations. This information is helpful for learners in their course learning choices.

We have three principles to judge whether two courses are equivalent:

1) Courses with the same name are equivalent.
2) If the concepts extracted from two course names are the same, then the two courses are equivalent.
3) If the key concepts extracted from two course descriptions are the same, then the two courses are equivalent.

4   Application

To fully explore the potential usage of open course data, we build a web application based on our knowledge graph, aiming to provide online learners with better learning guidance. In this section, we will show the system structure and two main application examples, namely concept query and learning path recommendation.

4.1   System Structure

The system structure of our web application is shown in Fig. 3. At the server back end, there are a concept graph and a course knowledge graph. At front end, learning guidance generated by the server according to user query are shown in different forms, including concept explanations, related courses, suggested learning path, etc.

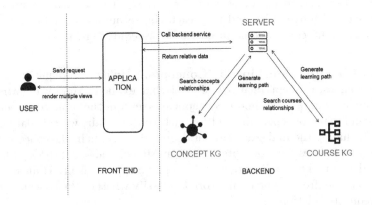

Fig. 3. System structure of the web application

4.2   Concept Query

After the user searches for certain concept or certain field in computer science, our system uses full-text matching and group matching methods to find the most suitable concept node in the concept graph. We then show the explanation and related concepts of this concept [3] to the user. Moreover, we display top courses where the concept is mostly taught in, and also give information on course introduction and homepage, which will help learners to quickly determine how and where to learn this concept.

4.3   Learning Path Recommendation

Learning path is an ordered sequence of learning activities. The learning object in the front of a learning path is the basis of the learning objects after it, and help students to understand following-up knowledge better. Learning path serves as primary guidance on the learning order of materials, and reduces the confusion of what to learn and what not to learn when students first get into unfamiliar fields.

Concept Learning Path. For concepts, we intend to demonstrate the relation network and suggested learning path based on the prerequisite and other relations between concepts.

The learning path of a concept is formed by the extension of concept pairs with a prerequisite relation. When a concept is being queried, a two-direction traverse is made on the concept graph. We firstly do forward breadth-first search starting from the queried node to discover its prerequisite concepts, then search backward to discover its subsequent concepts. This will form a concept network which clearly illustrates the learning order of these concepts.

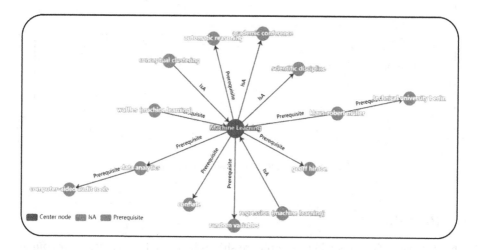

Fig. 4. Example of concept learning path

As an instance, Fig. 4 shows the learning path of concept "Machine Learning". The description on the edges are important. When it comes to *IsA*, subclass or superclass information are expressed according to the direction of the arrow. For instance, "conceptual clustering" is a method about "Machine Learning" and "Machine Learning" is a "scientific discipline". Similarly, when the description on the edge is *Prerequisite*, it means this neighbor concept is the prerequisite concept of "Machine Learning", or the other way around.

We also introduce word vectors to measure the relatedness between concepts, so as to improve the efficiency and precision while generating concept learning path. When doing traverse on each concept, we calculate the cosine similarity between concept and its neighbors, select several concepts with the highest score and add them to the traversing queue. Thus, we can obtain a concept learning path with a limited number of nodes and a high degree of correlation.

Courses Learning Path. For each concept, we show its related courses. Since the learning process of courses also needs an appropriate order, we present the course learning path to be additional guidance.

Course learning path is based on course graph and the prerequisite knowledge obtained from course description extracted before. We use tree diagram to present the prerequisite relationship of the courses, and courses on the right are prerequisite courses to those on the left.

For example, Fig. 5 illustrates that in order to learn the course "Network Control Systems", students have to first have basic understanding of course "Robust Control Theory"; and to grasp "Robust Control Theory", students would better know "Linear Systems Theory", which in turn requires a lot of math knowledge.

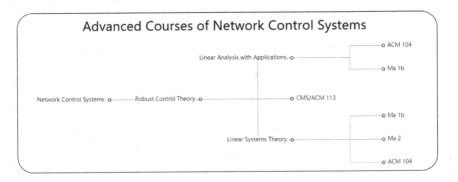

Fig. 5. Example of course learning path.

5   Results

The statistics of our knowledge graph are shown in Table 1. Currently, our course data is collected only from four universities and part of the MOOC platforms.

Once obtain more course data from the web, we can use the same method to process the data and enrich our course knowledge graph.

Table 1. Ontology statistics

Classes		Number	Relations	Number
Course	MOOC	1988	offeredBy	2371
	University course	560	openedOn	1988
University		405	taughtIn	12369
Concept		15397041	prerequisiteTo	15615270
Platform		37	equivalentTo	942

5.1 Concept Linking Results

Through the method mentioned in Sect. 3.2, we obtained 9940 key concepts from MOOCs and 2429 key concepts from university courses. After removing duplicated concepts, there are 11096 concepts linked to our concept graph.

Table 2 shows some result examples. Colored words in course description are "target concepts" annotated by human. Red color means the concept is also extracted by our approach, and blue color represents either the concept is missed or wrongly identified into key concepts.

In order to measure the performance of the concept linking process, we randomly select 100 courses and perform manually annotation. Among the selected samples, the precision, recall and F1-score are shown in Table 3.

5.2 Course Linking Results

As for course linking, we finally obtain 942 equivalentTo relations between courses offered by different universities or those opened on various platforms. Table 4 illustrates example results of course linking. Courses with equivalentTo relations teach roughly the same topics, and can be substitutions to each other. With this information, learners can easily choose alternative courses when learning.

6 Related Work

Related works can be generally divided into two categories: first is the construction of course graphs or course repositories, and second is to solve course related tasks.

Table 2. Concept linking examples

Course name	Course description	Key topics
Machine Learning	This course provides a broad introduction to machine learning and statistical pattern recognition.	machine learning, pattern recognition
Computer Security	Foundations of modern computer security, including software security, operating system security, network security, applied cryptography, human factors, authentication.	computer security, software security, network security, cryptography, operating system, human factors, authentication
Mining Massive Data Sets	The course will discuss data mining and machine learning algorithms for analyzing very large amounts of data. The emphasis will be on MapReduce and Spark as tools for creating parallel algorithms that can process very large amounts of data.	data mining, machine learning, MapReduce
Introduction to Computational Thinking	Provides a unifying introduction to computational thinking for technical and general-purpose applications. The class is intended to break down traditional walls between computation in science, engineering and computer science with applications from these disciplines. As an example, we will see how to apply differentiable programming to machine learning. The class uses Julia to provide students with a complementary programming language, with an emphasis on modifiable codes and performance. We assume no prior Julia experience, but we will build on Python from 6.0001.	computational thinking, machine learning, computer science, programming language, Julia, Python, differentiable

Table 3. Results of concept linking

	Precision	Recall	F1-score
MOOCs	0.782	0.688	0.732
University courses	0.713	0.692	0.702

Table 4. Course linking examples

Course name, University	Equivalent courses
Machine Learning, Stanford	Machine Learning, UW
	Introduction to Machine Learning, UW
	Modeling with Machine Learning: From Algorithms to Applications, MIT
Reinforcement Learning, Stanford	Reinforcement Learning: Foundations and Methods, MIT
	Advanced Survey of Reinforcement Learning, Stanford
Mining Massive Data Sets, Stanford	Machine Learning & Data Mining, Caltech

6.1 Course Graphs

Open online learning resources like MOOCs and university courses have great value for both learners and researchers. There exist some works that build datasets of such resources.

MOOC Datasets. MOOCCube [17] provides a publicly available large-scale data repository for NLP applications in MOOCs. Its dataset centers on both the course content and student behaviors like enrollment. MOOC-KG [5] constructs a MOOC knowledge graph to supply users with integrated MOOC information among several platforms through knowledge modeling, knowledge acquisition, conflict resolution and knowledge storage.

Web Resources. Websites like Class Central[5] and Course Graph[6] enable users to discover free online courses from top universities around the world. They also select the highest rated online courses and provide user comments.

6.2 Course Related Tasks

Several research topics related to open online learning resources are as follows: course concept extraction, prerequisite relation learning, and course learning guidance.

---

[5] https://www.classcentral.com/.
[6] http://coursegraph.com/.

Course Concept Extraction. In efforts to extract important concepts from online courses, Jiang et al. [1] propose a semi-supervised learning framework to identify concepts in unstructured MOOC data, Lu et al. [11] extract domain-specific concepts from sequential learning materials using a graph-based raking method, and Pan et al. [13] extract concepts from course video captions by embedding-based graph propagation.

Prerequisite Relation Learning. Yang et al. [16] extract course prerequisite relations from university syllabi to predict more prerequisite relations, while others focus on prerequisite relations between course-related concepts. For example, Liang et al. [10] discovered concept prerequisite relations from the university curriculum and Pan et al. [12] achieved this using MOOC video subtitles.

Learning Order Computing. Course learning guidance could be recommendations on related concepts and new courses like Zhang et al. [18]'s work, or proper learning paths generated by multidimensional data [15]. These recommendations guide students throughout their learning process, and greatly help students enhance their learning efficiency.

7   Conclusion

This paper presents a knowledge graph of open courses containing structured information about MOOCs and open courses. We define an open course ontology, and perform concept linking, prerequisite extraction and course linking to fully explore these open course data. We also build a web application as an example of the potential utilization of our course knowledge graph, enabling online learners to efficiently find suitable courses they are looking for. In future work, more data of multiple types will be included into our knowledge graph, and we will try to enrich the functions of our application to help learners better organize their learning process.

Acknowledgements. The work is supported by the National Key R&D Program of China (No. 2017YFC0804004).

References

1. Jiang, Z., Zhang, Y., Li, X.: MOOCon: a framework for semi-supervised concept extraction from MOOC content. In: Bao, Z., Trajcevski, G., Chang, L., Hua, W. (eds.) DASFAA 2017. LNCS, vol. 10179, pp. 303–315. Springer, Cham (2017). https://doi.org/10.1007/978-3-319-55705-2_24
2. Bojanowski, P., Grave, E., Joulin, A., Mikolov, T.: Enriching word vectors with subword information. Trans. Assoc. Comput. Linguist. 5, 135–146 (2017)
3. Cao, Y., Wang, X., He, X., Hu, Z., Chua, T.: Unifying knowledge graph learning and recommendation: towards a better understanding of user preferences. CoRR abs/1902.06236 (2019). http://arxiv.org/abs/1902.06236

4. Daiber, J., Jakob, M., Hokamp, C., Mendes, P.N.: Improving efficiency and accuracy in multilingual entity extraction. In: Proceedings of the 9th International Conference on Semantic Systems (I-Semantics) (2013)
5. Dang, F., Tang, J., Li, S.: MOOC-KG: a MOOC knowledge graph for cross-platform online learning resources. In: ICEIEC 2019 - Proceedings of 2019 IEEE 9th International Conference on Electronics Information and Emergency Communication, pp. 596–603 (2019). https://doi.org/10.1109/ICEIEC.2019.8784572
6. Fabbri, A.R., et al.: TutorialBank: a manually-collected corpus for prerequisite chains, survey extraction and resource recommendation. CoRR abs/1805.04617 (2018). http://arxiv.org/abs/1805.04617
7. Gordon, J., Aguilar, S., Sheng, E., Burns, G.: Structured generation of technical reading lists. In: Proceedings of the 12th Workshop on Innovative Use of NLP for Building Educational Applications, pp. 261–270. Association for Computational Linguistics, Copenhagen, September 2017. https://doi.org/10.18653/v1/W17-5029. https://www.aclweb.org/anthology/W17-5029
8. Gordon, J., Zhu, L., Galstyan, A., Natarajan, P., Burns, G.: Modeling concept dependencies in a scientific corpus. In: ACL 2016 (2016)
9. Jiang, L., et al.: MAssistant: a personal knowledge assistant for MOOC learners. In: EMNLP-IJCNLP 2019–2019 Conference on Empirical Methods in Natural Language Processing and the 9th International Joint Conference on Natural Language Processing, Proceedings of System Demonstrations, pp. 133–138 (2020)
10. Liang, C., Ye, J., Wu, Z., Pursel, B., Giles, C.L.: Recovering concept prerequisite relations from university course dependencies. In: 31st AAAI Conference on Artificial Intelligence, AAAI 2017, pp. 4786–4791 (2017)
11. Lu, W., Zhou, Y., Yu, J., Jia, C.: Concept extraction and prerequisite relation learning from educational data. In: Proceedings of the AAAI Conference on Artificial Intelligence, vol. 33, pp. 9678–9685 (2019). https://doi.org/10.1609/aaai.v33i01.33019678
12. Pan, L., Li, C., Li, J., Tang, J.: Prerequisite relation learning for concepts in MOOCs. In: ACL 2017–Proceedings of the Conference on 55th Annual Meeting of the Association for Computational Linguistics (Long Papers), vol. 1, pp. 1447–1456 (2017). https://doi.org/10.18653/v1/P17-1133
13. Pan, L., Wang, X., Li, C., Li, J., Tang, J.: Course concept extraction in MOOCs via embedding-based graph propagation. In: Proceedings of the Eighth International Joint Conference on Natural Language Processing, IJCNLP 2017, Taipei, Taiwan, 27 November–1 December 2017, Volume 1: Long Papers, pp. 875–884 (2017). https://aclanthology.info/papers/I17-1088/i17-1088
14. Roy, S., Madhyastha, M., Lawrence, S., Rajan, V.: Inferring concept prerequisite relations from online educational resources (2018)
15. Shi, D., Wang, T., Xing, H., Xu, H.: A learning path recommendation model based on a multidimensional knowledge graph framework for e-learning. Knowl.-Based Syst. **195** (2020). https://doi.org/10.1016/j.knosys.2020.105618
16. Yang, Y., Liu, H., Carbonell, J., Ma, W.: Concept graph learning from educational data. In: WSDM 2015 - Proceedings of the 8th ACM International Conference on Web Search and Data Mining, pp. 159–168 (2015). https://doi.org/10.1145/2684822.2685292

17. Yu, J., et al.: MOOCCube: a large-scale data repository for NLP applications in MOOCs. In: ACL (2020)
18. Zhang, J., Hao, B., Chen, B., Li, C., Chen, H., Sun, J.: Hierarchical reinforcement learning for course recommendation in MOOCs. In: Proceedings of the AAAI Conference on Artificial Intelligence, vol. 33, pp. 435–442 (2019). https://doi.org/10.1609/aaai.v33i01.3301435

Improving Question Answering over Knowledge Base with External Linguistic Knowledge

Peiyun Wu[1,2] and Xiaowang Zhang[1,2(✉)]

[1] College of Intelligence and Computing, Tianjin University, Tianjin 300350, China
{wupeiyun,xiaowangzhang}@tju.edu.cn
[2] State Key Laboratory of Communication Content Cognition, Beijing, China

Abstract. Semantic parsing is an important method to question answering over knowledge base (KBQA), which transforms a question into logical queries to retrieve answers. Existing works largely focus on fine-grained relations representation while ignoring the latent semantic information behind the implicit meaning of relations. In this paper, we leverage an external linguistic knowledge (ELK) to enhance relation semantics where ELK is used to remove the ambiguity of words occurring in a relation. Moreover, we present a sense-based attention for word-level relation representation and a Graph-Attention-Network (GAT)-based question encoder. Experiments evaluated on two data sets show that our model outperforms existing approaches.

Keywords: Question answering · Semantic parsing · Knowledge base · External linguistic knowledge

1 Introduction

KBQA is the task of accurately and concisely answering a natural language question over knowledge base (KB). As a critical branch of KBQA, semantic parsing based approaches [4–6] construct semantic parsing trees or equivalent query structures to represent the semantics of the given question, and then ranking them by calculating the semantic similarity with the que stion.

Most of the existing semantic parsing methods in KBQA focus on capturing the semantics of questions and relations in query graphs. The label of the relation in the first edge of the query graph is taken as core relations for measuring similarity together with some manual features [4,5]. Another approach to detecting individual relation with word-level is presented in [8] to improve the performance of matching from questions to query graphs where each word of a relation is independently represented, and relation is directly represented by integrating of its words' representations. [6] extends [8] by improving the representation of questions and processing more complex questions.

© Springer Nature Singapore Pte Ltd. 2021
H. Chen et al. (Eds.): CCKS 2020, CCIS 1356, pp. 225–236, 2021.
https://doi.org/10.1007/978-981-16-1964-9_18

However, current works that encode the relation semantics in a query graph mainly focus on tow aspects: *relation-level* and *word-level* (we denote their combination as fine-grained relation), the latent semantic information behind the meaning of relations is ignored. For *relation-level*, each relation is considered as a whole and individual token, which pays more attention to the overall information of the relation. The problem with this aspect is that it suffers from data sparsity in the training process. For *word-level*, the relation is treated as a sequence of words with the lack of global information of the original relation. In this aspect, word meanings are diversified without considering the word-level relation that might bring noises. For example, for relation *"be subject to"*, it means *"to be made to undergo an unpleasant experience"*, the *word-level* word *"subject"* might lose that meaning due to its polysemy.

In this paper, to cope with the above limitations, we propose a fined-grained relation with leveraging *WordNet* to enhance relations representation. *WordNet* [7] as a lexical database for English, contains synset (the set of synonyms) for each word, which can instruct word sense representation learning and word disambiguation. We both consider the original global relation and its word-level relations with word disambiguation. Our work has three major contributions:

- We propose a sense-based attention method to represent fined-grained relation semantics via introducing an external linguistic knowledge (ELK).
- We leverage syntactic dependency parsing and present a Graph-Attention-Network [10] (GAT)-based representation to encode questions.
- We introduce a hidden-relation-based attention mechanism to strengthen relation semantics by applying hidden relations in questions to suggest candidate relations in query graphs for better characterizing the intention of questions.

2 Query Graphs Generation

In this section, we introduce the query graph that we employ to store the semantic information of the question. And meanwhile, we give the generation process of candidate query graph space.

2.1 Query Graphs Definition

We use the query graph to represent a question, which will be transformed into SPARQL for querying on KB later. The query graph G is treated as a directed graph, and we can store it as a set of tuples U. We define the basic elements of a tuple in U as follows:

The vertex can be divided into three categories: question variable($?q$), intermediate variables(v) and KB entities. The $?q$-node always represents the answer to a question. A v-node can be unknown entities or values.

The edge represents KB relations between two vertices, we also consider the direction of an edge. Due to the existence of ternary relation in question such as *"clinton was president in 1993"*, we store such a ternary relation as $(starttime(positionheld(clinton, president), 1993))$.

The filter type is an additional element attached to a special constraint. For example, the tuple ($?q$, *birth_day*, 2000, >) requires for entities $?q$ whose birthday is greater than 2000.

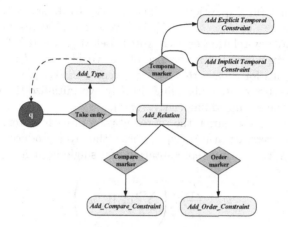

Fig. 1. Query graph generate process

2.2   Graphs Space Generation

Our approach to generating query graphs is an iterative process. Inspired by query graphs generation of [4], we consider 5 kinds of semantic constraints: **entity, type[1], temporal (explicit and implicit time), order[2]** and **compare[3]** to process more complex questions. Given a natural language question, we split it into a set of tokens T.

Firstly, we extract all possible entities from T with the help of the existing entity linking tool [11]. We can obtain the (mention, entity) pairs to get the entity linking list, which will be filtered with the scores evaluated by the entity linking tool. The set of candidate entities E will be finally returned.

And then, we construct new candidate graphs with a set of predefined operations set OP: Take_entity, Add_Type, Add_Temporal (explicit and implicit), Add_Compare and Add_Order. Figure 1 shows a flow path of our candidate semantic graph generation process.

3   Our Approach

We introduce our approach in the following two modules.

---

[1] The answer type of the question, often expressed by nouns in questions.

[2] We use order marker (predefined superlative word list) to detect ordinal questions.

[3] We use compare marker (such as more than, less than, etc.) to detect comparison questions.

3.1 Question Representation Encoding

In this module, we encode the syntactic dependency tree of a question to a vector as the representation of the given question via GAT. The module is used to provide the condition of calculating the similarity of a question with candidate query graphs.

We first transform a question $q = (x_1, \ldots, x_n)$ to its syntactic dependency tree and denote as (V_q, E_q), where V_q is a vertex set (words in q), E_q is an edge set (dependency-relation of each word). Let $W_{\text{glove}} \in \mathbb{R}^{|V| \times d}$ be the word embedding matrix where $|V|$ is the size of the vocabulary, d is the dimension of each word. Thus for each vertex $x_i \in V_q$, vector in W_{glove} is assigned to the corresponding vertex as its embedding. Finally, we initialize the embedding of V_q via all its vertices' embedding $(w_1, \ldots, w_n)$.

Now, we use GAT to enrich the representation of each vertex. The input to a layer is a set of vertex features $(w_1, \ldots, w_n)$, the layer produces a new output set $(w_1', \ldots, w_n')$. The output representation of a single layer in GAT is:

$$w_i' = \bigoplus_{k=1}^{K} \sigma \left(\sum_{j \in N_i} a_{ij}^k W_v^k w_j \right), \tag{1}$$

Here, $\oplus$ represents concatenation operation, K represents K-head attention, which can stabilize the learning process, and σ represents any non-linear function. N_i is the set of neighbor indices of vertex i in V_q. W_v^k represents k-th attention mechanism that corresponds to a linear transformation matrix. The attention coefficient a_{ij}^k is computed by a *softmax* function over all the values in the vertex i's neighborhood of the k-th attention mechanism. The single-head attention coefficient is expressed as:

$$a_{ij} = \frac{\exp\left(LeakyReLU\left(a^T[W_v w_i \oplus W_v w_j]\right)\right)}{\sum_{k \in N_i} \exp\left(LeakyReLU\left(a^T[W_v w_i \oplus W_v w_k]\right)\right)}, \tag{2}$$

where $.^T$ represents transposition, and a is a weight vector, LeakyReLU is a nonlinearity function. In the final layer of GAT, we use *averaging* to achieve multi-head attention and calculate the final output embedding as below:

$$w_i' = \sigma \left(\frac{1}{K} \sum_{k=1}^{K} \sum_{j \in N_i} a_{ij}^k W_v^k w_j \right), \tag{3}$$

Finally, we introduce a fully-connected layer and the ReLU non-linearity to obtain the final representation h_q of question q:

$$h_q = \text{ReLU} \left(W_q \cdot \sum_{i=1}^{n} w_i' + b_1 \right). \tag{4}$$

where W_q denotes the linear transformation matrix.

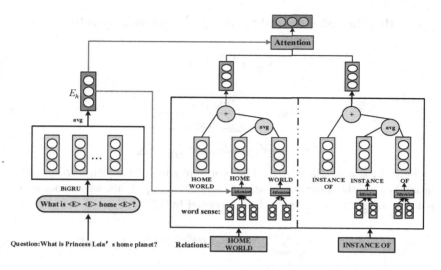

Fig. 2. Diagram for our relation representation model.

3.2 Relation Representation

In this module, we obtain the semantic representation of relations in two steps, namely, *sense-based attention* and *hidden-relation-based attention* (see Fig. 2). We use relations representation as corresponding query graph representation and to calculate similarity with the given question.

Sense-Based Attention. In this step, we obtain the representation of each fined-grained relation occurring in a query graph with leveraging *WordNet*.

Firstly, we define a *relational embedding matrix* as follows: let d_r be the dimension of relations, $|V_r|$ is the vocabulary size of relations,

$$W_{\text{rel}} \in \mathbb{R}^{|V_r| \times d_r}, \tag{5}$$

For *relation-level* relations, we directly take each relation as a whole unit and map it into vector representation by using W_{rel}, we denote $\mathcal{R}^{\text{whole}}$ as a vector sequence of relations in a query graph, as $\mathcal{R}^{\text{whole}}_{1:n} := \{r_1, \ldots, r_n\}$. We define *word-level* of the relation r_i, denoted by $\mathcal{R}^{\text{word}}$, as a sequence of vectors by applying $W_{glove} \in \mathbb{R}^{|V| \times d}$ to map each word w_{ij} in a relation r_i to its word embedding w_{ij} as $\mathcal{R}^{\text{word}}_{r_i} := \{w_{i1}, \ldots, w_{im}\}$. In *WordNet*, a word may have multiple senses, we denote $\mathcal{R}^{\text{sense}}$ as a set of sense vectors of word w_{ij} in $\mathcal{R}^{\text{word}}_{r_i}$ as $\mathcal{R}^{\text{sense}}_{w_{ij}} := \{s_{ij1}, \ldots, s_{ijk}\}$.

Next, we replace the entities in question $q = (x_1, \ldots, x_n)$ with a special $\langle E \rangle$ token, analogously, based on the word embeddings matrix W_{glove}, we can obtain the vector set of this sequence: $\{l_1, \ldots, l_n\}$. Then, we use $\{l_1, \ldots, l_n\}$ as input and obtain the average result of Bidirectional Gated Recurrent Unit (BiGRU) output as E_h:

$$E_h = \frac{1}{n} \sum_{i=1}^{n} \text{BiGRU}(l_1, \ldots, l_n), \tag{6}$$

Through this, the vector of word w_{ij} in $\mathcal{R}_{r_i}^{\text{word}}$ is represented as below:

$$w_{ij} = \sum_{y=1}^{k} a_{ijy} s_{ijy}, \tag{7}$$

$$a_{ijy} = \frac{\exp(w_{s_y})}{\sum_{x=1}^{k} \exp(w_{s_x})}, \tag{8}$$

$$w_{s_y} = W_s \cdot \tanh(E_h \cdot s_{ijy}^T) + b_2, \tag{9}$$

where W_s is a weight matrix. The representation of the j-th word in the i-th relation is a weighted sum of its senses $\{s_{ij1}, \ldots, s_{ijk}\}$. Thus, we can obtain each fined-grained relation representation, denoted by $\mathcal{R}^{\text{fine}}$:

$$\mathcal{R}^{\text{fine}}[i] = \frac{1}{m} \sum_{j=1}^{m} \mathcal{R}_{r_i}^{\text{word}}[j] + \mathcal{R}^{\text{whole}}[i], i \in (1, \ldots, n). \tag{10}$$

Hidden-Relation-Based Attention. In this step, we apply hidden relation in questions to assign weights of all relations in query graphs via attention mechanism.

First, we apply hidden relations in questions to assign weights of all relations as follows:

$$a_{r_i} = \frac{\exp(w_{r_i})}{\sum_{k=1}^{n} \exp(w_{r_k})}, \tag{11}$$

$$w_{r_i} = \tanh(W_h^T(E_h \oplus \mathcal{R}^{\text{fine}}[i]) + b_3), \tag{12}$$

Thus, we get the semantics of all relations R^{all} as follows:

$$R^{all} = \sum_{i=1}^{n} a_{r_i} \mathcal{R}^{\text{fine}}[i], \tag{13}$$

Finally, we introduce a fully-connected layer and the ReLU non-linearity to obtain the final relations representation h_r of all relations in a query graph:

$$h_r = \text{ReLU}\left(W_r \cdot R^{all} + b_4\right). \tag{14}$$

And we can compute the semantic similarity score between the query graph G and question q as following:

$$S(q, G) = cos\left(h_q, h_r\right). \tag{15}$$

4   Experiments and Evaluations

In this section, we evaluate our approach on the following experiments. We first compare with state-of-the-art model results and then perform analysis in detail to discuss how different modules of our approach improve KBQA task.

4.1 Experiments Setup

Due to Freebase [1] no longer up-to-date, including unavailability of APIs and new dumps, we use the full Wikidata [2][4] dump as our KB and host it with Virtuoso engine[5]. Besides, we adopt *SpaCy*[6] as syntactic dependency parser and use Deep Graph Library (DGL[7]) to transfer query graphs into DGL graph objects.

We conduct our experiments on the following two popular data sets, namely, *WebQuestionSP* [12], *QALD-7* [13].

- WebQuestionSP (WebQSP): It is a corrected version of the popular WebQuestions [14] data set. We slightly revise it that we map answers to Wikidata answer IDs instead of Freebase IDs since we use Wikidata.
- QALD-7: It was developed for Task 4 of the QALD-7 Shared Task, *"English question answering over Wikidata"*. The QALD-7 data set contains 80 questions, 42 of them are complex.

4.2 Training the Model

In training process, we use weak supervision to get positive pairs of questions and query graphs, as suggested in [14]. To maximize the margin between positive graphs g^+ and negative graphs g^-, for each training epoch, we adopt hinge loss as follows:

$$L = \max \sum_{g \in C} (0, (\lambda - cos(q, g^+) + cos(q, g^-))),$$

where C is the candidate query graph set of the question q.

Specifically, we generate 1620 positive query graphs from WebQSP and take 500 of 1620 as the validation set to get answers to all questions from Wikidata and overtake the gap between Freebase and Wikidata.

For all experiments, we initialize word embeddings using GloVe [15] word vectors with dimensions of 100. We use Adam optimizer [16] and train the model with a batch size of 64 at each epoch. When calculating the representation of questions, we use three-layer GAT and fully-connected layer with dropout $= 0.2$.

4.3 Baselines

We introduce five models for our experiments as baselines as follows:

1. *STAGG* (2015) (an re-implementation of [4])–questions are encoded by Deep Convolutional Neural Network (DCNN). The query graphs are scored with some manual features, and the representation of the core relation (the relation in the first edge) is encoded by DCNN.

---

[4] https://www.wikidata.org/wiki/Special:Statistics.
[5] http://virtuoso.openlinksw.com/.
[6] https://spacy.io/.
[7] https://www.dgl.ai/pages/about.html.

2. *Pooled Edges*–We use the DCNN to encode the question and apply a pooling operation over the representation of the individual relations.
3. *HR-BiLSTM* (2017) [8]–The model encodes questions that are encoded by residual BiLSTM and compute similarity with the pooling of *relation-level* and *word-level* relation. We re-implement it by adding a graph generating process in order to support complex questions.
4. *Luo et al.* (2018) [6]–Syntactic dependency path feature together with sequence information question encoder and sum fined-grained relation representation.
5. Our (w/o ELK)–A variant of our model without applying external linguistic knowledge in relation representation of a query graph.

4.4   Results

Now we perform our experiments on three datasets. We use precision, recall, and F1-score as metrics, where all results are average scores. Note that our results are different from those original results released in baselines over Freebase since we use Wikidata.

Table 1. WebQSP and QALD-7 results over Wikidata

Model	WebQSP			QALD-7		
	Precision	Recall	F1	Precision	Recall	F1
STAGG (2015)	0.1911	0.2267	0.1828	0.1934	0.2463	0.1861
Pooled Edges	0.2094	0.2553	0.2032	0.1904	0.1800	0.1605
HR-BiLSTM (2017)	0.2131	0.2573	0.2087	0.2172	0.2299	0.2099
Luo et al. (2018)	0.2255	0.2618	0.2162	0.2176	0.2451	0.2131
Our (w/o ELK)	0.2285	0.2702	0.2203	0.2316	0.2571	0.2254
Our	**0.2421**	**0.2843**	**0.2301**	**0.2549**	**0.2674**	**0.2472**

WebQSP and QALD-7. We compare the results on the WebQSP and QALD-7 dataset in Table 1. By Table 1, we show that our model outperforms all datasets and all metrics. We use QALD-7 solely for evaluation and train on WebQSP.

1. Our model achieves 25.9%, 13.2%, 10.3%, 6.4% higher F1-score compared to STAGG, Pooled Edges, HR-BiLSTM, Luo et al. (2018) on WebQSP. Analogously, we achieve 32.8%, 54.1%, 17.8%, 16.0% higher F1-score on QALD-7. So we can conclude that our relation representation method performs better than all baselines.
2. STAGG and Pooled Edges perform worse than other baselines on WebQSP and QALD-7, the main reason being that STAGG focuses on manual features and part of relations that lack flexibility, and Pooled Edges ignores the

word-level relations. In short, we can show that fined-grained relations representation performs better than models with manual features and relation-level semantics.

3. Our model improves the F1-score of WebQSP and QALD-7 by 4.4% and 9.7% over Our (w/o ELK). Hence, we can conclude that introduce ELK to remove word-level ambiguity can improve the performance of KBQA. Note that even without considering ELK, Our (w/o ELK) still performs better than other baselines, it shows the robustness of our model.

4.5 Ablation Study

In this subsection, we explore the contributions of various components in our model. For the ablation study, we use F1-score as our metrics.

Question Representation. To demonstrate the effectiveness of our question representation encoder, we apply different question encoding of each baseline to our model, and the results are shown in Table 2, where Our (GCN) denotes a variant of our question encoder that substitutes the GAT to Graph Convolutional Network (GCN) [23].

Table 2. Ablation results on question representation

QR of baselines	WebQSP	QALD
DCNN (STAGG, Pooled Edges)	0.2223	0.2363
HR-BiLSTM	0.2208	0.2299
Luo et al. (2018)	0.2241	0.2411
Our (GCN)	0.2276	0.2448
Our	**0.2301**	**0.2472**

The experimental results show that our question representation adopting GAT outperforms other models over all datasets.

– Firstly, our question encoder performs best compared to the other models. Specifically, Our question encoder betters Luo et al. (2018), with a 2.7%, 2.5% improvement on two datasets.
– Secondly, Luo et al. (2018) and our question encoder performs better compared to the other models that without considering syntactic dependency path information. Luo et al. (2018) and our question encoder both consider the syntactic dependency of question, it shows syntactic information is useful to improve the performance.
– Thirdly, GAT performs better compared to GCN in question presentation. Specifically, the GAT-based approach betters Our (GCN) with a 1.1%, 1.0% improvement on two datasets. The main reason is that GAT pays attention to the importance of different words in the question.

In short, we conclude that our GAT-based question encoder can better extract the semantics of questions.

Relation Representation. To demonstrate the effectiveness of our relation representation, we consider two types of using ELK ("−" is without using ELK, "+" is apply ELK to relation representation) and two kinds of operation over all relations in query graph ("pool" is max-pooling over relations, "HRA" is our hidden-relation-based attention mechanism), the results are shown in Table 3.

The experimental results show that our approach processing the fined-grained relation semantics with applying ELK and hidden-relation-based attention over relations performs best. "−" with "HRA" achieves 0.7%, 2.0% higher F1-score compared to "−" with "pool" on two datasets, and "+" with "HRA" achieves 0.9%, 2.4% higher F1-score compared to "+" with "pool" on two datasets. The comparison demonstrates that our hidden-relation-based attention is more efficient than pooling operation. Our method improves less on the WebQSP dataset because WebQSP contains more simple questions and fewer relations in the query graph. Analogous, "+" with "pool" achieves 4.3%, 9.2% higher F1-score compared to "−" with "pool", and "+" with "HRA" achieves 4.4%, 9.7% higher F1-score compared to "−" with "HRA" on two datasets, it demonstrates that use ELK can greatly improve performance in KBQA.

Table 3. Ablation results on relation representation

ELK	Operation	WebQSP	QALD
−	pool	0.2187	0.2210
−	HRA	0.2203	0.2254
+	pool	0.2281	0.2413
+	HRA	0.2301	0.2472

5   Related Works

Approaches that proposed for KBQA can be roughly divided into two classes: *semantic parsing, information retrieval.*

Semantic Parsing. Semantic parsing (SP) based approaches use a formalized representation to encode the semantic information of questions [14,17,18]. Most of the traditional semantic parsing styled methods [19–22] mainly rely on schema matching or hand-crafted rules and features.

Recently, neural networks (NN) based approaches have shown great performance on KBQA. SP+NN employed neural network to encode the semantics of questions and query graphs, and then select the correct query graphs of given questions in the candidate set for querying later [4–6,8]. SP+NN based approaches mostly encode relation semantics to represent the vector of query graphs. Different from these methods, we introduce ELK to enhance relation semantics, where ELK is used to remove the ambiguity of word-level relation. In addition, we take a hidden-relation-based attention mechanism to strengthen relation semantics. A few works pay more attention to modeling the whole query graph or structure [9,24].

Information Retrieval. Information retrieval (IR) based approaches retrieve the candidate entities (as answers) from KB using semantic relation extraction [12,25]. Those approaches map answers and questions into the same embedding space, where one could retrieve the answers independent of any grammar, lexicon, or other interior semantic structure. Belonging to the SP-based method, our approach encoded structured semantic information of question into formalized semantic graphs, which allows us to handle more complex problems.

6 Conclusion

In this paper, we utilize ELK to improve relation representation in KBQA. The knowledge is useful to extract word-level relation semantics with disambiguation. Our approach pays more attention to relational semantics of query graphs, which is an important semantic part of questions in KBQA. In this sense, we provide a new method of leveraging extra knowledge in KBQA. In future work, we are interested in extending our model for more complex practical questions.

Acknowledgment. This work is supported by the National Key Research and Development Program of China (2017YFC0908401) and the National Natural Science Foundation of China (61972455). Xiaowang Zhang is supported by the Peiyang Young Scholars in Tianjin University (2019XRX-0032).

References

1. Bollacker, K.D., Evans, C., Paritosh, P., Sturge, T., Taylor, J.: Freebase: a collaboratively created graph database for structuring human knowledge. In: SIGMOD, pp. 1247–1250 (2008)
2. Vrandecic, D., Krötzsch, M.: Wikidata: a free collaborative knowledgebase. CACM **57**(10), 78–85 (2014)
3. Cui, W., Xiao, Y., Wang, H., Song, Y., Hwang, S., Wang, W.: KBQA: learning question answering over QA corpora and knowledge bases. PVLDB **10**(5), 565–576 (2017)
4. Yih, W., Chang, M., He, X., Gao, J.: Semantic parsing via staged query graph generation: question answering with knowledge base. In: ACL, pp. 1321–1331 (2015)
5. Bao, J., Duan, N., Yan, Z., Zhou, M., Zhao, T.: Constraint-based question answering with knowledge graph. In: COLING, pp. 2503–2514 (2016)
6. Luo, K., Lin, F., Luo, X., Zhu, K.Q.: Knowledge base question answering via encoding of complex query graphs. In: EMNLP, pp. 2185–2194 (2018)
7. Miller, G.A.: WordNet: a lexical database for English. Commun. ACM **38**(11), 39–41 (1995)

8. Yu, M., Yin, W., Hasan, K.S., Santos, C.N., Xiang, B., Zhou, B.: Improved neural relation detection for knowledge base question answering. In: ACL, pp. 571–581 (2017)
9. Zafar, H., Napolitano, G., Lehmann, J.: Formal query generation for question answering over knowledge bases. In: Gangemi, A., et al. (eds.) ESWC 2018. LNCS, vol. 10843, pp. 714–728. Springer, Cham (2018). https://doi.org/10.1007/978-3-319-93417-4_46
10. Velickovic, P., Cucurull, G., Casanova, A., Romero, A., Liò, P., Bengio, Y.: Graph attention networks. In: ICLR (2018)
11. Yang, Y., Chang, M.: S-MART: novel tree-based structured learning algorithms applied to tweet entity linking. In: ACL, pp. 504–513 (2015)
12. Yih, W., Richardson, M., Meek, C., Chang, M., Suh, J.: The value of semantic parse labeling for knowledge base question answering. In: ACL, pp. 201–206 (2016)
13. Usbeck, R., Ngomo, A.-C.N., Haarmann, B., Krithara, A., Röder, M., Napolitano, G.: 7th open challenge on question answering over linked data (QALD-7). In: Dragoni, M., Solanki, M., Blomqvist, E. (eds.) SemWebEval 2017. CCIS, vol. 769, pp. 59–69. Springer, Cham (2017). https://doi.org/10.1007/978-3-319-69146-6_6
14. Berant, J., Chou, A., Frostig, R., Liang, P.: Semantic parsing on freebase from question-answer pairs. In: EMNLP, pp. 1533–1544 (2013)
15. Pennington, J., Socher, R., Manning, C.D.: GloVe: global vectors for word representation. In: EMNLP, pp. 1532–1543 (2014)
16. Kingma, D.P., Ba, J.: Adam: a method for stochastic optimization. In: ICLR (2015)
17. Reddy, S., et al.: Transforming dependency structures to logical forms for semantic parsing. TACL **4**, 127–140 (2016)
18. Zhang, H., Cai, J., Xu, J., Wang, J.: Complex question decomposition for semantic parsing. In: ACL, pp. 4477–4486 (2019)
19. Kwiatkowski, T., Zettlemoyer, L.S., Goldwater, S., Steedman, M.: Lexical generalization in CCG grammar induction for semantic parsing. In: EMNLP, pp. 1512–1523 (2011)
20. Hakimov, S., Unger, C., Walter, S., Cimiano, P.: Applying semantic parsing to question answering over linked data: addressing the lexical gap. In: Biemann, C., Handschuh, S., Freitas, A., Meziane, F., Métais, E. (eds.) NLDB 2015. LNCS, vol. 9103, pp. 103–109. Springer, Cham (2015). https://doi.org/10.1007/978-3-319-19581-0_8
21. Krishnamurthy, J., Mitchell, T.M.: Weakly supervised training of semantic parsers. In: EMNLP, pp. 754–765 (2012)
22. Cai, Q., Yates, A.: Large-scale semantic parsing via schema matching and lexicon extension. In: ACL, pp. 423–433 (2013)
23. Kipf, T.N., Welling, M.: Semi-supervised classification with graph convolutional networks. In: ICLR (2017)
24. Sorokin, D., Gurevych, I.: Modeling semantics with gated graph neural networks for knowledge base question answering. In: COLING, pp. 3306–3317 (2018)
25. Xu, K., Reddy, S., Feng, Y., Huang, S., Zhao, D.: Question answering on freebase via relation extraction and textual evidence. In: ACL, pp. 2326–2336 (2016)

Effectively Incorporating Knowledge in Open-Domain Dialogue Generation

Wenji Zhou[1,2,4], Tingting He[1,2,4(✉)], Miao Zhang[1,3,4], and Rui Fan[1,2,4]

[1] Hubei Provincial Key Laboratory of Artificial Intelligence and Smart Learning,
Central China Normal University, Wuhan 430079, Hubei, People's Republic of China
[2] School of Computer, Central China Normal University, Wuhan 430079,
Hubei, People's Republic of China
[3] National Engineering Research Center for E-Learning, Central China Normal University,
Wuhan 430079, Hubei, People's Republic of China
[4] National Language Resources Monitor and Research Center for Network Media,
Central China Normal University, Wuhan 430079, Hubei, People's Republic of China
{wenjizhou,zmzhangmiao,fanrui}@mails.ccnu.edu.cn,
tthe@mail.ccnu.edu.cn

Abstract. Dialogue generation is one of the most important parts in the dialogue system. Generating useful and informative responses in conversation has become a research hotspot. Previous work has proved that incorporating external knowledge is conducive to generating meaningful responses. But how to make full use of the existing information to select the most appropriate knowledge is a current research difficulty. In this paper, we propose a dialogue generation model with a lightweight knowledge routing module to sample knowledge needed for the conversation. In the knowledge routing module, we not only use the interactive information between the knowledge and the dialogue context utterance, but also the interactive information between each piece of knowledge for knowledge selection. Subsequently, the selected knowledge is incorporated into the dialogue generation model to generate responses. The experimental results show that our model tends to generate more meaningful and informative responses compared with baseline models.

Keywords: Dialogue generation · External knowledge · Deep learning

1 Introduction

Intelligent dialogue system is a research hotspot in the field of artificial intelligence, and dialogue generation is one of the most important parts in the dialogue system. In recent years, the Seq2Seq model [1–3] based on the encoder-decoder framework has been widely used in natural language generation. The dialogue generation model based on Seq2Seq can be trained in an end-to-end manner with large-scale training corpus. Although this method is very succinct, it tends to generate "secure" responses such as "I don't know" and "I agree with you" that contain useless information. These generic responses are acceptable, but we prefer informative and meaningful responses.

© Springer Nature Singapore Pte Ltd. 2021
H. Chen et al. (Eds.): CCKS 2020, CCIS 1356, pp. 237–249, 2021.
https://doi.org/10.1007/978-981-16-1964-9_19

In order to address this problem, many scholars proposed to improve the quality of dialogue generation by integrating knowledge. For example, Long et al. [4] tried to incorporate the retrieved knowledge into dialogue generation to produce more informative and meaningful responses. Zhang et al. [5] treated the persona profiles as knowledge to direct the generation of dialogue. Also, Zheng et al. [6] utilized persona-sparse data to build personalized dialogue models. Commonsense knowledge was incorporated in Zhou et al. [7], demonstrating that large-scale commonsense knowledge is able to facilitate dialogue understanding and generation. Both Dinan et al. [8] and Lian et al. [9] proposed effective mechanisms to select knowledge. Dinan et al. [8] used ground-truth knowledge to assist knowledge selection and Lian et al. [9] employed posterior knowledge distribution to guide knowledge selection. Nevertheless, most of the existing researches focus on how use the dialogue context to guide the knowledge selection. They only exploited the interactive information between the knowledge and the dialogue context utterance, without realizing that each piece of knowledge can also interact with each other. There is a certain degree of semantic similarity and logical correlation between some knowledge, which will also affect the weight distribution of each piece of knowledge. As shown in Table 1, we can find that in the first conversation, the dialogue context utterance is most similar to the knowledge K1 in semantic. So the knowledge K1 is more likely to be incorporated into the response. But knowledge K1 and knowledge K2 are semantically similar. The interaction between these two pieces of knowledge may also help to incorporate the knowledge K2 and generate more informative response R2. Similarly, in the second conversation, knowledge K1 and knowledge K2 are logically correlated. This correlation is contributed to selecting the knowledge K2 to generate more accurate response R2.

Table 1. Comparing the responses after using the semantic similarity and logical relevancy between each piece of knowledge or not.

Dialogue context utterance	Knowledge	Response1 (without knowledge interaction) Response2 (with knowledge interaction)
Do you like to watch comedies?	K1: I like comedies K2: I watched a lot of action movies	R1: Yes, I like comedies R2: Yes, I both like comedies and action movies
Where is Jackie Chan. from?	K1: The Best Actor award went to Jackie Chan K2: The winner of the Best Actor award was a Hong Kong actor	R1: He was awarded the best actor R2: He was awarded the best actor, and he is from Hong Kong

To overcome the above deficiencies, we propose a lightweight knowledge routing module to make full use of the interactive information between the knowledge and the dialogue context utterance, and the interactive information between each piece of knowledge to guide knowledge selection in open-domain dialogue generation. After knowledge selection, the influence of the knowledge most relevant to the current dialogue is enhanced, and the interference of irrelevant knowledge is reduced. The experimental results show that our model is more inclined to generate meaningful and informative responses.

Our main contributions are shown as follows:

- We incorporate external knowledge to generate more meaningful and informative responses in dialogue generation.
- The proposed knowledge routing module makes full use of the interactive information between knowledge and dialogue context utterance, and the interactive information between each piece of knowledge. In this way, the relevant knowledge is able to be selected and incorporated into response generation more properly.

2 Related Work

In recent years, the Seq2Seq model, which has caused an upsurge in academia, has been successfully applied to dialogue generation and achieved good results. However, the problem of generating short, meaningless and generic words arises in many studies.

In the last few years, scholars find that knowledge incorporation is an effective way to overcome the aforementioned deficiencies [10, 11]. Some studies regard unstructured [3, 5, 10] or structured text information [7, 11] as external knowledge and incorporate it into dialogue generation models. Long et al. [1] introduces the text knowledge information processed in convolution operation to the decoder together with dialogue history information to generate a reply. Ghazvininejad et al. [10] uses the memory network that stores textual knowledge to generate informative responses. A knowledge dialogue model using copy network is proposed by Zhu et al. [11]. In addition, some domain-specific knowledge bases are used to build dialogue generation model [12, 13]. Zhou et al. [7] is the first one to incorporate large scale commonsense knowledge into the dialogue generation process. Further, the posterior distribution is employed by Lian et al. [9] to guide knowledge selection.

However, most of existing models only consider the interactive information between the knowledge and the dialogue context utterance during the knowledge selection procedure, ignoring that there may be some semantic similarities or logical correlations between some knowledge. This part of knowledge can also interact with each other to influence knowledge selection. In comparison, our model has the ability to make full use of all this interactive information to guide knowledge selection and incorporate the knowledge properly to generate meaningful and informative responses.

3 Model

In this paper, we focus on how to select knowledge and to incorporate knowledge properly into dialogue generation model. Given a dialogue context utterance (including both the dialogue history and the current utterance) $X = x_1 x_2 \cdots x_n$ (where x_t represents the t-th word in X, and n is the number of words in X) and a collection of knowledge $\{K_1, K_2, \cdots, K_N\}$ (where N is the number of knowledge), we need to select relevant knowledge and incorporate the selected knowledge properly to generate a suitable response $y = y_1 y_2 \cdots y_m$ (where y_j represents the j-th word in y, and m is the number of words in y).

3.1 Architecture Overview

As shown in Fig. 1, there are three major components in the whole architecture of our dialogue generation model:

- **Encoder:** The encoder encodes the dialogue context utterance X into a vector x. Similarly, each piece of knowledge K_i is encoded into the vector k_i.
- **Knowledge Routing Module:** The knowledge routing module takes as input each piece of knowledge k_i and dialogue context utterance x. Then, it will output a proper knowledge representation k' relevant to the current dialogue. That is, the knowledge routing module has the ability to select the most important knowledge in $\{k_1, k_2, \cdots, k_N\}$, and represent the relevant knowledge into a single vector k'. Subsequently, the selected knowledge k' will be fed into the decoder together with the dialogue context utterance. Apparently, the knowledge routing module is the most important part in our dialogue model.
- **Decoder:** At last, the decoder generates a response Y based on the dialogue context utterance and the selected knowledge k'.

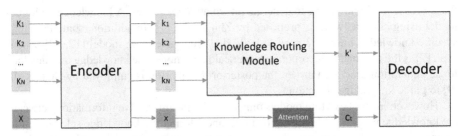

Fig. 1. The architecture overview of our model.

3.2 Encoder

In this paper, we use Bidirectional Gated Recurrent Unit (BGRU) [14] to encode. Given a dialogue context utterance $X = x_1 x_2 \cdots x_n$, the encoder encodes it into the corresponding hidden state $H = h_1 h_2 \cdots h_n$, which is defined as below:

$$h_t = BGRU(h_{t-1}, e(x_t)) \tag{1}$$

where $e(x_t)$ is the embedding of the word x_t and h_t is the hidden state of the encoder at time t. The hidden state of the last word in X is able to represent the semantic information of the whole sentence. This is the characteristic of RNN. Thus, we define $x = h_n$ to represent the dialogue context utterance X. Similarly, each piece of knowledge K_i in $\{K_1, K_2, \cdots, K_N\}$ is regarded as a sentence and encoded into vector k_i in the same way. The dialogue context utterance x and the knowledge k_i will be used later in knowledge routing module.

3.3 Knowledge Routing Module

We feed the dialogue context utterance x and each piece of knowledge in $\{k_1, k_2, \cdots, k_N\}$ into the knowledge routing module at the same time. The module is able to select proper knowledge in $\{k_1, k_2, \cdots, k_N\}$, that is to assign appropriate weights to each piece of knowledge, and output a single knowledge vector k' which can represent the selected knowledge. The detailed architecture of knowledge routing module is shown in Fig. 2.

Specifically, before the dialogue context utterance x and each piece of knowledge in $\{k_1, k_2, \cdots, k_N\}$ interact with each other, we perform global average pooling operations on the dialogue context utterance x and each piece of knowledge $\{k_1, k_2, \cdots, k_N\}$. This operation is inspired by the work of [15–17]. In this way, the vector representations of the dialogue context utterance x and each piece of knowledge $\{k_1, k_2, \cdots, k_N\}$ are compressed into corresponding scalar representations s_x and $\{s_{k_1}, s_{k_2}, \cdots, s_{k_N}\}$ respectively. This operation makes use of the various dimensional information of the vector so that the scalar is able to represent the semantic information of the vector. In addition, after the vector is converted into the scalar, the complexity of feature representation is reduced, and the calculation complexity in the model is reduced. Thus, the knowledge routing module is a lightweight module.

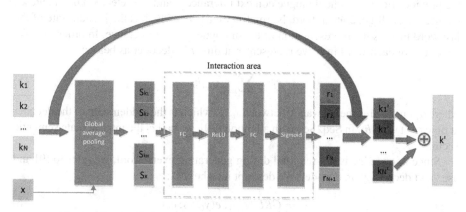

Fig. 2. The detailed structure of our proposed knowledge routing module.

In order to make the dialogue context utterance x and each piece of knowledge interact with each other, we design the following operation:

$$\mathbf{r} = \sigma\left(W_2\delta\left(W_1\left(s_{k_1}s_{k_2}\cdots s_{k_N}s_x\right)\right)\right) \tag{2}$$

where W_1 and W_2 are learnable parameters, δ is the ReLU [18] function and σ is the sigmoid function [19]. This process is equivalent to a two-layer fully connected operation on each piece of knowledge $\{s_{k_1}, s_{k_2}, \cdots, s_{k_N}\}$ together with the dialogue context utterance s_x. The first layer is activated with ReLU, and the second layer is activated with sigmoid. The two-layer full connection improves the generalization ability of the model, so that more interactive information can be learned between each piece of knowledge and the dialogue context utterance. The obtained $r = \{r_1, r_2, \cdots, r_N, r_{N+1}\}$ is the weight of each piece of knowledge and the dialogue context utterance respectively (where $\{r_1, r_2, \cdots, r_N\}$ is the weight of each piece of knowledge and r_{N+1} is the weight of the dialogue context utterance). Then, we can get the overall representation of the selected knowledge k' by performing the following operations:

$$k'_i = r_i \cdot k_i, \quad i = 1, 2, \cdots, N \tag{3}$$

$$k' = \sum_{i=1}^{N} k'_i \tag{4}$$

where k'_i is the weighted representation of knowledge. We take the sum of the weighted knowledge as the overall representation of the knowledge related to the current dialogue. Then k' is obtained and it is the output of the knowledge routing module. k' will be later used to generate responses in decoder.

3.4 Decoder

In decoder, we use a standard GRU as the basic unit. Given the word y_{t-1} generated at the previous time step, the dialogue context utterance x and the selected knowledge k', the decoder will generate a word. Normally, we use h_n, which is the hidden state of the last word in x, as the representation of the dialogue context utterance. In order to make better use of each word in x, we represent x at time t in decoder as below:

$$c_t = \sum_{i=1}^{n} \alpha_{t,i} h_i \tag{5}$$

where $\alpha_{t,i}$ measures the relevancy between s_{t-1} (which is the hidden state of the decoder at time $t-1$) and h_i. In Seq2Seq with attention [20], we call c_t the intermediate semantic vector.

Since knowledge is not required during generating every word, we refer to [9] and use "soft decoding" in decoder. We descript it as belows:

$$s_t^y = GRU(s_{t-1}, e(y_{t-1}), c_t) \tag{6}$$

$$s_t^k = GRU(s_{t-1}, k', c_t) \tag{7}$$

$$s_t = r \cdot s_t^y + (1-r)s_t^k \tag{8}$$

where $e(y_{t-1})$ is the embedding of y_{t-1} and r is a learnable coefficient. s_t^y is obtained by using $e(y_{t-1})$ to update the hidden state of the decoder, while s_t^k is obtained by using k'

to update the hidden state of the decoder. And the coefficient r balances the contribution of s_t^y and s_t^k to the final hidden state s_t of the decoder.

In the end, the decoder generates a response $y = y_1y_2 \cdots y_m$ sequentially. Each token is generated by sampling from the output probability distribution. Specifically,

$$y_t \sim p_t = P(y_t|y_{<t}, c_t, k') = softmax(W_1s_t) \tag{9}$$

where $y_{<t} = y_1y_2 \cdots y_{t-1}$ and W_3 is a learnable parameter.

3.5 Loss Function

We use NLL loss and BOW loss [21] in our model. NLL loss is to measure the difference between the target response Y and the generated response y, so that the model can continuously generate better responses. BOW loss is to quantity the relevancy between the selected knowledge k' and the target response Y. In this way, we can ensure the accuracy of the selected knowledge. The optimization methods of these two loss functions are as follows:

$$l_{NLL}(\theta) = -\frac{1}{m}\sum\nolimits_{t=1}^{m} logp_\theta(Y_t|y_t) \tag{10}$$

$$l_{BOW}(\theta) = -\frac{1}{m}\sum\nolimits_{t=1}^{m} logp_\theta(Y_t|k') \tag{11}$$

Thus, the total loss of our model can be calculated as below:

$$l(\theta) = l_{NLL}(\theta) + l_{BOW}(\theta) \tag{12}$$

4 Experiments

4.1 Dataset

In this paper, we conduct experiments on two datasets, which are Baidu competition dataset and Persona-chat dataset [5], to verify the validity of our model.

Baidu Competition Dataset: Baidu opened 120,000 turns of Chinese dialogue data in the knowledge driven competition in 2019. In the Baidu competition dataset, the training set is organized in the form of session, and each session includes the dialogue goal, relevant knowledge information and the dialogue content. The training set includes 19,858 sessions, with an average of 14.15 knowledge sentences and 9.05 dialogue sentences per session. The test set is organized as a sample, and each sample includes dialogue goal, relevant knowledge information, and dialogue history. The test set includes 5,000 samples, each sample contains 14.19 knowledge sentences and 3.80 history sentences on average.

Persona-Chat Dataset: The Persona-chat dataset contains a total of 151,157 turns of dialogues. Each dialogue is between 6 and 8 turns with an average of 15 words per sentence. Each turn of dialogue is created around a given persona, and each persona is described with an average of 4.49 knowledge sentences. The dialogue is to imitate the two interlocutors to meet and understand each other for the first time. This makes the dialogue more personal and appealing. Compared with other datasets, Persona-chat is able to generalize to more difficult tasks because the sentences are revised to avoid word overlap problem. We divided the dataset into three parts: 122,499 turns for training, 14,602 turns for verification, and 14,056 turns for testing.

4.2 Hyper-parameters Setting

Table 2. Hyper-parameters of our model.

Hyper-parameters	Size
embedding size	300
hidden size	800
vocabulary size	30000
batch size	64
learning rate	0.00005

The setting of hyper-parameters will have a great impact on the model. In this model, we consider the parameters of word embedding size, learning rate, batch size, the hidden layer size, etc. The experimental parameter settings in this paper are shown in Table 2.

4.3 Experiments for Comparison

In order to verify the effectiveness of our model, we have compared our model with the following baselines.

Seq2Seq: The Seq2Seq model with attention mechanism without incorporating knowledge [2].

Seq2SeqK: The knowledge is incorporated into the Seq2Seq model without selection [4].

PostKS: The knowledge selection is guided by posterior probability distribution [9]. Then the selected knowledge is incorporated into the Seq2Seq model.

KRS (Knowledge Routing Selection): This is our model using knowledge routing to select knowledge and then incorporate the selected knowledge into dialogue generation.

4.4 Automatic and Human Evaluation

Automatic Evaluation. In this paper, BLEU-1/2 and Distinct-1/2, which are popular metrics for evaluating the quality of the generated responses, are adopted as the criteria for automatic evaluation. Specifically, BLEU [22] is to compare the co-occurrence number of n-gram phrases between the generated sentences y and the target sentences Y. BLEU-1/2 is able to better evaluate the similarity of two sentences through the co-occurrence frequency of 1-g phrase and 2-g phrase. Distinct [23] is commonly used to assess the diversity of the generated responses. It computes the proportion of distinct n-gram phrases in all n-gram phrases of the generated sentence.

Human Evaluation. In this paper, we also perform human evaluation to assess the quality of responses generated by different models. We randomly select 200 responses generated by each model for three reviewers to evaluate. The average of their scores serves as the final score for each model. The evaluation criteria adopted by the reviewers are as follows:

0: The response is not fluent or relevant.
1: The response is acceptable but uninformative or meaningless, such as "I don't know".
2: The response is fluent, meaningful and informative.

4.5 Result and Analysis

The automatic evaluation and human evaluation performances are shown in Table 3 and Table 4 respectively. It is obvious that our model outperforms the other three baselines in most automatic metrics, especially on Baidu competition dataset.

Table 3. The automatic evaluation performances for different models.

Models	Baidu competition dataset		Persona-chat dataset	
	BLEU1/2	Distinct1/2	BLEU1/2	Distinct1/2
Seq2Seq	0.317/0.188	0.038/0.085	0.190/0.092	**0.004/0.016**
Seq2SeqK	0.309/0.186	0.041/0.090	0.194/0.092	0.004/0.014
PostKS	**0.327**/0.183	0.034/0.087	**0.198**/0.091	0.003/0.014
KRS	0.324/**0.196**	**0.041/0.100**	0.196/**0.096**	0.003/0.012

Specially, for Baidu competition dataset, our model achieves higher scores in all automatic metrics compared with Seq2Seq model, which demonstrates that incorporating knowledge helps to generate informative responses compared with those without knowledge. When comparing with Seq2SeqK model, which incorporates unselected knowledge, our model also performs better. This is because that some of the unselected knowledge will be introduced into the Seq2SeqK model as noise. This also shows that incorporating selected knowledge is conducive to generating accurate and fluent

Table 4. The human evaluation performances for different models.

Models	Baidu competition dataset	Persona-chat dataset
Seq2Seq	1.067	0.376
Seq2SeqK	1.092	0.387
PostKS	1.142	0.423
KRS	**1.217**	**0.478**

responses. Especially, our model outperforms the PostKS model which utilizes posterior probability distribution to guide knowledge selection in most metrics. This is because that our model not only uses the interactive information between the knowledge and the dialogue context utterance, but also the interactive information between each piece of knowledge. In this way, the knowledge selected by our model is more accurate and richer and the generated responses are more meaningful and informative. Although our model scores slightly lower on the BLEU1 metric than PostKS model, ours performs better on the BLEU2 metric. BLEU1 only compares the co-occurrence number of 1-g phrases between the generated sentences y and the target sentences Y. By contrast, BLEU2 calculates the number of 2-g phrases, which also focuses on the fluency of sentences. Therefore, the responses generated by our model are more fluent than those generated by PostKS model.

However, the Distinct1/2 metric on Persona-chat is not so good in our model. The cause may be that in the Persona-chat dataset, most knowledge (which is called persona in the dataset) starts with "I am", "I like", "I have", etc. These similar and general words and phrases will increase the similarity between each piece of knowledge when the knowledge interacts with each other. Incorporating this knowledge into dialogue will increase the probability of generating such general words and then the scores of the Distinct1/2 metric may decrease to a certain extent. Nevertheless, the quality of the responses generated by our model has not been reduced and has improved a lot. During human evaluation, it's obvious to find that Seq2Seq model without incorporating knowledge tends to generate "safe" responses such as "I am sorry" or "I am not sure". By contrast, the responses generated by our model contain more meaningful information.

4.6 Case Study

As shown in Table 5, we exhibit two typical examples of the generated responses on both Baidu competition dataset and Persona-chat dataset. The knowledge related to the generated responses is sampled for illustration. For the first conversation from Baidu competition dataset, it's obvious to find that two pieces of knowledge are incorporated into the response generated by our model (KRS) properly. The response is meaningful and informative. The response generated by PostKS is acceptable, but only contains one piece of knowledge. In contrast, the responses generated by Seq2Seq and Seq2SeqK are not fluent or irrelevant. For the second conversation from Persona-chat dataset, we can find that three pieces of knowledge are incorporated into the response. The

Table 5. Sample responses generated by all models on Baidu competition dataset and Persona-chat dataset. For Baidu competition dataset, we concatenate each knowledge triple into a sentence and encode it into a vector.

Baidu competition dataset	Dialogue Context Utterance	Person a: The movie directed by Thomas C. Grane is really well publicized. （Thomas C. Grane 导演的电影宣传的真到位。） Person b: Isn't it commonplace to promote movies? （电影做宣传不是常事吗？） Persona a:
	Knowledge	K1: ["Avatar: Create Pandora Planet"," director"," Thomas C. Grane"] （["阿凡达：打造潘多拉星球", "导演", "Thomas C. Grane"]） K2: ["Avatar: Create Pandora Planet"," domain","movie"] （["阿凡达：打造潘多拉星球", "领域", "电影"]） K3:["Avatar: Create Pandora Planet"," reputation"," good reputation"] （["阿凡达：打造潘多拉星球", "口碑", "口碑很好"]）
	Seq2Seq	Yeah, I watched a movie recently and it was highly rated. （是啊，我最近看了一部电影，评分很高的。）
	Seq2SeqK	Yeah, I watched a movie, which was adapted from a . （是啊，我看了一部电影，是根据一本改编的。）
	PostKS	Yeah, **Avatar: Create Planet Pandora**. This movie is really wonderful. （是啊，**阿凡达：打造潘多拉星球**这部电影，真的是很精彩。）
	KRS	Yes, **Avatar: Create Planet Pandora**. This movie has a **good reputation**. You can watch it. （是的，**阿凡达：打造潘多拉星球**，这部电影，**口碑不错**，你可以看看。）
Persona-chat dataset	Dialogue Context Utterance	Person a: Do you like the beach? Person b: I do enjoy the sounds of waves yes. Person a:
	Knowledge	K1: I enjoy my job of helping the youth of our city. K2: I need to lose weight. K3: I enjoy pizza.
	Seq2Seq	I like the <unk>.
	Seq2SeqK	I like to **read** and listen to music.
	PostKS	I am not sure that is a good thing.
	KRS	I love to **read**. I love to **eat too much**.

response sentence "*I love to read*" is related to the knowledge K1 which means that he may be a teacher. The response sentence "*I love to eat too much*" comes from the knowledge K2 and K3. However, the responses generated by the other three models are unacceptable. Hence, our proposed model tends to perform better by generating meaningful and informative responses.

5 Conclusion and Future Work

In this paper, we propose a knowledge routing module which can select knowledge more properly and then the selected knowledge is incorporated into dialogue generation model to generate meaningful and informative response. In our model, we make it possible not only for knowledge to interact with the dialogue context utterance, but also for each piece of knowledge to interact with each other. Automatic and manual evaluation results show that the model proposed in this paper is superior to the classical baselines and demonstrate the effectiveness of our model. Our work has significant reference value for how to select external knowledge in dialogue generation.

As for future work, we plan to take the advantage of Pre-training methods into our model for better performance.

Acknowledgments. This research is supported by the National Natural Science Foundation of China (61532008, 61932008, 61572223), the Key Research and Development Program of Hubei Province (2020BAB017), Wuhan Science and Technology Program (2019010701011392), Scientific Research Center Program of National Language Commission (ZDI135-135) and the National Key Research and Development Program of China (2017YFC0909502).

References

1. Cho, K., van Merrienboer, B., Gulcehre, C., et al.: Learning phrase representations using RNN encoder-decoder for statistical machine translation. In: EMNLP 2014, pp. 1724–1734 (2014)
2. Shang, L., Lu, Z., Li, H.: Neural responding machine for short-text conversation. In: ACL, vol. 1, pp. 1577–1586 (2015)
3. Sutskever, I., Vinyals, O., Le, Q.V.: Sequence to sequence learning with neural networks. In: NIPS 2014, pp. 3104–3112 (2014)
4. Long, Y., Wang, J., Xu, Z., et al.: A knowledge enhanced generative conversational service agent. In: Proceedings of the 6th Dialog System Technology Challenges (DSTC6) Workshop (2017)
5. Zhang, S., Dinan, E., Urbanek, J., Szlam, A., et al.: Personalizing dialogue agents: i have a dog, do you have pets too? In: ACL, vol. 1, pp. 2204–2213 (2018)
6. Zheng, Y., Zhang, R., Huang, M., et al.: A pre-training based personalized dialogue generation model with persona-sparse data. In: AAAI 2020, pp. 9693–9700 (2020)
7. Zhou, H., Young, T., Huang, M., et al.: Commonsense knowledge aware conversation generation with graph attention. In: IJCAI 2018, pp. 4623–4629 (2018)
8. Dinan, E., Roller, S., Shuster, K., et al.: Wizard of Wikipedia: knowledge-powered conversational agents. In: ICLR (Poster) (2019)
9. Lian, R., Xie, M., Wang, F., et al.: Learning to select knowledge for response generation in dialog systems. In: IJCAI 2019, pp. 5081–5087 (2019)
10. Ghazvininejad, M., Brockett, C., Chang, M.-W., et al.: A knowledge-grounded neural conversation model. In: AAAI 2018, pp. 5110–5117 (2018)
11. Zhu, W., Mo, K., Zhang, Y., et al.: Flexible end-to-end dialogue system for knowledge grounded conversation. CoRR abs/1709.04264 (2017).
12. Gu, J., Lu, Z., Li, H., et al.: Incorporating copying mechanism in sequence-to-sequence learning. In: ACL, vol. 1 (2016)
13. Xu, Z., Liu, B., Wang, B., et al.: Incorporating loose-structured knowledge into conversation modeling via recall-gate LSTM. In: IJCNN 2017, pp. 3506–3513 (2017)
14. Cho, K., van Merrienboer, B., Gülçehre, Ç., et al.: Learning phrase representations using RNN encoder-decoder for statistical machine translation. In: EMNLP 2014, pp. 1724–1734 (2014)
15. Yang, J., Yu, K., Gong, Y., et al.: Linear spatial pyramid matching using sparse coding for image classification. In: CVPR 2009, pp. 1794–1801 (2009)
16. Sánchez, J., Perronnin, F., Mensink, T., Verbeek, J.: Image classification with the fisher vector: theory and practice. Int. J. Comput. Vis. **105**(3), 222–245 (2013). https://doi.org/10.1007/s11263-013-0636-x

17. Shen, L., Sun, G., Huang, Q., et al.: Multi-level discriminative dictionary learning with application to large scale image classification. IEEE Trans. Image Process. **24**(10), 3109–3123 (2015)
18. Nair, V., Hinton, G.E.: Rectified linear units improve restricted Boltzmann machines. In: ICML 2010, pp. 807–814 (2010)
19. Han, J., Moraga, C.: The influence of the sigmoid function parameters on the speed of backpropagation learning. In: Mira, J., Sandoval, F. (eds.) IWANN 1995. LNCS, vol. 930, pp. 195–201. Springer, Heidelberg (1995). https://doi.org/10.1007/3-540-59497-3_175
20. Bahdanau, D., Cho, K., Bengio, Y.: Neural machine translation by jointly learning to align and translate. In: ICLR 2015 (2015)
21. Zhao, T., Zhao, R., Eskénazi, M.: Learning discourse-level diversity for neural dialog models using conditional variational autoencoders. In: ACL, vol. 1, pp. 654–664 (2017)
22. Papineni, K., Roukos, S., Ward, T., Zhu, W.-J.: BLEU: a method for automatic evaluation of machine translation. In: ACL 2002, pp. 311–318 (2002)
23. Li, J., Galley, M., Brockett, C., et al.: A diversity-promoting objective function for neural conversation models. In: HLT-NAACL 2016, pp. 110–119 (2016)

What Linguistic Information Does Reading Comprehension Require?

Yong Guan[1], Ru Li[1,2(✉)], and Shaoru Guo[1]

[1] School of Computer and Information Technology, Shanxi University,
Taiyuan, China
`liru@sxu.edu.cn`
[2] Key Laboratory of Ministry of Education for Computation Intelligence and Chinese
Information Processing, Shanxi University, Taiyuan, China

Abstract. Machine comprehension is one of the primary goals in Artificial Intelligence (AI) and Natural Language Processing (NLP). Accessing the difficulty level of machine reading comprehension (MRC) questions is important for building accurate MRC systems. In order to tackle this problem, we propose a novel idea to access the difficulty level of MRC questions, according to the amount of linguistic information required to answer them. Specifically, we systematically analyze and compare the performance for each BERT layer representation per question type on MRC datasets, and highlighted the characteristics of the datasets according to linguistic information of different layers. Our extensive analysis suggests that the superficial categories (or question types) of MRC questions do not directly reflect their difficulty levels and that it is possible to analyze the MRC questions' difficulty levels according to the amount of linguistic information required.

Keywords: Machine reading comprehension · Linguistic information · Neural network

1 Introduction

Machine reading comprehension (MRC) is a critical task in NLP [9], which aims to read and comprehend a given article or passage, and answer relevant questions based on its content.

Assessing the difficulty level of MRC questions and clarifying what linguistic information are required to answer them, are particularly important for building an accurate MRC system. Note MRC questions only have *superficial categories* [11,19], such as *question types*: *what, where, who*. The questions are as follows: 1) can these superficial categories tell what difficulty levels of MRC questions are? 2) If not, what is the best way to characterize the difficulty levels of MRC questions?

To answer these two questions, we systematically investigates the relation between the superficial categories (question types) and the difficulty level of

© Springer Nature Singapore Pte Ltd. 2021
H. Chen et al. (Eds.): CCKS 2020, CCIS 1356, pp. 250–260, 2021.
https://doi.org/10.1007/978-981-16-1964-9_20

MRC questions. We also measure what linguistic information are required for MRC questions with different difficulty levels. Note our focus in this paper is to answer these two important questions to facilitate machine reading comprehension task by employing state-of-the-art technologies, instead of proposing new but unproven technologies.

Our method is inspired by the recent analysis [4] about the compositional nature of BERT's representation, which concluded that BERT can capture a rich hierarchy of linguistic information, with *surface features in lower layers*, *syntactic features in middle layers* and *semantic features in higher layers*.

How to evaluate the difficulty level of a MRC question? It should reflect the difficulty of comprehending the question context when answering it. There should have positive relations between the difficulty level of MRC questions and amount of linguistic information required. In other words, the more linguistic information required to answer a question, the more difficult the question is.

To address the first question, for each *question type*, we systematically analyze and compare the performance for each BERT layer on MRC benchmark data MCTest. We adopted the methods to partition questions in nine different *question types* that [19] proposed for the fine-grained analysis of MRC capability.

To address the second question, we use linguistic information to classify MRC questions. According to the different levels of linguistic information that are needed to answer the questions, we classify the questions into four categories: surface features, syntactic features, semantic features and other. This not only enables us to understand the characteristics of the MRC questions, but also helps us understand how to build accurate MRC models.

The key contributions of this work are summarized as follows:

1. We propose a novel idea to measure the difficulty level of MRC questions: the more linguistic information required to answer a question, the more difficult the question is.
2. For each question type, we analyze the performance of each BERT layer and observe that question types do not directly reflect the question difficulty.
3. We propose to employ different levels of linguistic information to classify MRC questions into four different categories, which enables accurate MRC analysis and model building.

2 Relate Work

To facilitate the explorations and innovations in MRC area, many datasets have been established over years [5, 8, 12, 16, 17, 24].

With the increasing availability of MRC datasets and the development of deep learning, numerous neural network models have been proposed to tackle the challenging MRC task [14, 18, 22, 26], and good results have been achieved in last few years. Recently, pre-trained language model [10, 13, 15] has caused a stir in MRC community, especially after the release of BERT [2].

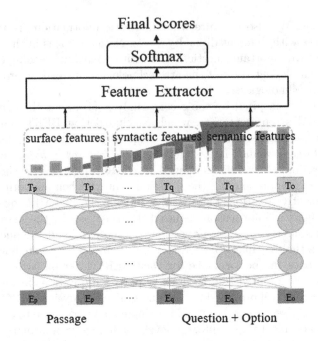

Fig. 1. Architecture of the model.

While there are various neural networks were carefully designed and evaluated in terms of accuracy on whole data by training end to end models, there seems to be a lack of work to study what information is required to solve the MRC questions. [7] finding that instead of using the whole information (question and passage) to train the models, only apply question or passage often perform surprisingly well. [20] split MRC dataset into easy and hard subsets according to the simple heuristics, such as entity type recognition and attention. In this work, we propose a novel idea to access the difficulty level of MRC questions, according to the amount of linguistic information required to answer them.

3   Method

BERT is based on a multi-layer bidirectional Transformer networks. We omit rather extensive formulations of BERT and recommend readers to get the details from [21]. In this paper, based on various observation [3,4] that BERT can capture a rich hierarchy of linguistic information representations, we investigate the relation between the difficulty level MRC questions and the amount of linguistic information required to answer them. Figure 1 overviews our framework, which includes three key modules: Encoder, Feature Extractor and Output Layer.

3.1 Encoder

In multiple-choice MRC task, there are three components in each example: passage, question and answer with multiple choices, shown in Table 5. However, the input sequence of the BERT model is either a single text sequence or a pair of text sequences. As such, we construct our input as: the passage as sequence A, and the concatenation of the question and one choice of answer as sequence B. The input sequence $X = \{x_1, ..., x_n\}$ can be denoted as:

$$[[CLS] \text{ passage } [SEP] \text{ question } + \text{ option } [SEP]]$$

Where n is the sequence length. Then the BERT model captures the contextual information for each token via self-attention and produces a sequence of contextual embeddings.

3.2 Feature Extractor

As the original pre-trained BERT is based on the last layer representations to predict the answers, this module fuses different layers to extract salient features. The input to the Feature Extractor is $F = \{H_i, ..., H_j\}, (i < j)$, where i and j represent the i-th and j-th layer in the BERT encoder respectively. In order to filter unnecessary information in the input, we apply k-max pooling [6] over F to select k most important weights.

$$F^* = MaxPooling(F) \tag{1}$$

Where MaxPooling is a column-wise maximum function. We use the Bidirectional Gated Recurrent Unit (GRU) [1] to read F^*, and then reweight the representation of F^* through the attention mechanism [23], which use each hidden state to get an importance socre. The importance socres are then normalized to get the new representation by weighted sum:

$$C_i = GRU(F_i^*) \tag{2}$$

$$\alpha_i = W^T C_i \tag{3}$$

$$\bar{C} = softmax(\alpha) \tag{4}$$

$$C^* = \sum_{i=0}^{k} \bar{C}_i F_i^* \tag{5}$$

Where W^T is learned parameters and k is the number of layers. $\bar{C}$ is weight matrix when computing the representation.

3.3 Output Layer

Finally, we combine the preliminary representation C and current feature vector C^* to construct the final state. The final state is then passed through a MLP to predict answer with a softmax layer.

$$Z = MLP([C; C^*; C + C^*; C - C^*]) \tag{6}$$

$$Z^* = ReLU(W_z Z + b_z) \tag{7}$$

$$P = softmax(W_o Z^*) \tag{8}$$

Where W_z and W_o are learned parameters. MLP denotes the Multilayer Perceptron.

4   Model Implementation

4.1 Analyze What Linguistic Information Required per Question Type on MRC Questions

We systematically analyze and compare the performance for each BERT layer per question type on MCTest. Specifically,

(1) We adopt the question types proposed in [19], that divided questions into 9 categories;
(2) We extract the representation of each layer of BERT, and use the first input token ([CLS]) representation, which summarizes the information from the actual tokens at every layer to predict answers.

4.2 Classify MRC Questions According to Linguistic Information

The amount of linguistic information could reveal the difficulty level of the MRC questions. Thus, we investigate the MRC questions based the linguistic structure implicitly learned by BERT's representations.

In particular, on the first pass, motivated by [4], we devide the 12 layers of BERT into 3 parts on average, where the three parts represent surface features (lower layers), syntactic features (middle layers) and semantic features (higher layers) respectively. Then, we input each of the three parts into Feature Extractor as shown in Sect. 3.3 and use each part to predict the answer separately. Finally, we classify the questions according to which linguistic information are sufficient to answer the question. More specifically, MRC questions are classified into 4 categories: surface features, syntactic features, semantic features and other category. The other category refers to the questions that can not be correctly answered in the previous three categories[1].

---

[1] For the same question, if the first two or three parts can answer it, we think it belongs to the first category, because the linguistic information in the first category is sufficient to solve this question. Other situations are similar to this.

Table 1. Performance for each BERT layer per question type of MCTest160.

Layer	Which	Where	What	When	Who	Whose	How	Why	Other	Total
1	16.67	16.67	30.95	20.0	32.14	–	28.57	37.84	40.0	30.83
2	16.67	33.33	33.33	0.0	25.0	–	33.33	27.03	60.0	30.83
3	16.67	16.67	34.92	40.0	25.0	–	19.04	32.43	20.0	30.42
4	16.67	58.33	50.79	60.0	42.86	–	47.62	45.95	0.0	47.50
5	16.67	58.33	54.76	40.0	50.0	–	52.38	51.35	20.0	51.67
6	16.67	66.67	57.14	60.0	50.0	–	61.90	54.05	20.0	55.0
7	16.67	66.67	57.94	60.0	50.0	–	66.67	54.05	20.0	55.83
8	0.0	75.0	55.56	40.0	42.86	–	71.43	40.54	40.0	52.08
9	0.0	66.67	52.38	40.0	53.57	–	42.86	43.24	0.0	48.33
10	66.67	66.67	69.05	60.0	75.0	–	52.38	45.95	100.0	65.0
11	100.0	83.33	81.75	80.0	82.14	–	76.19	78.38	80.0	81.25
12	100.0	91.67	81.75	80.0	67.86	–	76.19	72.97	60.0	78.75

Table 2. Performance for each BERT layer per question type of MCTest500.

Layer	Which	Where	What	When	Who	Whose	How	Why	Other	Total
1	24.0	25.86	29.34	42.86	27.27	33.33	32.56	24.19	25.0	28.33
2	36.0	20.69	29.02	14.29	19.48	33.33	23.26	32.26	25.0	27.0
3	44.0	29.31	31.86	42.86	27.27	0.0	32.56	35.48	37.5	32.0
4	28.0	39.66	44.16	85.71	38.96	0.0	34.88	45.16	25.0	41.83
5	32.0	48.28	43.22	71.43	41.56	33.33	30.23	56.45	25.0	43.5
6	36.0	46.55	44.16	71.43	38.96	0.0	39.53	61.29	37.5	44.83
7	32.0	46.55	45.43	85.71	37.66	33.33	41.86	54.84	37.5	45.0
8	32.0	44.83	46.37	85.71	41.56	33.33	34.88	46.77	37.5	44.5
9	28.0	43.10	46.06	85.71	36.36	0.0	37.21	41.94	37.5	42.83
10	56.0	63.79	66.56	71.43	46.75	33.33	53.49	59.68	50.0	61.33
11	68.0	75.86	77.29	100.0	68.83	66.67	76.74	82.26	37.5	75.83
12	72.0	77.59	83.91	85.71	67.53	66.67	79.07	85.48	50.0	80.0

5 Experiments

5.1 Benchmark Datasets

We focus on multiple-choice questions, as different answer candidates vary in abstraction level and usually do not directly appear in the document [25], which is suitable for measuring questions' difficulty level. We employ the method described in [19], to classify questions into different types on MCtest. Note MCTest is a reading comprehension task with small vocabulary limited to what young children would understand. It has two subsets: MCTest160 and MCTest500.

Table 3. Percent (%) of MCTest datasets.

MCTest	Surface features	Syntactic features	Semantic features	Other
160	47.5	16.67	25.0	10.83
500	41.83	15.83	30.33	12.0

Table 4. Percent (%) of RACE datasets.

RACE	Surface features	Syntactic features	Semantic features	Other
Middle	28.69	29.46	26.46	15.39
High	27.79	25.93	24.50	21.78

We also use Large-scale ReAding Comprehension Dataset From Examinations (RACE) for our experiments. RACE consists of two subsets: *RACE middle* and *RACE high*, corresponding to middle school and high school difficulty level respectively.

5.2 Experimental Results

We use PyTorch implementation of BERT, which hosts the models trained by [2]. All our experiments are trained on RACE and based on the *bert-base-uncased* variant. We employ MCTest data in our experiments as our test set, as it is too small to train a neural network model based on it.

(1) Table 1 and 2 show the performance for each BERT layer per question type of MCTest160, MCTest500, respectively. As can be seen from both tables, the accuracy of MCTest are getting higher when the number of layers increase across all nine question types (Columns 2–10), indicating superficial categories of MRC datasets do not directly affect the question difficulty level. It makes sense, as question words or answer words should not reveal what linguistic information that questions require.

(2) Table 3 shows the statistics of MCTest data based on observation of hierarchy of linguistic information. We can see from the table, MCTest160 requires 47.5% surface features, which is 5.67% more than that in MCTest500. On the other hand, MCTest160 requires less syntactic and semantic features than MCTest500, signifying MCTest500 is harder than MCTest160, which aligns very well with the principles their data construction.

(3) Table 4 shows our experimental results on RACE datasets. Overall, *RACE high* requires more sophisticated linguistic information than *RACE middle*. The most noteworthy phenomenon is that questions with different difficulty levels are almost evenly distributed. These results are consistent with the original intention in dataset settings that *RACE high* comes from high school exams while *RACE middle* comes from middle school exams.

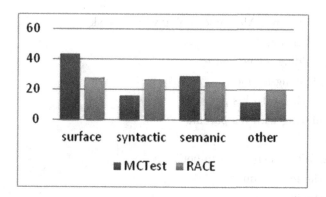

Fig. 2. Comparison between RACE datasets and MCTest datasets.

(4) Figure 2 shows MCTest is easier than RACE, as RACE requires more semantic features. This is consistent with the principles of data construction that MCTest is targeted to those young children, while RACE targets middle and high school students. We can conclude that the more complex the dataset is, the deeper semantic information is needed, which clearly shows our dataset partition method can reveal the difficulty level of datasets.

In addition, we also conduct our experiments on the *bert-large-uncased variant*, and observe that their experimental results are similar.

5.3 Case Study

Consider the example of the questions shown in Table 5. Q1 belongs to surface features category and Q2 belongs to syntactic features category. Answering these two questions requires different linguistic information. However, the two questions belong to the same category according to question type word *what*. To solve Q1, a MRC system only needs surface features to find *Mortamer was looking for food*. Q2, however, requires understanding of relations among the events, i.e. *the black sky* and *strong winds* are the evidence to *scared Mortamer*. This example shows that different features are required though the question words are the same, demonstrating superficial categories of MRC datasets does not directly reflect question difficulty levels.

Table 5. An example from MCTest with the answers marked *.

Passage:

Mortamer and his parents were outside hunting for food and the sky turned black! The sun was gone, hidden behind giant black clouds!.... scaring poor little Mortamer ...

Q1: What was Mortamer doing when the sky turned black?

A) building a tree house

B*) looking for food

C) playing with his pet snail

D) playing with his parents

Q2: What scared Mortamer?

A) the bright sun

B) the jungle birds

C) his friend, Johnson playing a trick

D*) The black sky and strong winds

---

6   Conclusion

In this paper we investigate how to access the difficulty level of MRC questions, according to the types of linguistic information required to answer them. Particularly, we conduct extensive experiments to evaluate the performance for each BERT layer per question type on MCTest datasets, and conclude that the superficial categories of MRC datasets do not directly reflect the question difficulty level. Finally, we propose an effective method which leverages linguistic information to classify questions' difficulty level more accurately. For future work, we plan to use the analysis from the present study in constructing a system that can be applied to multiple datasets and further improve the performance of MRC system.

Acknowledgement. We thank the anonymous reviewers for their helpful comments and suggestions. This work is supported by the National Natural Science Foundation of China (No. 61936012, No. 61772324).

References

1. Cho, K., et al.: Learning phrase representations using RNN encoder-decoder for statistical machine translation. In: Proceedings of the 2014 Conference on Empirical Methods in Natural Language Processing (EMNLP), pp. 1724–1734. Association for Computational Linguistics, Doha, October 2014. https://doi.org/10.3115/v1/D14-1179. https://www.aclweb.org/anthology/D14-1179

2. Devlin, J., Chang, M., Lee, K., Toutanova, K.: BERT: pre-training of deep bidirectional transformers for language understanding. CoRR abs/1810.04805 (2018). http://arxiv.org/abs/1810.04805

3. Goldberg, Y.: Assessing BERT's syntactic abilities. arXiv abs/1901.05287 (2019)

4. Jawahar, G., Sagot, B., Seddah, D.: What does BERT learn about the structure of language? In: ACL 2019-57th Annual Meeting of the Association for Computational Linguistics (2019)

5. Joshi, M., Choi, E., Weld, D.S., Zettlemoyer, L.: TriviaQA: a large scale distantly supervised challenge dataset for reading comprehension. CoRR abs/1705.03551 (2017). http://arxiv.org/abs/1705.03551

6. Kalchbrenner, N., Grefenstette, E., Blunsom, P.: A convolutional neural network for modelling sentences. In: Proceedings of the 52nd Annual Meeting of the Association for Computational Linguistics (Volume 1: Long Papers), pp. 655–665. Association for Computational Linguistics, Baltimore, June 2014. https://doi.org/10.3115/v1/P14-1062. https://www.aclweb.org/anthology/P14-1062

7. Kaushik, D., Lipton, Z.C.: How much reading does reading comprehension require? A critical investigation of popular benchmarks. arXiv preprint arXiv:1808.04926 (2018)

8. Lai, G., Xie, Q., Liu, H., Yang, Y., Hovy, E.: RACE: large-scale ReAding comprehension dataset from examinations. In: Proceedings of the 2017 Conference on Empirical Methods in Natural Language Processing, pp. 785–794. Association for Computational Linguistics, Copenhagen, September 2017. https://doi.org/10.18653/v1/D17-1082. https://www.aclweb.org/anthology/D17-1082

9. Levesque, H.J.: On our best behaviour. Artif. Intell. 212, 27–35 (2014). https://doi.org/10.1016/j.artint.2014.03.007. http://www.sciencedirect.com/science/article/pii/S0004370214000356

10. Mikolov, T., Chen, K., Corrado, G.S., Dean, J.: Efficient estimation of word representations in vector space. CoRR abs/1301.3781 (2013)

11. Narasimhan, K., Barzilay, R.: Machine comprehension with discourse relations. In: Proceedings of the 53rd Annual Meeting of the Association for Computational Linguistics and the 7th International Joint Conference on Natural Language Processing (Volume 1: Long Papers), pp. 1253–1262. Association for Computational Linguistics, Beijing, July 2015. https://doi.org/10.3115/v1/P15-1121. https://www.aclweb.org/anthology/P15-1121

12. Nguyen, T., et al.: MS MARCO: a human generated machine reading comprehension dataset. CoRR abs/1611.09268 (2016). http://arxiv.org/abs/1611.09268

13. Peters, M.E., et al.: Deep contextualized word representations. CoRR abs/1802.05365 (2018). http://arxiv.org/abs/1802.05365

14. Qiu, L., et al.: Dynamically fused graph network for multi-hop reasoning. In: Proceedings of the 57th Annual Meeting of the Association for Computational Linguistics, pp. 6140–6150. Association for Computational Linguistics, Florence, July 2019. https://doi.org/10.18653/v1/P19-1617. https://www.aclweb.org/anthology/P19-1617

15. Radford, A.: Improving language understanding by generative pre-training (2018)
16. Rajpurkar, P., Zhang, J., Lopyrev, K., Liang, P.: SQuAD: 100,000+ questions for machine comprehension of text. CoRR abs/1606.05250 (2016). http://arxiv.org/abs/1606.05250
17. Richardson, M., Burges, C.J., Renshaw, E.: MCTest: a challenge dataset for the open-domain machine comprehension of text. In: Proceedings of the 2013 Conference on Empirical Methods in Natural Language Processing, pp. 193–203. Association for Computational Linguistics, Seattle, October 2013. https://www.aclweb.org/anthology/D13-1020
18. Seo, M.J., Kembhavi, A., Farhadi, A., Hajishirzi, H.: Bidirectional attention flow for machine comprehension. CoRR abs/1611.01603 (2016). http://arxiv.org/abs/1611.01603
19. Smith, E., Greco, N., Bošnjak, M., Vlachos, A.: A strong lexical matching method for the machine comprehension test. In: Proceedings of the 2015 Conference on Empirical Methods in Natural Language Processing, pp. 1693–1698. Association for Computational Linguistics, Lisbon, September 2015. https://doi.org/10.18653/v1/D15-1197. https://www.aclweb.org/anthology/D15-1197
20. Sugawara, S., Inui, K., Sekine, S., Aizawa, A.: What makes reading comprehension questions easier? arXiv preprint arXiv:1808.09384 (2018)
21. Vaswani, A., et al.: Attention is all you need. In: Guyon, I., et al. (eds.) Advances in Neural Information Processing Systems, vol. 30, pp. 5998–6008. Curran Associates, Inc. (2017). http://papers.nips.cc/paper/7181-attention-is-all-you-need.pdf
22. Wang, W., Yang, N., Wei, F., Chang, B., Zhou, M.: Gated self-matching networks for reading comprehension and question answering. In: Proceedings of the 55th Annual Meeting of the Association for Computational Linguistics (Volume 1: Long Papers), pp. 189–198. Association for Computational Linguistics, Vancouver, July 2017. https://doi.org/10.18653/v1/P17-1018. https://www.aclweb.org/anthology/P17-1018
23. Yang, J., Zhao, H.: Deepening hidden representations from pre-trained language models for natural language understanding. ArXiv abs/1911.01940 (2019)
24. Yang, Z., et al.: HotpotQA: a dataset for diverse, explainable multi-hop question answering. In: EMNLP (2018)
25. Yin, W., Ebert, S., Schütze, H.: Attention-based convolutional neural network for machine comprehension. CoRR abs/1602.04341 (2016). http://arxiv.org/abs/1602.04341
26. Yu, A.W., et al.: QANet: combining local convolution with global self-attention for reading comprehension. CoRR abs/1804.09541 (2018). http://arxiv.org/abs/1804.09541

Enhancing Embedding via Two-Level Features for Machine Reading Comprehension

Shuyi Wang, Hui Song[✉], Bo Xu, and Hongkuan Zhang

School of Computer Science and Technology, Donghua University, Shanghai 201620, China
{2181754,2181729}@mail.dhu.edu.cn, {songhui,xubo}@dhu.edu.cn

Abstract. Machine reading comprehension (MRC) is a challenging task in natural language processing domain. Mainstream methods such as BERT or XLNet create Language Model (LM) with rich contextual information, but suffer from lacking semantic and syntactic message for MRC. We present a novel model TLE-BERT to enhance text encoding with the token-level and sentence-level features. In token level, we add semantic representation to the BERT based embedding. The desired features such as POS, entity type, noun phrase and synonyms of tokens are chosen by the sentinel mechanism. In sentence level, we introduce the grammatical structure to text encoding with dependency relation of the sentence, which is converted into m-hop matrices and merged into the embedding with a multi-head attention transformer layer. Each feature improves the performance of the model in varying degrees, and experiments show that TLE-BERT which combines all features organically strengthen the expressive of text representation, achieving a significant improvement on the popular MRC datasets.

Keywords: Feature enhancing · Two-level integration · Attention mechanism

1 Introduction

Machine Reading Comprehension (MRC), which mainly asks a machine to answer questions based on a given context, aims to make the machine understand and process human language. Common MRC tasks can be divided into four categories [22] according to the form of their answers: cloze tests, multiple choice, span extraction and free answering, and the last one is the ultimate goal of MRC research.

In this paper we focus on the third one, span extraction. This technology selects the correct continuous sub-sequence from an article to answer the question, which is very popular in most scenarios of question answering systems, such as the domain specific search engines, service consulting, dialogue-type QA and etc.

The origin of this task can be traced back to the 1970s [20] and people's solutions to MRC go through three stages. The early methods are rule-based. They can perform shallow language processing, but need to manually construct features. Machine learning-based methods [22] mainly depend on the existing language feature tools. Now the mainstream deep learning-based technology [27] performs deep feature extraction and obtains context-aware representation, making up the deficiencies of the traditional

© Springer Nature Singapore Pte Ltd. 2021
H. Chen et al. (Eds.): CCKS 2020, CCIS 1356, pp. 261–273, 2021.
https://doi.org/10.1007/978-981-16-1964-9_21

models. Also benefited from the capacity of deep learning, a series of large-scale MRC datasets are gradually released, such as CNN & Daily Mail [25], SQuAD [26], MS MARCO [11], which greatly promote the development of MRC techniques. Nowadays MRC has been a rising concern, not only a research hot spot in academia, but also a new approach worth trying out in industry.

The latest deep learning methods such as BERT [4] and XLNet [18] have achieved good language representation by training on large corpora, and have made break-through progress in various NLP tasks including MRC. But compared with human reading, they lack semantics and factual knowledge to implement complex inference and understanding.

To make up for the weakness, SemBERT [7] combined explicit contextual semantics of Semantic Role Labeling (SRL) with BERT embedding. KT-NET introduces external knowledge, entity and synonym, through knowledge bases (KBs). When integrating the external knowledge, it employs a mechanism to select the desired knowledge, and merges them to meet knowledge-aware prediction.

However, there are still some problems with these models although they improved performance. Sometimes engaged knowledge introduce extra entities to the predicted answer, which leads to the wrong boundary of answer (shown in Fig. 1a). On the other hand, if segments more than one share the same syntax form with the answer in the context, the model may be confused and lead to pick the incorrect one (shown in Fig. 1b).

Context: " The game was played on February 7, 2016, at Levi's Stadium in the San Francisco Bay Area at Santa Clara, California."
Question: " What city did Super Bowl 50 take place in?"
Wrong answer: Santa Clara, California
Correct answer: Santa Clara

(a)

Context: " On May 21, 2013, NFL owners at their spring meetings in Boston voted and awarded the game to Levi's Stadium."
Question: " Where was the meeting held when the NFL owners voted on the location for Super Bowl 50?"
Wrong answer: Levi's Stadium
Correct answer: Boston / in Boston

(b)

Fig. 1. (a) is an example containing redundant words, (b) is an example of making wrong prediction in spans with the same form.

In this paper, we explore the problems of the existing models and propose a two-level representation enhancing approach TLE-BERT for MRC. At the lower level of the model, we add the semantic information of entity type, noun phrases and POS to avoid the predicted answer containing redundant words, extra synsets such as **WordNet** are also included to get link keywords in context and questions. At the higher level, to identify the phrases with similar grammatical form, the dependencies of tokens in the sentence are transformed into multi-head attention coefficients to make the representation reflect the syntactic relation. Experiments on popular datasets prove the effectiveness of our method.

The contributions of this paper are summarized as follows:

- Through the algorithm analysis and case study of MRC task, we confirm the issues of existing models and propose a two-level approach for integrating token features and grammatical structure of token-pairs into sentence encoding.
- For token-level, we select the valid features from the candidates for different tokens according to the actual context, sentinel vector is added. And for sentence level, we transform the dependency tree of sentence into m-hop matrixes, and then integrate the information through multi-head attention mechanism.
- We combine different features that need to be introduced to conduct experiments, find that the impact of features varies on different datasets. We choose DocQA, pre-trained model Google BERT_base, KT-NET_base as baseline methods. Our model achieves **+0.63 *EM*/+ 0.68 *F*₁** improvements in SQuAD1.1 dataset and **+6.43 *EM*/+ 6.79 *F*₁** improvements in ReCoRD dataset than the best result of baseline method, better than the current state-of-the-art models.

2 Background and Related Work

The early reading comprehension model is mainly rule-based matching and machine learning-based methods [22], and recently the deep learning-based models improve the MRC performance in two ways, train a rich language representation, or build an adaptive model for MRC task.

2.1 Pre-trained Language Representation Model

Natural language processing (NLP) tasks require machines to have the ability of understanding language and reasoning like humans. In order to have these capabilities, the embedding representation of tokens has evolved from traditional distributed representation such as Word2Vector [6] and Glove [17] to the pre-trained context-aware representation. These context-aware embedding are obtained by language models (LMs) relying on unique pre-training tasks and large-scale corpora. The pre-trained LMs have become hotspot in research.

ELMo [1] is a bidirectional LM through the LSTM, and GPT [2, 19] composed of transformer [3]. GPT can capture longer semantic structure information than ELMo but neither ELMo nor GPT have truly realized bidirectional context-aware.

By pre-training tasks of the mask language model (MLM) and the next sentence prediction (NSP), BERT encodes the text information into the token embedding to achieve a deeper contextual semantic awareness. And then XLNET [18] replaces the transformer in BERT with transformer-xl to improve the ability to process long texts.

However, the parameters of XLNET and BERT are enormous, and the lightweight ALBERT [5] solved this problem. It also verifies and resolves the ineffectiveness of BERT in NSP tasks. SpanBERT [10] proposed the task of segment boundary target (SBO), which works together with BERT's MLM task to make the model better pre-dict the fragments in the text and mine the relationship between the text fragments.

However, the features that the LMs can obtain are still limited to the context. ERNIE-THU [12] added entity knowledge through an external knowledge base (KB), enabling the model to better understand the semantics. Baidu-ERNIE [13, 14] added a new mask pre-training task for entities and phrases, which helps the model to learn potential phrases and entity information, so do WKLM [16]. LIBERT [15] uses synonyms and hyponym-hypernym words in the KB to enrich the embedding of the generated model linguistic information.

2.2 Machine Reading Comprehension Model

At present, BERT is one of widely used pre-trained LMs. Many workers have explored natural language understanding tasks such as MRC based on it. Although BERT has been trained on large corpus to make the model obtain deep semantic perception, which are learned by the model during the training process. However, when solving existing NLP tasks, semantic information is often not enough, so workers have made some explorations when applying the model to downstream tasks.

The authors of SemBERT believe that the existing pre-trained model BERT only obtains simple semantic features and does not have very clear semantic blocks, so they introduce semantic role labels for each word through the semantic role labeling model (SRL), and work with BERT. However, semantic role labels still belong to the underlying information in the context, so the SemBERT model still has limitations in the task of reading comprehension.

K-BERT [8] argues that recently models cannot answer the question through external knowledge like humans, so they embedded external knowledge from the KBs for BERT. They also solved problems of knowledge noise and heterogeneous embedding space, make it play a better effect on the knowledge-driven problems.

KT-NET [9], as well as K-BERT, uses the external KBs to introduce knowledge to enrich the embedding representation, but they judge and filter all the relevant knowledge retrieved, and set the sentinel to judge whether the current external knowledge is needed. KT-NET allows the model to possess and apply knowledge through the introduction of external knowledge, but it is only for the enrichment of knowledge, and the deeper context information is still shallow like BERT.

3 Method

Our work concentrates on span extraction of MRC task, which predicts the answer by locating the start and the end position in a context. Given a question $Q = \{q_1, q_2, \ldots, q_l\}$ with l tokens and a context $C = \{c_1, c_2, \ldots, c_m\}$ with m tokens, the extracted answer is a span in C. In this paper, English tokens are all sub-token level.

We construct the answer extraction model based on BERT, and enhance the BERT embedding in two levels: token-level and sentence-level (as shown in Fig. 2).

As shown in Fig. 2, the token-level, for each token $t \in Q \cup C$, we add external knowledge Syn from the synonym dictionary and the semantic features F of the token. The features are introduced by sentinel mechanism. In sentence-level, the dependency relations R of a sequence is transformed into a matrix, and fused by multi-head attention

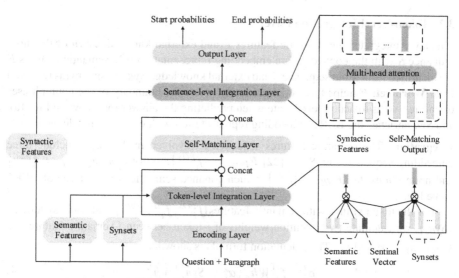

Fig. 2. Structure of TLE-BERT model

mechanism with the token embedding. More detail of the TLE-BERT model is followed in later sections.

3.1   Encoding Layer Based BERT Model

The input sequence is represented as $S = \{s_1, s_2, \ldots, s_n\}$, where n is the number of tokens. It contains the question sequence Q, the context sequence C and three tags $[CLS]$, $[SEP]$, $[SEP]$, which is depicted as :

$$S = \{[CLS], Q, [SEP], C, [SEP]\} \tag{1}$$

among them $[CLS]$ is the classification label used in the classification task and the label $[SEP]$ is used to distinguish the question and the context sequence.

For each token s_i, three embeddings, the initial text embedding s_i^{token}, position embedding s_i^{pos} and segment embedding s_i^{seg} are added together to obtain the initial input h_i^0.

$$h_i^0 = s_i^{token} + s_i^{pos} + s_i^{seg} \tag{2}$$

Input embedding h_i^0 passes through the N-layer *encoder* of BERT to get context-aware representation h_i^N, and the output embedding sequence H of this layer is expressed as:

$$H = \{h_0^N, h_1^N, \ldots, h_n^N\}, h_i^N = encoder\left(h_i^{N-1}\right) \tag{3}$$

3.2 Token-Level Integration Layer

This layer integrates the semantic features F and external knowledge Syn of the input sequence S with the context-aware embedding representation h_i^N. Semantic features F are identified from input sequence S and external knowledge Syn are synonyms retrieved from **WordNet**. F helps the model gradually notice the boundary of the noun phrases or entities, so as to accurately locate when predicting the answer (as shown in Fig. 1a), and Syn enriches the BERT embedding representation with external knowledge.

Integration. The semantic features sequence F involved in this layer include three embedding sequences, POS tag [22] $F_{POS} = \{f_i^{POS}\}$, entity type tag $F_{ET} = \{f_i^{ET}\}$ and noun phrase tag $F_{NP} = \{f_i^{NP}\}$, which enhance semantics on the basis of BERT embedding.

For the token s_i in S with semantic features $\{f_i^{POS}, f_i^{ET}, f_i^{NP}\}$ and some synonyms $\{Syn_i\}$, we calculate two attention weights of i-token between j-semantic feature and k-Synonym, α_i^j and α_i^k. The calculation formula is shown as:

$$\alpha_i^j = f_i^{jT} W h_i^N, \alpha_i^k = Syn_i^{kT} W h_i^N \tag{4}$$

where $j = 0, 1, 2$ represent POS, entity type and noun phrase feature respectively, while the k is the k-th synonym of the i-token h_i^N.

We employ sentinel attention mechanism [23] to organically integrate these features and knowledge. The sentinel vector $\bar{f}$ for semantic features and $\overline{Syn}$ for knowledge are added to autonomously learn whether it is necessary to retain the semantic features or knowledge for this token. The calculation of the attention weight for sentinel vectors $\bar{f}$ and $\overline{Syn}$ is shown as:

$$\beta_i^1 = \bar{f}^T W h_i^N, \beta_i^2 = \overline{Syn}^T W h_i^N \tag{5}$$

The calculation of the final semantic feature f_i' and knowledge Syn_i' based on previous attention is shown as:

$$f_i' = \sum_j \alpha_i^j f_i^j + \beta_i^1 \bar{f}, Syn_i' = \sum_k \alpha_i^k Syn_i^k + \beta_i^2 \overline{Syn} \tag{6}$$

while $\sum_j \alpha_i^j + \beta_i^1 = 1$ and $\sum_k \alpha_i^k + \beta_i^2 = 1$. When β is 1, the token's attention of the semantic features or knowledge is 0.

We concatenate the h_i^N, f_i' and Syn_i' as the output of i-token in this layer, the output h_i^t is shown as:

$$h_i^t = \left[h_i^N, f_i', Syn_i' \right] \tag{7}$$

3.3 Self-matching Layer

This layer takes the token-level enriched representation $\{h_i^t\}_{i=1}^n$ as input and then use self-attention mechanism to interact context-aware embedding $\{h_i^N\}$, semantic features $\{f_i'\}$ and external knowledge $\{Syn_i'\}$. This operation mainly discovers the direct and indirect interactions between tokens [9]. The output of this layer we represent as $H^s = \{h_i^s\}_{i=1}^n$.

3.4   Sentence-Level Feature Integration Layer

In reading comprehension, people reason the matter out with grammatical structure information. For example, in Fig. 1b, we come to the answer ***Boston***, for it modifies ***meetings***, not ***Levi's***. To get the structure message of the tokens, we introduce the dependencies of sentence-level into token embedding.

The grammatical structure of a sentence is expressed as a dependency relation set of token pairs, each of them is expressed as a triple, for example $\langle in, prep, meetings \rangle$, $\langle Boston, pobj, in \rangle$. The words in these triples are regarded as having 1-hop relationship. The prepositional phrase ***inBoston*** modifies the ***meetings***, so that ***Boston*** and ***meetings*** establish a 2-hop relationship through the middle word ***in***, and so on. Through m-hop, each word in S gets in touch with others. Here, m could be set through experimental study.

We convert the pair-wise dependency of tokens in the sentence S into a 1-hop dependency matrix R^1. R^1 is defined as:

$$r_{ij} = \begin{cases} 0, s_i \circ s_j \\ 1, s_i \cdot s_j \end{cases} \tag{8}$$

r_{ij} represents whether there is a dependency relationship between the i-th token and the j-th token, 1 indicates there exist relation between them, and 0 indicates no. And m-hop dependency matrix can be gotten through R^1:

$$R^m = \left(R^1\right)^m \tag{9}$$

The example of multi-hop matrices is shown in Fig. 3.

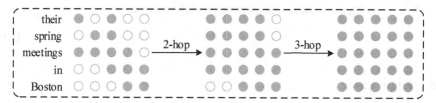

Fig. 3. An example of dependency matrices

We construct a multi-head attention transformer layer to integrate m-hop matrices which represent the sentence-level dependencies into the embedding. Here the previous layer H^s works as this layer's input sequence, as shown in Fig. 4.

We get three input-based tensor representations *Query*, *Key*, *Value* through the input of this layer H^s, the calculation for k-th head's results is shown as:

$$\left\{ Query^k, Key^k, Value^k \right\} = \left\{ H^s W^k_{Query}, H^s W^k_{Key}, H^s W^k_{Value} \right\} \tag{10}$$

where W^k_{Query}, W^k_{Key}, W^k_{Value} are k-th head weights randomly initialized, and the value of k is between 1 and m.

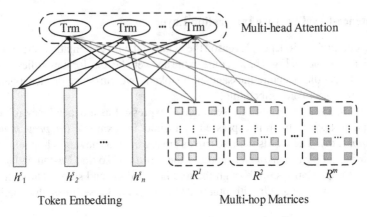

Token Embedding   Multi-hop Matrices

Fig. 4. The structure of dependency attention layer

When calculating the k-th head self-attention matrix $\boldsymbol{M}^k$, we add $\boldsymbol{R}^k$ to $Query^k Key^k$ to introduce the token dependency information, shown as following:

$$\boldsymbol{M}^k = softmax\left(\frac{Query^k Key^k + \boldsymbol{R}^k}{\sqrt{d_{Key^k}}}\right) \qquad (11)$$

where d_{Key^k} is the dimension of Key^k, $\boldsymbol{V}^k$, the k-th head self-attention embedding based on k-hop matrix of the sentence is calculated as:

$$\boldsymbol{V}^k = \boldsymbol{M}^k Value^k \qquad (12)$$

Then we concatenate all heads to get the embedding representation $\boldsymbol{V} = \left[\boldsymbol{V}^1, \boldsymbol{V}^2, \ldots, \boldsymbol{V}^m\right]$, and product the output $\boldsymbol{O} = \{o_i\}$ of this layer:

$$O = WV \qquad (13)$$

3.5 Output Layer

In the output layer, we add a fully connected layer and the softmax method to get the start and end probability of each token in the context in order to predict the final answer span.

The start probability $\boldsymbol{p}_i^1$ and end probability $\boldsymbol{p}_i^2$ for each token is calculated as:

$$\boldsymbol{p}_i^1 = \frac{w_1^T o_i}{\sum_j w_1^T o_i}, \boldsymbol{p}_i^2 = \frac{w_2^T o_i}{\sum_j w_2^T o_i} \qquad (14)$$

The objective function is the log-likelihood:

$$L = -\frac{1}{N}\sum_{i=1}^N \left(\log p_{y_i^1}^1 + \log p_{y_i^2}^2\right) \qquad (15)$$

while N is the number of samples in the train process, and y_i^1 and y_i^2 are i-th sample's truly start and end position.

4   Experiments

4.1   Datasets

The datasets we used in this article is SQuAD1.1 (StandFord Question Answering Datasets) and ReCoRD (Reading Comprehension with Commonsense Reasoning Dataset). SQuAD1.1 is a popular reading comprehension data set launched by Stanford University in 2016. And it constructs 10,000+ questions based on 500 articles in Wikipedia. The train-set contains 87,599 samples, and the dev-set contains 10,570 samples. Questions in the SQuAD1.1 appear in the form of interrogative sentences, and the 11 types of samples are summarized in Table 1. The ReCoRD contains 110,730 samples with a large portion of queries requiring commonsense reasoning, which is larger than SQuAD1.1. Questions in the ReCoRD are a sentence with @placeholder instead of the missing text span that needs to be predicted.

Table 1. Types of questions in SQuAD1.1.

Question type	Number	Question type	Number
What	56905	Where	4090
Who	9900	Be/Do/etc.	1671
Which	6620	Why	1353
When	6258	Whom	394
How many	5735	Whose	350
How	4893		

4.2   Data Preprocessing

The features of the sentences we used are mainly obtained through the spaCy tool, which is an industrial-grade NLP tool that supports multiple NLP functions such as tokenization, POS tagging, stemming, named entity recognition, noun phrase extraction, and so on.

Since the tokenization mechanism of the pre-trained language model BERT is different from the spaCy's, which leads to different tokenized results from two systems, as shown in Fig. 5.

Original: For whos glory did Father O'Hara believed that the Notre Dame football team played ?
BERT: For who s glory did Father O'Hara believed that the Notre Dame football team played ?
spaCy: For whos glory did Father O ' Hara believed that the Notre Dame football team played ?

Fig. 5. The difference between BERT and spaCy tokenization results.

We modify the tokenized method of spaCy to make it consistent with BERT. Based on this word tokenization set, the dependency relationship of the dataset is calculated.

4.3 Evaluation Standard

MRC task usually needs F_1 value and Extract Match (**EM**) for comprehensive evaluation.

F_1 calculates the degree of overlap between the predicted and the correct answer. The *recall* is the ration of the number of overlapped characters to the number of correct answer characters. The *precession* is the ration of the number of overlapped characters to the predicted answer characters. F_1 is calculated according to *precession* and *recall*. The maximum value of F_1 is 1 and the minimum value is 0.

$$F_1 = 2 * \frac{precession*recall}{precession+recall} \tag{16}$$

EM indicates precise matching, the value of **EM** is 1 means that the predicted answer is exactly the same as the correct answer, otherwise is 0. According to the characteristics of the dataset design, in the SQuAD dataset, each question may have multiple candidate answers. Therefore, the predicted answer only needs to exactly match one of the candidate answers.

4.4 Result Analysis

Our experiment uses the parameters of the BERT_base of the PaddlePaddle version. The trained model has 12 layers with 768 hidden layer dimension, and the maximum sequence length is set to 384. We set the token-level feature's dimension as 20, and the tokenized mechanism is WordPiece.

The learning rate (Lr) directly affects the convergence of the model. As learning rate increases, the model may change from under-fitting to over-fitting, as well as Epoch. We test on hyper-parameters and find when the value of Lr is set to 3e-05 and Epoch equals to 3, the TLE-BERT performance reaches the best on SQuAD1.1 (as shown in Fig. 6). And for ReCoRD, we pre-train the model with 10 epochs, and then fine-tune 4 epochs to get the best F_1.

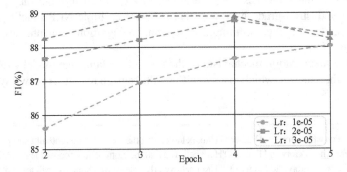

Fig. 6. Results under different hyper-parameters on SQuAD1.1.

On SQuAD1.1 and ReCoRD, we compared TLE-BERT with three models DocQA, Google BERT_base and KT-NET (based on BERT_base) in our environment, the results

Table 2. Results on SQuAD1.1 and ReCoRD.

Model	SQuAD1.1		ReCoRD	
	EM	F_1	*EM*	F_1
DocQA (ELMo)	–	–	44.13	45.39
Google BERT_base (ours)	81.25	88.41	55.99	57.99
KT-NET_base (ours)	81.56	88.75	60.79	62.91
TLE-BERT (ours)	**81.88**	**89.09**	**62.42**	**64.78**

are shown in Table 2. Both token-level and sentence-level features are capable of improving BERT_base. On SQuAD1.1, TLE-BERT get a gain of **+0.63 *EM*/+ 0.68 F_1** over the BERT_base. And on ReCoRD, it achieves **+6.43 *EM*/+ 6.79 F_1** improvements.

We also explore which semantic feature is most helpful to the model. Therefore, we try the model with different feature on SQuAD1.1 and ReCoRD, the results show in Table 3.

Table 3. TLE_BERT with different features.

Features	SQuAD1.1		ReCoRD	
	EM	F_1	*EM*	F_1
NAN	81.56	88.75	60.79	62.91
+POS	81.91	**89.06**	62.06	64.55
+ET	81.64	88.91	62.33	**64.67**
+NP	81.69	88.90	61.70	64.21
+DEP	81.83	**89.05**	62.23	**64.69**
ALL	**81.88**	**89.09**	**62.42**	**64.78**

As shown in Table 3, the performance of the model is improved when those features are introduced. NAN in Table 3 is the KT-NET_base model containing synsets knowledge, without any token or sentence level features we mention in this paper. We found that, the degree of improvement of each feature is different on various dataset. At the same time, the performance of single feature is also different for dataset.

The token-level feature POS achieve better performance in SQuAD1.1 than in ReCoRD, while ET is the opposite. POS provides grammatical information that the SQuAD dataset lacks, and ReCoRD needs entity information for reasoning. The sentence-level feature dependency relation improves the performance of the model on both datasets, which also confirms our previous conjecture that the pre-trained language model does not obtain the sentence dependency as well.

5 Conclusion

This paper proposes a new method of enhancing language representation through token-level and sentence-level features, which selects appropriate features as external information according to the characteristics of the dataset or the application background to improve the model performance. We employ an attention mechanism to fuse multiple token-level features and make the model learn to pick a valid feature set by sentinel vectors. The sentence-level features enhance the language representation with syntactic information of the sentence. Experiments have shown that our method is better than the current SOTA BERT_base models.

References

1. Peters, M.E., et al.: Deep contextualized word representations. In: North American Chapter of the Association for Computational Linguistics (2018)
2. Radford, A., Narasimhan, K., Salimans, T., Sutskever I.: Improving language understanding by generative pre-training. Technical Report, OpenAI (2018)
3. Vaswani, A., et al.: Attention is all you need. In: Advances in Neural Information Processing Systems, vol. 30, pp. 5998–6008 (2017)
4. Devlin, J., Chang, M.W., Lee, K., Toutanova, K.: BERT: pretraining of deep bidirectional transformers for language understanding. arXiv:1810.04805 (2018)
5. Lan, Z., Chen, M., Goodman, S., et al.: ALBERT: a lite BERT for self-supervised learning of language representations. In: Eighth International Conference on Learning Representations, ICLR 2020, pp. 1–17 (2020)
6. Mikolov, T., Sutskever, I., Chen, K., Corrado, G.S., Dean, J.: Distributed representations of words and phrases and their compositionality. In: Advances in Neural Information Processing Systems, vol. 26, pp. 3111–3119 (2013)
7. Zhang, Z., Wu, Y., Zhao, H., et al.: Semantics-aware BERT for language understanding. In: Thirty-Fourth AAAI Conference on Artificial Intelligence, pp. 1–8 (2020)
8. Liu, W., Zhou, P., Zhao, Z., et al.: K-BERT: enabling language representation with knowledge graph. In: Thirty-Fourth AAAI Conference on Artificial Intelligence, AAAI 2020, pp. 1–8 (2020)
9. Yang, A., Wang, Q., Liu, T., et al.: Enhancing pre-trained language representations with rich knowledge for machine reading comprehension. In: Proceedings of the 57th Annual Meeting of the Association for Computational Linguistics, ACL 2019, pp. 2346–2357 (2019)
10. Joshi, M., Chen, D., Liu, Y., et al.: SpanBERT: improving pre-training by representing and predicting spans. In: Topology, Algebra, and Categories in Logic. arXiv:1907.10529 (2019)
11. Nguyen, T., Rosenberg, M., Song, X., Gao, J.F., Tiwary, S., Majumder, R., Deng, L.: Ms marco: a human generated machine reading comprehension dataset. arXiv:1611.09268 (2016)
12. Zhang, Z., Han, X., Liu, Z., et al.: ERNIE: enhanced language representation with informative entities. In: 57th Annual Meeting of the Association for Computational Linguistics. ACL 2019, pp. 1–11 (2019)
13. Sun, Y., Wang, S., Li Y., et al.: ERNIE: Enhanced Representation through Knowledge Integration. arXiv:1904.09223 (2019)
14. Sun, Y., Wang, S., Li, Y., et al.: ERNIE 2.0: A Continual Pre-training Framework for Language Understanding. arXiv:1907.12412 (2019)
15. Lauscher, A., Vulic, I., Ponti, E.M., Korhonen A., Glavas G.: Informing Unsupervised Pretraining with External Linguistic Knowledge. arXiv:1909.20339 (2019)

16. Xiong, W., Du J., Wang, W.Y., et al.: Pretrained encyclopedia: weakly supervised knowledge-pretrained language model. In: Eighth International Conference on Learning Representations, ICLR 2020, pp. 1–13 (2020)
17. Pennington, J., Socher, R., Manning, C.: Glove: global vectors for word representation. In: Proceedings of the 2014 Conference on Empirical Methods in Natural Language processing, EMNLP 2014, pp 1532–1543 (2014)
18. Yang, Z., Dai, Z., Yang, Y., Carbonell, J., Salakhutdinov, R., Le Q.: Xlnet: generalized autoregressive pretraining for language understanding. arXiv:1906.08237 (2019)
19. Radford, A., Narasimhan, K., Salimans, T., Sutskever, I.: Language models are unsupervised multitask learners. Technical Report, OpenAI (2019)
20. http://www.research-covers/languageunsupervised/languageunderstandingpaper.pdf (2018)
21. Lehnert, W.G.: The Process of Question Answering. PhD thesis, Yale University (1977)
22. Liu, S., Zhang, X., Zhang, S., Wang, H., Weiming, Z.: Neural Machine Reading Comprehension: Methods and Trends. arXiv:1907.01118 (2019)
23. Dai, L., Xu, B., Song, H.: Feature-level attention based sentence encoding for neural relation extraction. In: Tang, J., Kan, M.-Y., Zhao, D., Li, S., Zan, H. (eds.) NLPCC 2019. LNCS (LNAI), vol. 11838, pp. 184–196. Springer, Cham (2019). https://doi.org/10.1007/978-3-030-32233-5_15
24. Yang, B., Mitchell, T.: Leveraging knowledge bases in LSTMs for improving machine reading. In: 55th Annual Meeting of the Association for Computational Linguistics. ACL 2017, vol. 1, pp. 1436–1446 (2017)
25. Hermann, K.M., et al.: Teaching machines to read and comprehend. In: Advances in Neural Information Processing Systems, pp. 1693–1701 (2015)
26. Rajpurkar, P., Zhang, J., Lopyrev, K., Liang, P.: Squad: 100,000 + questions for machine comprehension of text. In: Proceedings of the 2016 Conference on Empirical Methods in Natural Language Processing, EMNLP 2016, pp. 2383–2392 (2016)
27. Yu, A.W., Dohan D., Luong, M.-T., Zhao, R., Chen, K., Norouzi, M., Le, Q.V.: Qanet: combining local convolution with global self-attention for reading comprehension. arXiv:1804.09541 (2018)

Obstetric Diagnosis Assistant via Knowledge Powered Attention and Information-Enhanced Strategy

Kunli Zhang[1,2], Xu Zhao[1,2(✉)], Lei Zhuang[1], Hongying Zan[1,2], and Qi Xie[1]

[1] School of Information Engineering, Zhengzhou University, Zhengzhou, China
{ieklzhang,ielzhuang,iehyzang,ieqxie}@zzu.edu.cn
[2] Peng Cheng Laboratory, Shenzhen, China

Abstract. The obstetric Electronic Medical Records (EMRs) contain a large amount of medical data and health information. The obstetric EMRs play a vital role in improving the quality of the diagnosis assistant service. In this paper, we treat the diagnosis assistant as a multi-label classification task and propose a **K**nowledge powered **A**ttention and **I**nformation-**E**nhanced (KAIE) model for the obstetric diagnosis assistant. In order to make most of the information in EMRs, we propose to utilize the numerical information and chief complaint information to enhance the text representation. In addition to the use of information in EMRs, we integrate external knowledge from the COKG medical knowledge graph into the model. Specifically, we propose a multi-way attention mechanism for the generation of knowledge-aware representations based on text representations. Experiment results show that our approach is able to bring about +1.37 F1 score improvements upon the strong BERT baseline in the diagnosis assistant task.

Keywords: Diagnosis assistant · Knowledge graph · Obstetric EMRs · Multi-label classification

1 Introduction

In the last several decades, with the rapid development in economy and raising living standards, the demand for medical services has grown rapidly. Specifically in China, which has a huge population, the total amount of medical resources is still insufficient. The imbalance between the supply and demand for medical services is still the focus of China's healthcare industry. Although the implementation of China's Universal Two-child Policy in 2016 achieved many benefits, it also leads to an increase in the proportion of older pregnant women and the incidence of dystocia, fetal malformation, and other related complications [22]. Compared to the overall supply of the medical industry, the lack of obstetric medical resources is prominent. The available experts are also far from sufficient, which raises researches in diagnosis assistant.

H. Chen et al. (Eds.): CCKS 2020, CCIS 1356, pp. 274–286, 2021.
https://doi.org/10.1007/978-981-16-1964-9_22

For data-driven diagnosis approaches, medical institutions have accumulated many obstetric Electronic Medical Records (EMRs) since the issue of the Basic Norms of Electronic Medical Records (Trial) [5] by the National Health and Family Planning Medical Affairs Commission in 2010. EMRs are detailed records of medical activities, dominated by the semi-structured or unstructured texts. Obstetric EMRs mainly includes two parts: the course record and the discharge summary. The first course of the disease record in EMRs includes the patient's complaint, physical examination, auxiliary examination, and the preliminary diagnosis results, diagnosis basis, and further diagnosis plan.

In general, the admission diagnosis contains multiple diagnoses, including a normal diagnosis, pathological diagnosis, and description of complications, rather than a single diagnosis. As a result, the diagnosis assistant task based on the Chinese obstetric EMRs is usually regarded as a multi-label text classification problem. However, the doctor's diagnosis and treatment process are inseparable from the rich clinical experience and medical field knowledge accumulated over the years which is not available in EMRs. Therefore, we suggest it is important to introduce the medical domain knowledge in the process of using EMRs for diagnosis assistant. Since the introduction of domain knowledge requires formal expression before it can be used in the research of diagnosis assistant, we propose to integrate the Chinese Obstetric Knowledge Graph (COKG) into the EMRs-based diagnosis assistant.

In this paper, we adopt the knowledge graph embedding learning method to learn knowledge representations based on COKG, utilize the self-attention mechanism to gather information from all matched triplets in the COKG, and propose the multi-way attention mechanism to retrieve relevant information from knowledge representations. In order to make most of the large amount raw texts for language understanding, we choose the BERT (Bidirectional Encoder Representation from Transformers) [7] encoder to integrate our approaches. We modify the input part of the BERT model to introduce chief complaint information for information enhancement and use the multi-head attention [20] to incorporating the numerical features contained in the EMRs. Our experiment on the real-world Obstetric First Course Record Dataset supports the effectiveness of our approach.

The main contributions of this paper are as follows:

- In this paper, we propose the KAIE (Knowledge powered Attention and Information-Enhanced) model using knowledge powered attention and information enhancement strategy to integrate knowledge into the diagnosis assistant.
- We modify the input part of BERT to introduce chief complaint information for information enhancement, and propose to introduce numerical features through multi-head attention mechanism.
- We propose the multi-way attention mechanism to integrate knowledge representations into the text representations of EMRs.

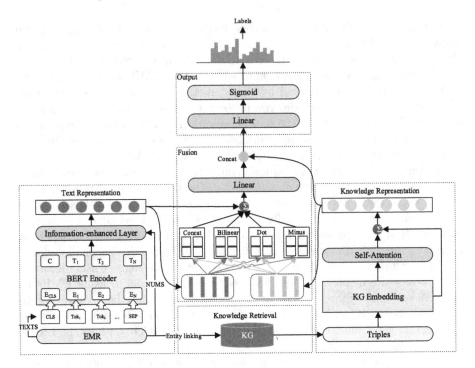

Fig. 1. The architecture of the KAIE model.

2   Related Work

In this paper, the obstetric EMRs diagnosis assistant task is treated as a
multi-label text classification problem. The multi-label classification in traditional machine learning is usually regarded as a binary classification problem
or adjust the existing algorithm to adapt to the multi-label classification task
[17,19,27,28]. With the adaptation of deep learning approaches, CNN and RNN
are used in multi-label classification tasks [1,3,9,13,23]. In recent years, ELMo
[14], OpenAI GPT [16], and BERT [7] model have achieved significant improvements in multi-label natural language processing tasks. They can be applied to
various tasks after fine-tuning.

However, due to the differences between the pre-training task and the fine-
tuning task, these models do not perform well on knowledge-driven tasks. For
example, Google BERT pre-trained on Wikipedia cannot fully realize its value
when processing EMRs analysis tasks in the medical field, and additional domain
knowledge is required to obtain improved performance. One way to solve this
problem is to pre-train the model on a specific domain instead of using the provided public domain model on the general domain. For example, ERNIE [18] and
BERT-WWM [6] adopt whole word masking rather than single character masking for pre-training BERT in Chinese corpora. Span-BERT [8] extend BERT by
masking contiguous random spans. RoBERTa [11] uses three methods to improve

the BERT performance by removing the next sentence prediction, dynamically changing the masking strategy, and using more and longer sentences during pre-training. XLNET [24] replace the transformer with transformer-XL to improve the ability in encoding long sentences. However, pre-training is time-consuming and computationally expensive, which is not affordable for many scenarios. An alternative is to introduce corresponding knowledge graph information when fine-tuning for specific tasks. Regarding the integration of external knowledge into the model, Chen J et al. [4] use BiLSTM to model the text and introduce external knowledge through C-ST attention and C-CS attention. Li M [10] et al. use BiGRU to extract word features, and use a similar matrix based on convolutional neural network and self-entity and parent-entity attention to introduce knowledge graph information. Yang A et al. [21] use knowledge base embedding to enhance the output of BERT for machine reading comprehension.

In terms of the diagnosis assistant based on Chinese obstetric EMRs, Zhang et al. [26] utilize four multi-label classification methods, backpropagation multi-label learning (BP-MLL), random k-labelsets (RAkEL), multi-label k-nearest neighbor (ML-KNN), and Classifier Chain (CC) to build the diagnosis assistant models. Ma et al. [12] fuse numerical features by employing the concatenated vector to improve the performance of the diagnosis assistant. Zhang et al. [25] encode EMRs with BERT, and propose an enhanced layer to enhance the text representation for diagnosis assistant.

3   Methodology

3.1   Model Architecture

Our proposed KAIE model is shown in Fig. 1. It can be divided into five parts: text representation generation module, knowledge retrieval module, knowledge representation module, fusion module, and the output module.

For the input EMRs, the model first uses the BERT model to extract context features, and then introduces the numerical features in the EMRs through the information-enhanced layer to obtain the text feature representation of EMRs. The model retrieves related triplet information from the obstetric knowledge graph through entity linking, and the obtained triplet set is converted into the corresponding feature vector through the pre-trained KG embeddings, and information between triplets is mixed by the self-attention layer. Knowledge representations are deeply integrated into text representations through the multi-way attention mechanism. The final diagnostic label set is produced by the output module.

3.2   Text Representation Generation Module

The function of the text generation representation module is to convert the input text sequence into the corresponding embedding representation. Compared to the BERT model, the input of KAIE model is composed of four parts: Token embedding, Position embedding, Segment embedding, and KeyInfo embedding which contains chief complaint information.

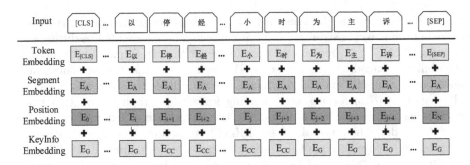

Fig. 2. The input embedding of the KAIE model.

Input to BERT Encoder. We add the chief complaint information important for the diagnosis results to the BERT input to enhance the text representation. The chief complaint contains the patient's pain or the most obvious symptoms or signs. It is also the main reason and duration of the disease. The chief complaint in the EMRs is generally a summarization (normally about 20 words) to explain the key symptoms of admission, which is the foundation for diagnosis.

For example, the chief complaint in one EMR is that "It is found that S/D value is high for 2 h after menses stops for more than 7 months." (停经 7月余，发现 S/D 值高 2 小时). The corresponding admission diagnosis is "fetal distress ...". The S/D value represents the resistance to the umbilical cord blood flow, and high resistance is a sign of fetal deficiency. That information is more critical for the diagnosis than other texts in the first course record. We suggest that if the chief complaint is distinguished from the ordinary text during input, it may help improve the performance of machine diagnosis.

Inspired by the idea of using history information embedding in [15], we propose to add KeyInfo embeddings which contain chief complaint information into the input of BERT. The characters in the input sequence are divided into general information and chief complaint information. As shown in Fig. 2, characters corresponding to the chief complaint are marked as E_{cc} in the KeyInfo embedding, and the remaining characters are indicated by E_G.

Information-Enhanced Layer. In addition to the chief complaint information described above, EMRs also contain certain examinations or indications characterized by numerical values, which are also important information for diagnosis. However, as BERT limits the length of the input text sequence, we separately extract the numerical value in EMRs to enhance the information, which not only can meet the limit of the input length, but also can better use the numerical information in the EMRs for diagnosis.

To enhance the text representation with numerical features. We employ the multi-head self-attention model proposed in Transformer [20] to fuse information, as shown in Eq. (1)–(4).

Where $[C]$ is the hidden layer state representation of $[CLS]$ which contains the representation information of the entire sequence. $[C']$ stands for the text representation after fusing the numerical information. $Num_{1......M}$ is a vector containing M values, which is obtained by standardizing and normalizing the numerical information in EMRs. W^S, W^Q, W^K, W^V and W^O are trainable parameters, where $Q \in d^{model}$.

$$Q = K = V = W^S Concat([C]; Num_{1.....M}) \qquad (1)$$

$$Attention(Q, K, V) = softmax(\frac{QK^T}{\sqrt{d_k}})V \qquad (2)$$

$$head_i = Attention(QW_i^Q, KW_i^K, VW_i^V) \qquad (3)$$

$$[C'] = Concat(head_1, ..., head_h)W^O \qquad (4)$$

3.3 Knowledge Retrieval Module

The Knowledge Retrieval Module (KR) aims to retrieve triplet information related to text sequences from the knowledge graph. For the given text sequence $S = \{w_0, w_1, w_2, ..., w_n\}$ and knowledge graph $\mathbb{K}$, the output of KR is a set $K = \{t_0, t_1, t_2, ..., t_m\}$ which contains m triples. This procedure can be divided into two steps: Entity-Link and Triplet-Query.

Entity-Link. In this paper, we use the bi-directional matching method to find all entities that may be included in the sentence. The dictionary we used contains predefined disease and symptom entities. Finally, a set $E = \{e_0, e_1, e_2, ..., e_l\}$ of l entities is obtained. It can be formulated as follows:

$$E = Entity - Link(S) \qquad (5)$$

Triplet-Query. After obtaining the E, we calculate the similarity between the entities in E and all entities in the knowledge graph through three methods: 1) Levenshtein Distance. 2) Jaccard coefficient. 3) The longest common substring. Specific calculation is shown in Eq. (6)–(9).

$$score_L = 1 - \frac{Levenshtein(e_i, k_j)}{max(len(e_i), len(k_j))} \qquad (6)$$

$$score_J = jaccard(bigram(e_i), bigram(k_j)) \qquad (7)$$

$$score_l = \frac{|longestcommonsubsequence(e_i, k_j)|}{max(len(e_i), len(k_j))} \qquad (8)$$

$$score_{final} = Average(score_L, score_J, score_l) \qquad (9)$$

Where e_i is the ith entity in the set E, and k_j is the jth entity in the knowledge graph $\mathbb{K}$.

Then we take their average as the final score and add the triplet with the highest score to set K. The overall process of triplet-query can be formalized in Eq. (10):

$$K = Triplet - Query(E) \tag{10}$$

3.4   Knowledge Representation Module

Given the extracted triplet (h, r, t) from the knowledge graph, we use the knowledge graph embedding method to map the triples into a continuous vector space. For any given triple (h, r, t), the purpose is to learn the vector representation of its head entity h, relationship r, and tail entity t.

We apply TransE [2] for knowledge graph embedding learning. Given the vector l_r of the relationship r between the head entity vector l_h and the tail entity vector l_t. For all triplets (h, r, t) in knowledge graph, TransE expects $l_h + l_r \approx l_t$, and its objective function is shown in Eq. (11), which minimizes the L_1 or L_2 distance between $l_h + l_r$ and l_t.

$$f_r(h, t) = |l_h + l_r - l_t|_{L_1/L_2} \tag{11}$$

The K obtained by the KR module is converted to KG embeddings K' by looking up TransE embeddings. To gather information of all extracted triplets in the set, we employ the self-attention mechanism to attend across triplets. As shown in Eq. (12) and (13).

$$Attention(Q, K, V) = softmax(QK^T)V \tag{12}$$

$$K'' = Attention(Q, K, V) \tag{13}$$

where $Q = K'W^Q$, $K = K'W^K$, $V = K'W^V$. W^Q, W^K and W^V are trainable parameters, K'' stands for the final knowledge representation.

3.5   Fusion Module

After obtaining the information-enhanced text representation $[C']$ and the final knowledge representation K'', we further merge these two representations using the multi-way attention mechanism, which ensembles several widely adopted attention mechanisms, as shown in Eq. (14)–(22).

Concat Attention

$$a_i = softmax(V_c^T(tanh(W_c^1[C']_i + W_c^2K_j''))) \tag{14}$$

$$[C'']_c = \sum_{i=1}^{N} a_i[C']_i \tag{15}$$

Bilinear Attention

$$a_i = softmax([C']_i^T W_b K_j'') \tag{16}$$

$$[C'']_b = \sum_{i=1}^{N} a_i [C']_i \tag{17}$$

Dot Attention

$$a_i = softmax(V_d^T tanh(W_d([C']_i \odot K_j''))) \tag{18}$$

$$[C'']_d = \sum_{i=1}^{N} a_i [C']_i \tag{19}$$

Miuns Attention

$$a_i = softmax(V_m^T tanh(W_m([C']_i - K_i''))) \tag{20}$$

$$[C'']_m = \sum_{i=1}^{N} a_i [C']_i \tag{21}$$

Final Output

$$[C'']_{final} = Concat(Concat([C'']_c; [C'']_b; [C'']_d; [C'']_m)W^{F1}; K'') \tag{22}$$

where V_c^T, W_c^1, W_c^2, W_b, V_d^T, W_d, V_m^T, W_m and W^{F1} are trainable parameters.

3.6 Output Module

Given the representation $[C'']_{final}$ of the fusion knowledge graph by fusion module, we calculate the probability representation P of the label via Eq. (23).

$$P = \sigma([C'']_{final} W^{F2}) \tag{23}$$

where W^{F2} is a trainable parameter, and sigmoid is used to classify each label as 0 or 1, the threshold of which is 0.5.

For the entire task, the objective function is to minimize the cross-entropy in Eq. (24).

$$\mathcal{L} = -\frac{1}{N} \sum_{i=1}^{N} [y_i log P_i + (1 - y_i) log(1 - P_i)] \tag{24}$$

where $y_i \in \{0, 1\}$, N is the number of labels.

4   Experiments

In this section, we present the results with the KAIE model on Chinese obstetric EMRs.

We used the traditional machine learning models CC [17] and RAkEL [19], the deep learning models BiGRU, CNN, and SGM [23] as baselines. These results are compared with the strong BERT model and our KIAE model. The F1 (F1-measure), P (Precision), R (Recall), and HL (Hamming Loss) were used as evaluation metrics.

4.1   Dataset

We conducted our experiments on the obstetric first course record dataset and COKG[1] (Chinese Obstetric Knowledge Graph).

Obstetric First Course Record Dataset. The first course records include 24,339 EMRs from multiple hospitals in China. They were pre-processed through the steps of anonymization, data cleaning, structuring, and diagnostic label standardization. 21,905 of them were used for training and 2,434 were used for testing.

COKG. COKG uses the MeSH-like framework as the knowledge ontology to define the entity and relationship description system with obstetric diseases as the core. It contains knowledge from various sources such as the professional thesaurus, obstetrics textbooks, clinical guidelines, network resources, and other multi-source knowledge. COKG includes a total of 10,674 types of entities. Among them, 4,888 kinds of entities are semi-automatically extracted, and 5,786 kinds of entities are automatically extracted.

For the Obstetric First Course Record Dataset, we used the main complaints, admission examination, obstetric examination, auxiliary examination, and other information in the first course record to get final diagnosis results in our experiments. We kept 180 diagnostic labels with the least frequency of 10 to filter the dataset.

4.2   Hyper-parameter Setting

The experimental settings of traditional machine learning methods are the same as [26], the LDA be used as the classification features, and the number of text topics is 120. For deep learning methods, the word embedding dimension of BiGRU and CNN is 100, the dropout of BiGRU is 0.1 and the word embedding dimension of SGM is 256.

---

[1] http://47.106.35.172:8088/.

Table 1. The results on Chinese obstetric EMRs.

Model	F1(%)	P(%)	R(%)	HL(%)
CC	54.83	60.21	50.34	0.0306
RAkEL	59.68	63.87	56.01	0.0254
BiGRU	74.51	79.48	70.12	0.0243
CNN	76.72	80.65	73.17	0.0198
SGM	60.40	62.67	58.29	0.0200
BERT	79.74	80.63	78.87	0.0056
KAIE	**81.11**	**83.22**	**79.12**	**0.0051**

Table 2. The results of ablation studies.

Model	F1(%)	P(%)	R(%)	HL(%)
KAIE	81.11	83.22	79.12	0.0051
-N	80.99	81.84	80.16	0.0053
-KeyInfo	80.73	81.93	79.55	0.0053
-KG	80.87	83.23	78.64	0.0052

The pre-trained BERT model we used is bert-base-chinese[2], and the learning rate is $5e-5$, the length of input is 512.

4.3   Results on Chinese Obstetric EMRs

Experimental results on the obstetric first course record dataset are shown in Table 1. It shows that the improvements of our model over the traditional machine learning methods and deep learning methods. Specifically, compared with the strong BERT, our model is significant and consistent with overall evaluation metrics. On one hand, unlike CNN, SGM and other methods, BERT can encode long-distance dependencies and uses multi-head attention to map features to different spaces, so the F1 value is higher than these methods. On the other hand, Domain knowledge also plays a vital role in the diagnosis process of doctors. After further introducing the relevant information in COKG, the chief complaint information and numerical information in EMRs, the results of our model have also been further improved.

We suggest that the improvements support the effectiveness of our KAIE approach to integrate the knowledge graph, especially when considering that the strong BERT is pre-trained on a large amount of data.

---

[2] https://github.com/google-research/bert.

4.4 Ablation Studies

In this subsection, we explore the impacts of the numerical features, chief complaint information, and the KG on performance, and results are shown in Table 2. It can be seen from the results in the table that removing the chief complaint information has the greatest impact on the results. It also illustrates the important position of the chief complaint information in the EMRs data. If we can make more effective use of the main complaint information in the EMRs, the effectiveness of the diagnosis may be further improved. The second is that the removal of knowledge graph information has a greater impact on the results, by analyzing the final prediction results, we found that labels with lower accuracy rates are not conventional obstetric diseases, and our cokg does not contain these diseases, which may be the reason for its lower accuracy rate. Another important point is that cokg is constructed from multiple sources of obstetric disease texts, and the details of the description of different diseases are different. And More than 60% of diseases have less than 10 triples in COKG, which is another reason for the lower accuracy of certain diseases.

From Table 2, we can find that: 1) In addition to the text information in the EMRs, numerical information also plays a vital role in diagnosis. 2) Among all texts of EMRs, the chief complaint which contains important information of the patient is also important for the diagnosis; 3) External knowledge shows a positive effect on improving the diagnosis quality.

5 Conclusion

This paper treats the obstetric EMRs diagnosis assistant as a multi-label text classification task. We propose a KAIE model for this task, which integrates the numerical information, chief complaint information, and external knowledge from COKG to improve the performance of diagnosis. Our experiment results on the obstetric EMRs support the effectiveness of our approach compared to the strong BERT model, and demonstrate that even though the pre-training of BERT involves a large number of corpora, knowledge graph of the specific domain can still provide useful information. In future research, we will integrate more public knowledge graphs and use more advanced knowledge graph embedding methods to further improve the diagnosis results.

Acknowledgements. This work has been supported by the National Key Research and Development Project (Grant No. 2017YFB1002101), Major Program of National Social Science Foundation of China (Grant No. 17ZDA138), China Postdoctoral Science Foundation (Grant No. 2019TQ0286), Science and Technique Program of Henan Province (Grant No. 192102210260), Medical Science and Technique Program Cosponsored by Henan Province and Ministry (Grant No. SB201901021), Key Scientific Research Program of Higher Education of Henan Province (Grant No. 19A520003, 20A520038), the MOE Layout Foundation of Humanities and Social Sciences (Grant No. 20YJA740033), and the Henan Social Science Planning Project (Grant No. 2019BYY016).

References

1. Baker, S., Korhonen, A.: Initializing neural networks for hierarchical multi-label text classification. In: BioNLP 2017, Vancouver, Canada, pp. 307–315. Association for Computational Linguistics (2017). https://doi.org/10.18653/v1/W17-2339, https://www.aclweb.org/anthology/W17-2339

2. Bordes, A., Usunier, N., Garcia-Duran, A., Weston, J., Yakhnenko, O.: Translating embeddings for modeling multi-relational data. In: Advances in Neural Information Processing Systems, pp. 2787–2795 (2013)

3. Chen, G., Ye, D., Xing, Z., Chen, J., Cambria, E.: Ensemble application of convolutional and recurrent neural networks for multi-label text categorization. In: 2017 International Joint Conference on Neural Networks (IJCNN), pp. 2377–2383. IEEE (2017)

4. Chen, J., Hu, Y., Liu, J., Xiao, Y., Jiang, H.: Deep short text classification with knowledge powered attention. In: Proceedings of the AAAI Conference on Artificial Intelligence, vol. 33, pp. 6252–6259 (2019)

5. China's Ministry of Health: Basic specification of electronic medical records (trial). Technical Report, vol. 3 (2010)

6. Cui, Y., et al.: Pre-training with whole word masking for Chinese BERT. arXiv preprint arXiv:1906.08101 (2019)

7. Devlin, J., Chang, M.W., Lee, K., Toutanova, K.: BERT: pre-training of deep bidirectional transformers for language understanding. In: Proceedings of the 2019 Conference of the North American Chapter of the Association for Computational Linguistics: Human Language Technologies, Minneapolis, Minnesota, vol. 1, pp. 4171–4186. Association for Computational Linguistics (2019)

8. Joshi, M., Chen, D., Liu, Y., Weld, D.S., Zettlemoyer, L., Levy, O.: SpanBERT: improving pre-training by representing and predicting spans. Trans. Assoc. Comput. Linguist. **8**, 64–77 (2020)

9. Kurata, G., Xiang, B., Zhou, B.: Improved neural network-based multi-label classification with better initialization leveraging label co-occurrence. In: Proceedings of the 2016 Conference of the North American Chapter of the Association for Computational Linguistics: Human Language Technologies, pp. 521–526 (2016)

10. Li, M., Clinton, G., Miao, Y., Gao, F.: Short text classification via knowledge powered attention with similarity matrix based CNN. arXiv preprint arXiv:2002.03350 (2020)

11. Liu, Y., et al.: Roberta: a robustly optimized BERT pretraining approach. arXiv preprint arXiv:1907.11692 (2019)

12. Ma, H., Zhang, K., Zhao, Y.: Study on obstetric multi-label assisted diagnosis based on feature fusion. J. Chin. Inf. Process. **32**(5), 128–136 (2018)

13. Ma, S., Sun, X., Wang, Y., Lin, J.: Bag-of-words as target for neural machine translation. In: Proceedings of the 56th Annual Meeting of the Association for Computational Linguistics, Melbourne, Australia, vol. 2, pp. 332–338. Association for Computational Linguistics (2018)

14. Peters, M., et al.: Deep contextualized word representations. In: Proceedings of the 2018 Conference of the North American Chapter of the Association for Computational Linguistics: Human Language Technologies, New Orleans, Louisiana, vol. 1, pp. 2227–2237. Association for Computational Linguistics (2018)

15. Qu, C., Yang, L., Qiu, M., Croft, W.B., Zhang, Y., Iyyer, M.: BERT with history answer embedding for conversational question answering. In: Proceedings of the 42nd International ACM SIGIR Conference on Research and Development in Information Retrieval, pp. 1133–1136 (2019)

16. Radford, A., Narasimhan, K., Salimans, T., Sutskever, I.: Improving language understanding with unsupervised learning. Technical Report, OpenAI (2018)
17. Read, J., Pfahringer, B., Holmes, G., Frank, E.: Classifier chains for multi-label classification. Mach. Learn. **85**(3), 333 (2011)
18. Sun, Y., et al.: Ernie: enhanced representation through knowledge integration. arXiv preprint arXiv:1904.09223 (2019)
19. Tsoumakas, G., Katakis, I., Vlahavas, I.: Random k-labelsets for multilabel classification. IEEE Trans. Knowl. Data Eng. **23**(7), 1079–1089 (2010)
20. Vaswani, A., et al.: Attention is all you need. In: Advances in Neural Information Processing Systems, pp. 5998–6008 (2017)
21. Yang, A., et al.: Enhancing pre-trained language representations with rich knowledge for machine reading comprehension. In: Proceedings of the 57th Annual Meeting of the Association for Computational Linguistics, pp. 2346–2357 (2019)
22. Yang, H.l., Yang, Z.: Effect of older pregnancy on maternal and fetal outcomes. Chin. J. Obstetric Emerg. (Electron. Edn.) **5**(3), 129–135 (2016)
23. Yang, P., Sun, X., Li, W., Ma, S., Wu, W., Wang, H.: SGM: sequence generation model for multi-label classification, pp. 3915–3926 (2018)
24. Yang, Z., Dai, Z., Yang, Y., Carbonell, J., Salakhutdinov, R.R., Le, Q.V.: Xlnet: generalized autoregressive pretraining for language understanding. In: Advances in Neural Information Processing Systems, pp. 5754–5764 (2019)
25. Zhang, K., Liu, C., Duan, X., Zhou, L., Zhao, Y., Zan, H.: Bert with enhanced layer for assistant diagnosis based on Chinese obstetric EMRs. In: 2019 International Conference on Asian Language Processing (IALP), pp. 384–389. IEEE (2019)
26. Zhang, K., Ma, H., Zhao, Y., Zan, H., Zhuang, L.: The comparative experimental study of multilabel classification for diagnosis assistant based on Chinese obstetric EMRs. J. Healthcare Eng. (2018)
27. Zhang, M.L., Zhou, Z.H.: Multilabel neural networks with applications to functional genomics and text categorization. IEEE Trans. Knowl. Data Eng. **18**(10), 1338–1351 (2006)
28. Zhang, M.L., Zhou, Z.H.: Ml-knn: a lazy learning approach to multi-label learning. Pattern Recogn. **40**(7), 2038–2048 (2007)

Emotion Role Identification in Social Network

Yakun Wang[1], Yajun Du[1($\boxtimes$)], and Chuan Xie[2]

[1] School of Computer and Software Engineering, Xihua University,
Chengdu 610039, China
wangyk@stu.xhu.edu.cn, duyajun@mail.xhu.edu.cn
[2] School of Computer Science and Technology,
Southwest University for Nationalities, Chengdu 610039, China

Abstract. Emotion is a status that combines people's feelings, thoughts, and behaviors and plays a crucial role in the communication between people. Considerable study suggests that human emotions can also be conveyed through online interactions. For a systematical literature review, we find that few studies focus on the influence of some special users on the process of emotional transmission in online social networks. To fill this gap, we first introduce the definition of emotion role, they are special users who play an important role in the process of emotion contagion of online social networks. We then propose an Emotion Role Mining approach (ERM) to detect users' emotion roles in social networks. A set of features and measures is proposed and calculated to identify and represent these users based on the analysis of their emotion influence and long-term emotional preferences. Experiments and evaluations are conducted to demonstrate the practicability and usefulness of the proposed approach using Micro-blog data. Comparison experiments indicate that the proposed approach outperforms several baseline methods.

Keywords: Online social networks · Emotion contagion · Emotion role

1 Introduction

With the development of the Internet, especially the rapid development of social networks, researchers found that people's emotions can also flow like information and pass it to others via virtual connections on the Internet. Various contributions advanced the hypothesis that emotions can also be conveyed through online interactions. In [8], Hancock and Gee examined emotional communication in CMC by inducing negative affect in one condition and neutral affect in another and revealed emotional contagion took place, in which partners interacting with participants in the negative affect condition had significantly less positive affect than partners in the control condition. Later, Guillory et al. [6] demonstrates interpersonal emotions can be shared in distributed groups via linguistic cues. Anna et al. [2] showed the collective character of affective phenomena on a large

© Springer Nature Singapore Pte Ltd. 2021
H. Chen et al. (Eds.): CCKS 2020, CCIS 1356, pp. 287–298, 2021.
https://doi.org/10.1007/978-981-16-1964-9_23

scale and proved that collective emotional states can be created and modulated via Internet communication. Kramer et al. [12,13] indicates not only that emotional contagion is possible via text-only communication and that emotions flow through social networks, but also that emotion spreads via indirect communications media, and emotions expressed by others on Facebook influence our own emotions.

Emotion is a status that combines people's feelings, thoughts, and behaviors, which correlates to people's social behavior, living habits and other characteristics. Nowadays, people are used to expressing their daily feelings directly in the online social networking space through texts. And what they express is often mixed with personal emotions. The content generated by individuals in daily social network communications, and the emotions therein expressed, may affect the emotional status of others. In online social networks, the change of an individual's emotional state can influence others in positive or negative ways and the spread of emotion may further trigger large cascade adoptions of happiness or depression [20]. Emotion role detection is crucial to many social network applications, including advertising targeting, marketing, personalization, recommendation, etc.

In social network analysis, various contributions to mining social roles according to different role definitions. Abnar et al. [1] proposed an SSRM framework to analyze social networks and classified each individual into four different categories based on their structural role in a community. Zhao et al. [25] investigated the social roles and statuses that people act in online social networks from the perspective of network structures. However, most jobs do not define roles from an emotional perspective.

In this paper, we propose an Emotion Role Mining approach (ERM) to define and identify these emotion roles. In detail, following a social networking model to describe and analyze user's emotional features, define and identify three special emotion users in the process of emotion contagion in social networks. A set of attributes and measures are then defined and calculated based on emotional influence and long-term emotional preference.

We summary our contributions as follows:

- We have defined three special emotional roles, combining six-element emotions with users' online network characteristics.
- We design some new emotional features and propose a approach to identify the emotion roles of online social network users.
- We conduct extensive experiments to validate the proposed approach over several baselines. We verify the effectiveness of our proposed approach.

The remainder of this article is organized as follows. Section 2 introduces related work. Section 3 describes the definition of emotion role. Section 4 formulates the features for emotion role detection and further elaborates the proposed ERM approach for emotion role identification. Section 5 discusses experimental evaluation followed by conclusions and future work in Sect. 6.

2   Related Work

2.1   Emotion Classification

Most research in this area focuses on sentiment analysis classifying text as positive or negative [19]. However, most of these tasks simply considered human emotions for positive or negative, simple happiness scores, did not take into account the richness of human emotions, fine-gained emotion status such as anger or sadness plays a vital role in the outbreak of social events such as catastrophes. Therefore, we mainly consider Ekman's six emotions [3] happiness, surprise, anger, disgust, fear, sadness. Recent years, as micro-blog has used more and more widely, emotion detection on micro-blog posts has attracted widespread attention. Li et al. [15] classified emotions in micro-blog posts based on extracted cause events. Yuan et al. [23] used emotions and smiles as noisy labels, and used SVMs for automatic emotion detection for micro-blog texts.

2.2   Role Detection in Online Social Network

This category focuses on methods of mining social roles according to different role definitions. Tchokni et al. [21] proposed to divide Twitter users into different levels according to their language usage. Kalia et al. [10] presented an approach for detecting organizational hierarchy that aimed to determine the degree of coordination among team members. In Reference [4], the authors put forward some theories about social influence networks and explored the issue of opinion change, which inspired us to further analyze online social networks. However, neither of these definitions or classification methods take into account user's long-term emotional status. In this article, we focus on user's emotional features and propose a new definition of emotion roles.

3   Emotion Role Definition

Emotion influence, which indicates one's ability to lead public emotion. In the process of emotion contagion in online social networks, some users have considerable emotion influence. They often share their personal emotions and attitudes with other network users, thereby affecting and changing the emotions and attitudes of these users. Therefore, these special users have an important influence on the emotions, behaviors and behaviors of others. We define this type of user as an emotion leader. By measuring and ranking the emotional influence of the emotion leader in the network, we select the user with the highest emotional influence as the emotion leader.

In addition, in online social networks, there is a phenomenon of emotional flocking. People who express similarly valenced emotions on specific political topics were more frequently connected in network clusters on Twitter [9]. More generally, online microblogging websites were argued to host emotion communities, which consist of interconnected users who are characterized by similar patterns of emotion expressions [11]. A structural hole appears to be an "empty space" between two sets of unconnected nodes in a social network [14]. Therefore, we divide emotional communities according to emotional preferences, and define such users who connect different emotional communities as emotional structure hole users. They have rich control capabilities and information resources in social networks, which play a role in the spread of emotions across communities.

Emotion leaders are outstanding individuals with great emotion influence on a network. They could make a strong impact on emotion propagation.

Emotion mediators are individuals who play an important role in connecting emotional communities in a network. They act as bridges between distinct communities.

Emotion followers are the most set of individuals in a network. Their emotions could hardly spread out and affect others.

4   Proposed Approach

In this section, we first introduce the preliminary steps of personal emotion recognition, and then explain the user emotion features for identifying emotion roles based on emotion influence and emotional preference. Finally, we will introduce our ERM approach and its corresponding methods.

4.1   Preliminaries

All feature extraction of emotion roles are based on emotion recognition of users' online behaviors, so emotion recognition is a crucial way to deeply understand users' emotion roles. In our study, we use CSR [22] method to classify user's emotion into six categories: {happiness, surprise, anger, disgust, fear, sadness}.

4.2   Emotional Feature

The emotion role is the concentrated expression of user's emotional features in social networks. Therefore, we mainly consider two emotional features (i.e., emotion influence, emotion preference).

Emotion Influence. Emotion influence is a measurement of a user's ability to guide online emotions. Regarding emotion influence, there is currently no standard evaluation method. We consider this feature from three perspectives. The first one is range factor (RF), which is related to the scale of information propagation caused by user. Similar to some existing studies, most studies summarized the problem as finding leaders and followers in online social networks. The second is emotion change factor (EF), aiming to measure the user's ability to affect the emotions of others. The last is repost ratio (RR), which is used to measure the ability to influence others. In order to quantify the user's emotion influence, we formally define three factors as follows:

Range Factor (RF). The emotional impact of the range of users in a social network is an important factor reflects their emotion influence. For a specific user v, range factor indicates the influence scope, caused by user v's emotions. The higher the RF value, the greater number of individuals whose emotions are affected. We define the structural factor RF_v as follows:

$$RF_v = \frac{1}{|P_v|} \sum_{p \in P_v} AvgD_p * \log N_p \tag{1}$$

where $P_v = \{p_{v1}, p_{v2}, \ldots, p_{vn}\}$ denotes a set of original posts of user v, $AvgD_p$ is the average influence depth of a specific emotion shared by user v, and N_p is number of users who have been influenced by content p posted by user v.

Furthermore, it is noted that the degree of change in emotional status also plays a crucial role in emotion contagion processes. By vectorizing the emotions expressed by the user, the degree of change of the user's emotional status can be quantified, which will finally beneficial to extract features that affect the emotion of others.

Emotion Change Factor. Given a specific user v, the emotion change factor indicates the change of other users after being affected by user v's emotions. It also shows the power of emotion influence of user v. We classify emotion of users' posts into six categories. It is worth mentioning that since the emotional state is represented by a vector, we use Euclidean distance to measure the difference between two different emotions. We define emotion change factor EF_v as follows:

$$EF_v = \frac{1}{|P_v|} \sum_{p \in P_v} \sum_{i \in IU_p} |B_i(p) - A_i(p)| \tag{2}$$

$P_v = \{p_{v1}, p_{v2}, \ldots, p_{vn}\}$ is a set of original posts of user v, and IU_p is a group of users affected by post p. We use $B_i(p)$to denote the emotional status of user i before influenced by post p and $A_i(p)$ to denote the emotional status after influenced by p. The Euclidean distance is used to measure the difference between above two distributions, which is normalized by exponential function.

Repost Ratio. Given a specific user v, the repost ratio indicates the rate at which user v's posts are reposted. We utilize repost ratio to quantify the user's ability to affect others. Higher repost ratio value means higher ability that one's emotions could affect others. This metric defined as:

$$RR_v = \frac{\#Number\_of\_posts\_reposted(v)}{|P_v|} \tag{3}$$

where $Number\_of\_posts\_reposted(v)$ is a function counting the number of user v's posts that have been reposted by others.

Emotion Preference. In online social networks, some people prefer to express positive emotions and share good experiences. However, there are other people who prefer to share negative feelings. The user's emotional preference reflects the user's online emotional features, which is related to the user's educational experience, living environment, etc. To quantitatively measure emotional preferences, we define EP_v to denote user v's emotional preference. Specifically, we assign a discrete score for each basic emotion category, and represent the emotional preference of an individual by a 6-tuple vector:

$$EP_v = \{ep_{happiness}, ep_{sad}, ep_{anger}, ep_{surprise}, ep_{fear}, ep_{disgust}\} \qquad (4)$$

where ep_i represents the proportion of i-th emotion (e.g., happiness, anger) in the user's historical posts, which can be quantified as follows:

$$ep_i = \frac{\#i_{th}\_emotion(P_v)}{|P_v|} \qquad (5)$$

where $i_{th}\_emotion(P_v)$ is a function counting the number of posts about i-th emotion from user v, P_v denotes a set of original posts of user v. EP_v can indicate the user v's long-term emotional preference.

4.3   Emotion Role Identification

Having the definition of three fundamental roles, we now describe how they can be identified in a given network. More formally, we identify each of the four fundamental roles defined in the previous section as follows:

Emotion Leaders members with great emotion influence. First, emotion influence EF is used to score the ability of users to guide public emotions. Then, rank and analyze these users, determines the leaders. More specifically, users who have greatest emotion influence, are identified as emotion leaders.

In our study, each user is characterized by two features, namely RF_v and EF_u, respectively. We select the Support Vector Machine (SVM) for emotion leader identification. We build a vector $V_v = (RF_v, EF_v, RR_v)$ as the input to SVM. Finally, we use SVM to classify the users into two different categories (i.e., emotion leaders and emotion followers).

Emotion Followers members are identified in contrast to the emotion leaders, i.e. users with low emotion influence in the network.

Emotion Mediators members are the ones that connect different emotional communities. First, we will divide different communities in the social network according to the similarity of the user's emotional preferences, the user's emotional preference similarity is defined as follows:

$$sim(v, u) = \frac{EP_v * EP_u}{|EP_v||EP_u|} \qquad (6)$$

Where $sim(v, u)$ represents the similarity of emotional preference between user v and user u.

Similar nodes are considered more likely to be connected than less similar ones. Hence we apply the proposed dissimilarity index to uncover the community structure in networks, which also verifies the proposed index. The modularity (Q) proposed by Newman [17] is widely adopted to indicate the strength of the community structure of a specific partition. The function of the modularity is calculated as follows:

$$Q = \frac{1}{2m} \sum_{ij} (A_{ij} - \frac{k_i k_j}{2m}) \delta(C_i, C_j) \tag{7}$$

where

$$\delta(C_i, C_j) = \begin{cases} 1 & if \quad C_i = C_j \\ 0 & otherwise \end{cases} \tag{8}$$

The idea of the emotional community detection algorithm is to first divide each user into a community, calculate the similarity of emotional preference between the two users through Eqs. 5, 6, and then merge the two emotional communities with the highest similarity. Through continuous merger, the final community division results are obtained. After dividing the emotional community, we use HIS method [16] to identify the emotional structure hole spanners

---

Algorithm 1: Find Emotion Mediators

Input: Graph $G = \{V, E\}$, User's posts
Output: top-k emotion mediators
1 Calculate the emotional preferences of all users by Eq.4
2 %Emotional preference similarity matrix M
3 $M \leftarrow$ according to the Eq.6
4 $maxQ \leftarrow 0$
5 Initialize each node in the network as a cluster
6 Set $S = \{v_1, v2, \ldots, v_n\}, n = |V|$
7 CommunityList $C \leftarrow \{\}$
8 %Merge community
9 **foreach** *cluser i and cluster j* **do**
10   **if** *they have most similarity* **then**
11   Merge them and update S

12   Update M according to the Eq.6
13   Compute modularity Q according to the Eq. 7
14   **if** $Q > maxQ$ **then**
15         $maxQ \leftarrow Q$
16         $C \leftarrow S$

17 %Find emotion mediators
18 Use HIS method to compute all users' important score
19 Sort each node according to important score

---

in the social network. The emotion mediator detection algorithm is summarized in Algorithm 1.

5   Experiments and Results

In this section, experiments and evaluations are conducted in classify a user's emotion role into three categories (i.e., emotion leader, emotion mediator and emotion follower), in order to demonstrate the practicability and usefulness of our proposed methods.

5.1   Dataset

As mentioned above, Micro-blog, one of the famous social media networking systems in China, has been employed for our experimental analysis. The dataset is from [24]. The complete dataset concludes the directed following networks and posts of $1,776,950$ users (from September 2012 to October 2012). We then filter the data according to two rules: (1) Users should continuously maintain a high activeness (e.g., posting at least one Micro-blog post every two days); (2) The content of posts should not contain Internet slang words. Finally, 5407 users with $596,782$ posts were used to conduct the experiments.

To generate the ground truth for emotion role identification, we have manually labeled the users as one of the three emotion roles according to their posts. To model and analyze social behaviors relating to emotion contagion in Micro-blog, posting contents, reposting contents and mentioning other users (e.i., "@name") are regarded as emotion behaviors.

5.2   Emotion Leader Identification

We first compare our method with four baselines to measure the performance of identifying emotion leaders. We apply support-vector-machine-based ERM approach and three other different classification methods including Logistic Regression (LR), Decision Tree (DT), and Naive Bayes (NB) to identify users into two categories (i.e., emotion leaders and emotion followers). All these tests were performed based on WEKA [7], and the results are shown in Fig. 1.

The effectiveness results in precision/recall/F-score on the Micro-blog data set are illustrated in Figs. 1(a), 1(b) and 1(c), respectively. We can infer that our approach outperforms the baselines on all the measures in Fig. 1. The proposed approach obtained an approximate 82% precision.

Figure 2 shows the number of users affected by top-k emotion leader nodes detected by different methods on Micro-blog dataset. As shown in Fig. 2, We can see in the micro-blog the proposed model significantly outperform the comparison methods.

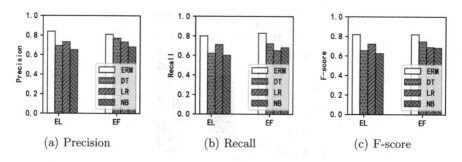

(a) Precision   (b) Recall   (c) F-score

Fig. 1. Results on the Micro-blog Data Set

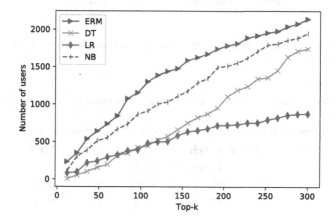

Fig. 2. Number of users affected by top-k emotion leader nodes detected by different methods on Micro-blog

Factor Contribution Analysis. Regarding emotional influence, we considered different features: range factor and emotion change factor. Here we show how these features contribute in identification task. Specifically, we first train a model without applying any features (referred to as Base). We then incrementally add the range factor (referred to as Base+RF) and emotion change factor (referred to as All) and evaluate their improvements in identification performance over that using only basic features.

The Fig. 3 shows the average F-score score after ignoring these features. We can see a clear drop in the performance when some features are ignored. This shows that by combining different features, our method works well, and that each factor in our method contributes to the performance.

5.3   Emotion Mediator Identification

We compare several methods for detecting emotion mediators. The baselines including PageRank [18], Pathcount [5], Constraint [14]. Table 1 lists the

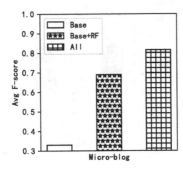

Fig. 3. Contribution of different features

performance of the different approaches on Micro-blog data sets with the following observations.

Table 1. Performance of emotion mediator identification by different approaches

Emotion role	Method	Precision	Recall	F-score
Emotion mediator	PageRank	36.3%	41.2%	38.6%
	Pathcount	57.3%	60.2%	58.7%
	Constraint	31.7%	33.5%	32.6%
	ERM	**68.6%**	**72.4%**	**70.4%**

From the results, we see that our method clearly outperforms the comparison algorithms by +10–30%. As expected, just choosing important nodes (by PageRank) is not a reasonable strategy. Pathcount achieve a better performance than PageRank. This is because the page is to find authoritative nodes in the network, which is different from looking for bridges between emotional network communities. This shows that there is a big difference between emotion network structure and social network structure. How to combine the two network structures for mining emotion mediators would be an interesting future work.

6   Conclusion and Future Work

In this paper, we study the novel problem of identifying emotion role in social network. We identified emotion leaders, emotion mediators and emotion followers. To identify emotion roles, we developed and quantified the emotion role metrics by introducing several features. Built on these analyses, we further developed the ERM approach to identify emotion roles. In order to evaluate our proposed approach, we applied it on the Micro-blog dataset. The identification of emotion roles can efficiently help public opinion detection and personalized recommendation.

In the future, we will introduce more features into our method, such as time evolution and neighbor relations, to improve the performance of our method. More evaluations and experiments will be conducted to improve the proposed methods with better identification results in more complex situations.

Acknowledgement. This work was supported by the National Natural Science Foundation (Grant Nos. 61872298, 61532009, 61802316, and 61902324), the Innovation Fund Of Postgraduate, Xihua University (Grant Nos. ycjj2019023).

References

1. Abnar, A., Takaffoli, M., Rabbany, R., Zaïane, O.R.: SSRM: structural social role mining for dynamic social networks. In: IEEE/ACM International Conference on Advances in Social Networks Analysis and Mining (2014)
2. Chmiel, A., Sienkiewicz, J., Thelwall, M., Paltoglou, G., Kevan, B.: Collective emotions online and their influence on community life. Plos One **6**, e22207 (2011)
3. Ekman, P.: An argument for basic emotions. Cogn. Emotion **6**(3), 169–200 (1992)
4. Friedkin, N., Johnsen, E.: Social influence networks and opinion change. Adv. Group Process. **16** (1999)
5. Goyal, S., Vega-Redondo, F.: Structural holes in social networks. J. Econ. Theory **137**(1), 460–492 (2007)
6. Guillory, J., Spiegel, J., Drislane, M., Weiss, B., Donner, W., Hancock, J.: Upset now?: emotion contagion in distributed groups, pp. 745–748 (2011)
7. Hall, M., Frank, E., Holmes, G., Pfahringer, B., Reutemann, P., Witten, I.H.: The weka data mining software: an update. SIGKDD Explor. Newsl. **11**(1), 10–18 (2009)
8. Hancock, J., Gee, K., Ciaccio, K., Lin, J.: I'm sad you're sad: emotional contagion in CMC, pp. 295–298 (2008)
9. Himelboim, I., Sweetser, K.D., Tinkham, S.F., Cameron, K., Danelo, M., West, K.: Valence-based homophily on twitter: network analysis of emotions and political talk in the 2012 presidential election. New Media Soc. **18**(7), 1382–1400 (2014)
10. Kalia, A., Buchler, N., Ungvarsky, D., Govindan, R., Singh, M.: Determining team hierarchy from broadcast communications (2014)
11. Kanavos, A., Perikos, I., Hatzilygeroudis, I., Tsakalidis, A.: Emotional community detection in social networks. Comput. Electr. Eng. **65**, 449–460 (2018)
12. Kramer, A.: The spread of emotion via facebook. In: Conference on Human Factors in Computing Systems - Proceedings (2012)
13. Kramer, A., Guillory, J., Hancock, J.: Experimental evidence of massive-scale emotional contagion through social networks. In: Proceedings of the National Academy of Sciences of the United States of America, vol. 111 (2014)
14. Lazega, E., Burt, R.S.: Structural holes: the social structure of competition. Revue Franaise de Sociologie **36**(4), 779 (1995)
15. Li, W., Xu, H.: Text-based emotion classification using emotion cause extraction. Expert Syst. Appl. **41**(4, Part 2), 1742–1749 (2014)
16. Lou, T., Tang, J.: Mining structural hole spanners through information diffusion in social networks. In: Proceedings of the 22nd International Conference on World Wide Web. WWW 2013, New York, pp. 825–836. Association for Computing Machinery (2013)

17. Newman, M.E.J., Girvan, M.: Finding and evaluating community structure in networks. Phys. Rev. E **69**(2) (2004)
18. Page, L., Brin, S., Motwani, R., Winograd, T.: The pagerank citation ranking: Bringing order to the web. Technical Report, Stanford InfoLab (1999)
19. Pang, B., Lee, L.: Opinion mining and sentiment analysis. Found. Trends Inf. Retr. **2**(1–2), 1–135 (2008)
20. Tang, J., et al.: Quantitative study of individual emotional states in social networks. IEEE Trans. Affect. Comput. **3**(2), 132–144 (2012)
21. Tchokni, S., eaghdha, D., Quercia, D.: Emoticons and phrases: status symbols in social media. In: Proceedings of the 8th International Conference on Weblogs and Social Media. ICWSM 2014, pp. 485–494 (2014)
22. Wen, S., Wan, X.: Emotion classification in microblog texts using class sequential rules. In: Proceedings of the Twenty-Eighth AAAI Conference on Artificial Intelligence. AAAI 2014, pp. 187–193. AAAI Press (2014)
23. Yuan, Z., Purver, M.: Predicting emotion labels for Chinese microblog texts. In: Gaber, M.M., Cocea, M., Wiratunga, N., Goker, A. (eds.) Advances in Social Media Analysis. SCI, vol. 602, pp. 129–149. Springer, Cham (2015). https://doi.org/10.1007/978-3-319-18458-6_7
24. Zhang, J., Liu, B., Tang, J., Chen, T., Li, J.: Social influence locality for modeling retweeting behaviors. In: Proceedings of the Twenty Third International Joint Conference on Artificial Intelligence. IJCAI 2013, pp. 2761–2767. AAAI Press (2013)
25. Zhao, Y., Wang, G., Yu, P.S., Liu, S., Zhang, S.: Inferring social roles and statuses in social networks. In: Proceedings of the 19th ACM SIGKDD International Conference on Knowledge Discovery and Data Mining KDD 2013, New York, pp. 695–703. Association for Computing Machinery (2013)

Academic Field and Future Influence Prediction for Scholar Profile Construction

Jiaying Tang, Hongying Jin, Daling Wang$^{(\boxtimes)}$, Shi Feng, and Yifei Zhang

School of Computer Science and Engineering, Northeastern University, Shenyang, China
{wangdaling,fengshi,zhangyifei}@cse.neu.edu.cn

Abstract. Collecting scholar information from massive academic resources to construct scholar profiles can provide a reference for various academic activities. For a scholar profile, except basic attributes such as age, gender, job title, some potential attributes need to be predicted for describing the scholar. The scholars academic field and future influence are two important attributes for constructing profile. How to predict the potential attributes precisely is a challenge. Towards that, we propose two models to predict academic field and influence respectively for constructing scholar profile in this paper. In detail, we propose a double-layer attention model of paper information and journal information representation model for predicting scholars academic field based on multiple data sources. We also propose a two-level fusion model based on feature combination for predicting scholars' future academic influence. Extensive experiments are conducted on AMiner dataset, and the results prove that our models are superior to the state-of-the-art methods.

Keywords: Scholar profile · Feature construction · Academic field · Future influence

1 Introduction

Scholar profiles depict scholars basic attributes and academic characteristics, and effective utilization of such information can provide important support for the scientific activities such as scholar retrieval, domain expert discovery, and academic resource recommendation.

In academic research, a scholar often publishes many papers. We can easily perceive the explicit academic field from the text content, or figure out other implicit field through the algorithms such as data mining and machine learning. The academic field as an important component of the scholar profile can reflect not only the scholar's research direction, but also the attention of different scholars on hot fields and scientific research trends.

In recent years, the explosive growth of academic papers makes it a challenge to find influential scholars. Academic papers can show the research results

© Springer Nature Singapore Pte Ltd. 2021
H. Chen et al. (Eds.): CCKS 2020, CCIS 1356, pp. 299–311, 2021.
https://doi.org/10.1007/978-981-16-1964-9_24

of scholars and reflect the influence and popularity to a certain extent. The academic influence of scholars can be measured by the citation frequency of papers, h-index and other methods. As an important attribute of scholar profile, future academic influence can not only reflect the contribution of scholars in a certain field in the next few years and the quality and popularity of published papers, but also discover the potential of scholars and the status of a field.

We hope that enough data can be used to portray a scholar in multiple dimensions and at all levels. For the requirement, we present a scholar profile in this paper. The basic structure of our scholar profile includes basic attributes and prediction attributes, as shown in Fig. 1. Basic attributes include age, gender, job title, and contact information. For predictive attributes, we focus on two important attributes which are scholars' academic field and future academic influence. We can directly extract or crawl basic attributes from scholar homepage. However, every predictive attribute is a combination of many attributes and usually not available directly. In order to better describe the profile of scholars, we predict such attributes based on our proposed two models.

Fig. 1. Basic structure of scholar profile

To sum up, our main contributions in this paper are as follows:

- We present a framework of scholar profile including scholars basic attributes and prediction attributes, and propose an approach of constructing the scholar profile by crawling the scholars basic information and predicting academic field and future academic influence of the scholars.
- We propose a double-layer attention model of paper information and journal information representation model for predicting scholars academic field based on multiple data sources. We also propose a two-level fusion model based on feature combination for predicting scholars' future academic influence.
- We conduct extensive experiments on AMiner dataset. The results show our models are superior to the state-of-the-art methods.

2   Related Work

Scholar profile can be viewed as a user profile in specific fields. At present, most attribute prediction algorithms constructed by user profile regard user profile inference as a classification task, and the goal is to predict unary predicates describing user attributes including gender [3], age [2] or political polarity [7]. The multi-dimensional attribute prediction has also been widely developed, but there are still two problems: on the one hand, the information is untrue or missing; on the other hand, most methods currently use a single data source to predict multiple attributes, without considering the influence of the correlation between multiple types of data on the prediction of profile attributes.

Many scholar profile construction systems have been designed, such as Aminer [9], which has achieved good results in promoting academic development. At the same time, it is also the key to large-scale think tanks to realize expert discovery and academic impact assessment.

The academic field mining is similar to user interest mining. Michael et al. [6] speculated on the user's preference for the website theme based on the user profile. Palacios-Núñez et al. [5] identified developmental tendencies in the intellectual property academic field through the identification of invisible colleges. For network mining, Zhang et al. [15] proposed a partially labeled paired factor graph model for social relationship discovery problems in large social networks. Wan et al. [10] searched and mined academic social networks.

Researches related to the measurement of scholars' influence, such as paper citation frequency and h-index. Bai et al. [1] measured the scholars' influence according to the research institutions and the changes in the paper over time. Yan et al. [14] studied the impact of a paper's new combinations and new components on its citation. Yan et al. [13] introduced the task of predicting the number of citations to make predictions. However, the peak of the papers cited by scholars is the first few years after the paper was published. Most of the current research on the influence of scholars only focuses on this stage, so it is impossible to predict the authoritativeness and popularity of scholars in the future.

Different from the related work, our proposed two predictive models fuse multiple data sources, are based on feature combination, and consider variational timing characteristics in academic circles.

3   Proposed Model

For establishing multi-dimensional profile attributes for scholars, after extracting the basic information of scholars, we predict scholars' academic fields and academic influence in the next decade.

3.1   Academic Field Prediction

Academic field prediction is a multi-label learning task. We convert the multi-label data set into a series of corresponding single-label data sets, determine whether a certain scholar data belongs to the label, and finally combine the single-label results. Our predictive model is shown in Fig. 2 and the details are as follows.

Scholar's Paper Information: For a paper, the semantic information of the title can best reflect the academic field of scholars. We use all the scholar's paper information as the data source of the model and adopt a double-layer attention model. The model named "PAC" is shown in the Fig. 2(a).

We sort each paper of scholars in the paper dataset D on descending order of publication time, aggregate all papers' titles into a document, and get the matrix W through the embedding layer. At the same time, W and the academic field label Y pass through a Bi-LSTM to get two vectors S_i, y_i. Pass S_i and y_i through the softmax layer to obtain α_i. Using S_i as the key and α_i as the query, the similarity score m_1 is calculated by the General Attention method.

A new vector v_s is obtained by stitching m_1 with y_i, $|m_1 - y_i|$ and $m_1 \odot y_i$. Input v_s to the fully connected layer to obtain the attention vector u_t, and the paper text representation d_1 is obtained through the attention layer. The vectors d_1 pass a layer of CNN to extract the local feature vector v_c related to the task, and the final result y is obtained through a fully connected layer.

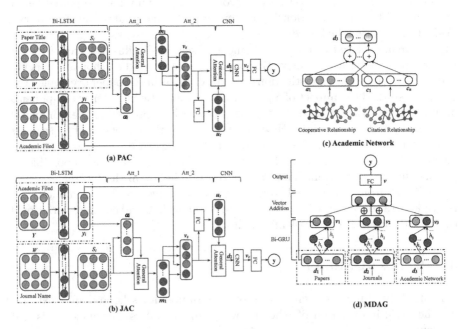

Fig. 2. The model of academic field prediction

Academic Journal Information: Scholars tend to submit papers to conferences or journals. All the journals are extracted from the data set to construct a collection of journals. The abbreviations of the journals are converted to full names and then redistributed to scholars. For the characterization of journal information, the method we adopt is the same as that of paper information. The model marked as "JAC" is shown in Fig. 2(b). We get the vector d_2.

Scholar Network Representation: The academic network of scholars in our paper includes two types of relationships: cooperative relationship and citation relationship. The network structure of scholars is defined as $G_s = (V_s, E_s)$, where V_s is the set of vertices. Every $v \in V_s$ represents a scholar. And E_s is the set of edges, which represents the relationship between the scholars. The weights indicate the number of collaborations and citations between the two scholars. The academic network relationship model is shown in Fig. 2(c).

We choose the network representation learning LINE model [8] to convert the node attributes and link relationships in the scholar's network relationship into feature vectors. Finally, the two feature vectors a, c are added together as a network relationship feature vector for scholars, $d_3 = a + c$.

Multiple Data Sources: We propose a model framework that connects three different data sources and considers the interaction between different data sources. This model takes scholar paper information, journal information, and scholar network relationships as the data source input, and obtains the high-level feature representation of the three data sources through the processing of the sub-model. We treat d_1, d_2, d_3 as a sequence, and obtain a vector to extract low-level features by sharing the Bi-GRU network g_1. The feature representation d_l of different data sources are spliced to obtain a feature representation v_l at different levels. Finally, input the scholar's vector v by adding to a fully connected layer to get the predicted label. The model named "MDAG" is shown in Fig. 2(d). The formal description of the process can be expressed by the following:

$$z_l = \sigma(W_z d_l + U_z h_{l-1} + b_z) \tag{1}$$

$$r_l = \sigma(W_r d_l + U_r h_{l-1} + b_r) \tag{2}$$

$$\widetilde{h_l} = \tanh(W_h d_l + U_h(r_l \odot h_{l-1}) + b_h) \tag{3}$$

$$h_l = (1 - z_l) \odot h_{l-1} + z_l \widetilde{h_i}) \tag{4}$$

$$g_l = [h_l, h_l] \tag{5}$$

$$v_l = [d_l, g_l] \tag{6}$$

$$v = W_v [v_1 \oplus v_2 \oplus v_3] + b_v \tag{7}$$

$$y = fully\_connected\_layer(v) \tag{8}$$

where z_l and r_i are the update gate and reset gate in the GRU network, $\widetilde{h_l}$ represents the candidate memory unit, and h_l represents the memory unit at the current moment. $\boldsymbol{W}_v$, $\boldsymbol{W}_z$, $\boldsymbol{W}_r$, $\boldsymbol{W}_h$, $\boldsymbol{U}_z$, $\boldsymbol{U}_r$, $\boldsymbol{U}_h$ are the weight matrix in the training process, $\boldsymbol{b}_z$, $\boldsymbol{b}_r$, $\boldsymbol{b}_h$, $\boldsymbol{b}_v$ are the deviation vector. All classification models use cross entropy loss function.

3.2 Future Academic Influence Prediction

In this paper, the future influence of a scholar is defined as: his all papers' citation number within a certain period in academic circles (in the next 10 years). The higher citations, the more attention his papers receive and the more recognized the scholar is.

Paper Timing Characteristics: Time has a great influence on paper citations. Paper citations will continue to increase over time, but the average annual citations will decrease. In addition, scholars prefer to cite newly published papers. Newly published and high-quality papers will be more popular. We use the PageRank algorithm with time perception to calculate the impact of the paper, and newly published papers will be given a higher score.

We establish a paper-citation network $G_p = (V_p, E_p)$, where V_p is the set of vertices, representing a scholar, and E_p is a collection of edges, indicating the citation relationship between papers. This graph is a directed graph. The out-degree represents the number of times a paper is cited, while the in-degree represents the number of times a paper cites other papers. After that, we use the PageRank with time series to calculate the PageRank score of each paper as follows:

$$PR\left(v_i\right) = (1-d)\frac{f\left(a_i\right)}{\sum_{k=1}^{N} f\left(a_k\right)} + d\sum_{j=1}^{m} \frac{PR\left(v_j\right) f\left(a_j\right)}{L(v_j)} \tag{9}$$

where v_j represents the citation of v_i , a_i and a_j represent the published time of two papers, $L(v_j)$ represents the out-degree of the node v_j of the paper. And the published time of paper a is defined as $a = T_c - T_p$, T_p is the published year, and T_c is end year. d is a damping coefficient, set to 0.85. $f\left(a\right)$ represents the time function of the paper, and the closer the year of publication, the greater the value of the time function. We get $f\left(a_i\right)$ as follows:

$$f\left(a_i\right) = \varepsilon e^{-b*a_i} + c \tag{10}$$

where ε represents the amplification rate of the function, which is used to adjust the value of the function, b represents the time decay rate. And c represents the offset of the function, which is used to adjust the function decay rate. Unlike the previous PageRank algorithm, we assign different weights to each node through the time when paper is published. Papers with larger values of $f\left(a_i\right)$ are given greater weight.

Each paper node $PR\left(v_i\right)$ undergoes a certain number of iterations until it converges, and the output is the influence score s, which is expressed as follows:

$$s_i = PR\left(v_i\right) \tag{11}$$

Time Series Characteristics of Journals: We use the PageRank score of a journal or conference as its influence. We build a heterogeneous network which includes two types of nodes: papers and journals. The edges set represents the publication relationship between papers and journals. The initial value of each journal is $1/N$, and that of the paper is calculated by Eq. 11. Equation 10 is also used to calculate the time series characteristics of the journal. The impact score of the journal $J(v_i)$ is as follows:

$$J(v_i) = \frac{1}{sum} \sum_{j=1}^{m} \frac{f(a_i) \sum_{k=1}^{m} score_k}{N} \tag{12}$$

where $score_k$ represents the PR value of the journal's papers, sum represents the sum of all journal values except the journal v_i, m represents the number of papers published in the journal, and N represents the total number of journals. We get the vector definition $j = [j_1, j_2, \cdots j_n]$ of the journal's influence.

Scholar Feature Space: We construct a scholar's feature space composed of scholars' professional characteristics, social attributes and personal basic attributes, and define them as $x = [x_1, x_2, x_3, x_4, x_5, x_6]$. The professional characteristics are composed of total number of papers, number of papers published as first or second author, number of papers published as * (here * means any position) author, number of papers published as first author, number of paper citations as first author, number of paper citations within two years.

We establish two characteristics on the scholar's cooperation network: the depth of cooperation in the network n_1 and the scholar's ability to publish papers independently of others n_2. The two characteristics are expressed as follows:

$$n_1 = \frac{1}{N_i} \sum_{j \in N_i} \frac{N_i \bigcap N_j}{N_i \bigcup N_j} \tag{13}$$

where N_i represents the number of neighbor nodes of scholars i, that is, scholars who have a cooperative relationship with it.

$$n_2 = \frac{T_{ij}}{2}(\frac{1}{T_i} + \frac{1}{T_j}) \tag{14}$$

where T_{ij} represents the number of papers published jointly by scholar i and scholar j, and T_i represents the number of papers published by scholar i.

Therefore, we define the social attributes of scholars as vector $n = [n_1, n_2]$. Scholars' basic personal attributes are defined as $c = [c_1, c_2]$, where c_1 is age, c_2 is job title. The professional characteristics, social attributes and basic personal attributes of scholars are stitched together to obtain the feature space representation of scholars as $a = [x, n, c]$.

Feature Combination: We propose a two-level catboost fusion regression model of feature combinations to predict the future influence of scholars. The

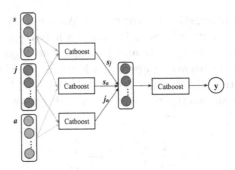

Fig. 3. The framework of CSCF model

first level takes different combinations of three features as input (paper timing features s, journal timing features j, and scholar feature space a), and uses catboost to learn the relationship between features. We obtain three results as the output of the first level. The second level merges the three results as input, and uses the second level catboost to learn the interaction between the three first level results to get the final result. We combine different feature sources in pairs. The model named "CSCF" is shown in Fig. 3.

We calculate the degree of difference between the predictor's influence and real influence in the next 10 years to measure the effectiveness of the model. We use mean square error as our loss function.

4   Experiments

4.1   Dataset

The dataset for our experiment comes from Aminer[1]. We extract information from the dataset and store them in a file for scholars. This file can be used as a dataset for scholars' academic field and academic influence prediction in this paper. In addition, the column attributes are "contact information", "sex", "age" and "title" as the basic information. Before extracting basic information and constructing a corpus, the raw data needs to be cleaned and processed. We delete the symbol and noise data, and use word2vec to map tokens to vectors. In order to avoid data imbalance, negative sampling is used for the academic field. For academic influence, the scholars whose influence index is lower than 5 and who have not published papers in 10 years are filtered out. The statistics information of the academic field prediction dataset and the academic influence prediction dataset are shown in the Table 1. All the task data are as of 2008, and the next 10 years (from 2009 to 2018) are used to test the prediction results.

---

[1] https://www.aminer.cn/data.

Table 1. Academic field dataset

Task	Data set	Scholar	Paper	Citations	Journal	Label
Academic	Raw data	1,715,170	3,074,071	7,038,195	23,640	1,467
Field	Task data	11,367	432,383	808,195	11,364	790
Academic	Raw data	1,715,170	3,074,071	7,038,195	23,640	–
Influence	Task data	101,458	372,348	808,195	18,564	–

4.2 Baselines

We conduct experiments in the academic field on other neural network methods and use them as baseline models. In addition, two methods from other papers are selected.

- **LDANE** [12]: Use the method based on LDA topic model to process.
- **MDAR (Multi-Data RNN)**: All places where Bi-LSTM is used are replaced with RNN, and the last Bi-GRU is also replaced with RNN.
- **MA (Multi-Attention)** [4]: Three types of text information are aggregated through the attention layer, which is vectorized by LSTM, and the output is summed with the network vector through another attention layer.
- **MDC (Multi-Data + Concatenation)**: For the fully connected layer, just connect d_1, d_2, d_3, and put them into the fully connected layer.

We conduct academic impact prediction experiments on multiple regression models and use them as baseline models.

- **XGBoost**: By building multiple regression trees, use the previously constructed features as input to XGBoost.
- **Light_GBM** [11]: Use a gradient−based unilateral GOSS sampling algorithm to improve accuracy.
- **CatBoost**: Replace the gradient estimation method with a sorting boosting method, thereby reducing the bias of gradient estimation.
- **CSXBF**: Replace CatBoost with XGBoost in the model.
- **CSLGBF**: Replace CatBoost with Light_GBM in the model.
- **SCF**: The features are directly input into the two-level CatBoost without being combined.

4.3 Results and Discussion

It can be observed from Table 2 that multiple data sources are more advantageous than single data sources in academic fields predicting tasks. The results of the MA model on Precision are better than the MDAG model, and the experimental results of other metric are not obvious. This shows that adding a layer of recurrent neural network to the high-level features of different data sources can learn the correlation between different data sources. On the other hand, not

Table 2. The results of academic field

Method		Recall	Precision	F1
Single data source	LDANE	0.498	0.442	0.468
	PAC	0.503	0.472	0.487
	JAC	0.483	0.458	0.470
Multiple data sources	MDC	0.598	0.596	0.596
	MA	0.603	**0.607**	0.604
	MDAR	0.613	0.605	0.608
	MDAG	**0.622**	0.606	**0.614**

every type of data source in this paper is suitable for attention. Our model only adopts an attention on the paper information and journal information.

We study whether the sequence order of high-level features of the three data sources before sharing the Bi-GRU layer will affect the experimental results. Table 3 shows the results of six different combinations. The experimental results show that there are significant differences in sequences of different orders. The reason is mainly due to the influence of the update state and reset state in the GRU. If it has nothing to do with the current state, it will reduce the information of the previously hidden state. The text information of the paper in the middle position can effectively capture the journal information and scholar network information from scholars, because the scholar network in the experiment is also constructed according to the paper data set.

Table 3. The results of different sequence orders

Order	Recall	Precision	F1
$d_1 - d_2 - d_3$	0.616	0.596	0.605
$d_2 - d_1 - d_3$	**0.622**	**0.606**	**0.614**
$d_3 - d_1 - d_2$	0.618	**0.606**	0.612
$d_3 - d_2 - d_1$	0.612	0.604	0.607
$d_2 - d_3 - d_1$	0.610	0.598	0.604
$d_1 - d_3 - d_2$	0.608	0.601	0.606

The experimental results of academic influence prediction are shown in Table 4. In addition to the slightly worse RMSE performance of the two-level fusion model CSCF of feature combination, the other two indicators are better than the results of the one-level model. The reason is that the second-level model can more fully use the first-level task results to learn the correlation between the first-level results, and can better learn the diversity between different feature types to achieve lower prediction errors.

Table 4. The results of academic influence

Method		MAE	RMSE	R^2
One−level model	LR	0.807	0.092	0.861
	XGBoost	0.622	**0.072**	0.838
	Light_GBM	0.992	0.294	0.913
	CatBoost	0.602	0.091	0.874
Two−level model	CSXBF	0.620	0.084	0.861
	CSLGBF	0.864	0.147	0.753
	SCF	0.673	0.082	0.849
	CSCF	**0.542**	0.077	**0.901**

From the experimental results of the two-level model, our CSCF is better than CSXBF and CSLGBF of other feature combinations. The reason is that Catboost can assign indicators to the classification features for processing, while the XGBoost and LGBM deal directly with classification features.

Unlike other work, our experiment incorporates the basic personal attributes of scholars. We believe that age, professional title and scholars' future academic influence are related. We analyze the experimental results of multiple two-level fusion models based on feature combinations to verify whether the scholar's personal basic attributes will affect the experimental results.

The scholars are divided according to different age ranges, and the average value of scholars' influence is taken as the influence of scholars of this age in the next 10 years. From Fig. 4(a), we can see that the higher the age, the greater the influence of scholars.

Similarly, the scholars are divided according to professional titles, and the average value of the influence in the collection is selected. From Fig. 4(b), except for the CSLGBF model, the higher the title, the greater the influence. The exception may be an error in the prediction result or an data set imbalance.

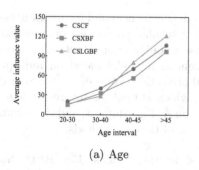

(a) Age

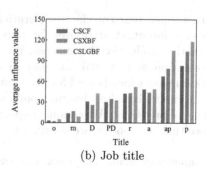

(b) Job title

Fig. 4. The effect on the experimental results. From left to right in (b) is other, master, Doctor, Post Doctorate, researcher, assistant, associate professor, professor.

4.4   Ablation Experiments

In order to better understand the actual effect of the model on learning between different data sources, we conduct relevant ablation experiments.

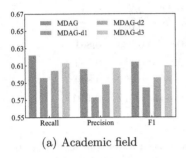

(a) Academic field

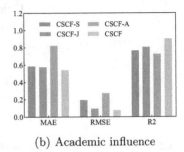

(b) Academic influence

Fig. 5. Evaluation index results of ablation experiments

The predicted results in the academic field are shown in Fig. 5(a). MDAG gets the best prediction results. This proves that the integration of diverse data types will provide more information for academic field prediction. After removing the paper information, the three evaluation indicators are the most decreased. Therefore, the paper's text information retains the most valuable information. The predicted results of academic influence are shown in Fig. 5(b). Our model CSCF results are the best, and once again proves the strong correlation between different types of features. In addition, the model effect after removing the scholar's feature space changes the most, which means that the scholar's feature space is the most influential in the experiment feature.

5   Conclusion

In this paper, based on the consideration that academic field and future influence are two important attributes for constructing scholar profile, we propose two predictive models for the purpose. We propose a double-layer attention of paper information and journal information representation model based on multiple data sources, which is used for academic field prediction. We propose a two-level fusion model based on feature combination, which is used for future academic influence prediction. We conduct experiments on AMiner dataset, and the results prove that our models are superior to the state-of-the-art methods.

Acknowledgement. The work was supported by the National Key R&D Program of China under grant 2018YFB1004700, National Natural Science Foundation of China (61772122, 61872074), and National Defense Basic Research Program (JCKY2018205C012).

References

1. Bai, X., Zhang, F., Ni, J., Shi, L., Lee, I.: Measure the impact of institution and paper via institution-citation network. IEEE Access **8**, 17548–17555 (2020)
2. Ciot, M., Sonderegger, M., Ruths, D.: Gender inference of twitter users in non-English contexts. In: EMNLP, pp. 1136–1145 (2013)
3. Farnadi, G., Tang, J., Cock, M.D., Moens, M.: User profiling through deep multi-modal fusion. In: WSDM, pp. 171–179 (2018)
4. Miura, Y., Taniguchi, M., Taniguchi, T., Ohkuma, T.: Unifying text, metadata, and user network representations with a neural network for geolocation prediction. In: ACL, pp. 1260–1272 (2017)
5. Palacios-Núñez, G., Vélez-Cuartas, G., Botero, J.D.: Developmental tendencies in the academic field of intellectual property through the identification of invisible colleges. Scientometrics **115**(3), 1561–1574 (2018)
6. Pazzani, M.J., Billsus, D.: Learning and revising user profiles: the identification of interesting web sites. Mach. Learn. **27**(3), 313–331 (1997)
7. Rao, D., Yarowsky, D.: Detecting latent user properties in social media. In: Proceedings of the NIPS MLSN Workshop, pp. 163–170 (2010)
8. Tang, J., Qu, M., Wang, M., Zhang, M., Yan, J., Mei, Q.: LINE: large-scale information network embedding. In: WWW, pp. 1067–1077 (2015)
9. Tang, J., Zhang, J., Yao, L., Li, J., Zhang, L., Su, Z.: ArnetMiner: extraction and mining of academic social networks. In: SIGKDD, pp. 990–998 (2008)
10. Wan, H., Zhang, Y., Zhang, J., Tang, J.: AMiner: search and mining of academic social networks. Data Intell. **1**(1), 58–76 (2019)
11. Weihs, L., Etzioni, O.: Learning to predict citation-based impact measures. In: JCDL, pp. 49–58 (2017)
12. Wen, A.: Study on the Key Technology of Scholarly User Profile based on Multi-Source and Heterogeneous Big Data. Ph.D. thesis, South China University of Technology (2018)
13. Yan, R., Tang, J., Liu, X., Shan, D., Li, X.: Citation count prediction: learning to estimate future citations for literature. In: CIKM, pp. 1247–1252 (2011)
14. Yan, Y., Tian, S., Zhang, J.: The impact of a paper's new combinations and new components on its citation. Scientometrics **122**(2), 895–913 (2020)
15. Zhang, Y., Tang, J., Yang, Z., Pei, J., Yu, P.S.: COSNET: connecting heterogeneous social networks with local and global consistency. In: SIGKDD, pp. 1485–1494 (2015)

Exploiting Knowledge Embedding to Improve the Description for Image Captioning

Dandan Song[✉], Cuimei Peng, Huan Yang, and Lejian Liao

Beijing Engineering Research Center of High Volume Language Information, Processing and Cloud Computing Applications, Beijing Key Laboratory of Intelligent Information Technology, School of Computer Science and Technology, Beijing Institute of Technology, Beijing, China
{sdd,pcm,yh,liaolj}@bit.edu.cn

Abstract. Most existing methods for image captioning are based on the encoder-decoder framework which directly translates visual features into sentences, without exploiting commonsense knowledge available in the form of knowledge graph. Inspired by the success of information retrieval and question answering systems that leverage prior knowledge, we explore a knowledge embedding approach for image captioning. In this paper, we propose a Knowledge Embedding with Attention on Attention (KE-AoA) method for image captioning, which judges whether or how well the objects are related and augments semantic correlations and constraints between them. The KE-AoA method combines knowledge base method (TransE) and text method (Skip-gram), adding external knowledge graph information (triplets) into the language model to guide the learning of word vectors as the regularization term. Then it employs the AoA module to model the relations among different objects. As more inherent relations and commonsense knowledge are learned, the model can generate better image descriptions. The experiments on MSCOCO data sets achieve a significant improvement on the existing methods and validate the effectiveness of our prior knowledge-based approach.

Keywords: Image captioning · Knowledge representation · Knowledge embedding · Multi-head attention

1 Introduction

Automatically generating a natural language description of an image is a fundamental and challenging task. It has been actively studied in recent Computer Vision (CV) and Natural Language Processing (NLP) research communities. The image caption generation task has been applied to many significant areas, such as image retrieval and visual aids.

Recent deep learning models primarily utilize an encoder-decoder framework to generate captions. In such a framework, an image is first encoded to a set

H. Chen et al. (Eds.): CCKS 2020, CCIS 1356, pp. 312–321, 2021.
https://doi.org/10.1007/978-981-16-1964-9_25

of visual feature vectors via a CNN and then decoded into words via an RNN. However, a common problem has never been substantially resolved: when facing a complex scene, the model only outputs a trivial and inaccurate caption. In our observation, human beings can easily recognize and describe images that contain complex scenes and details with a glance. Even if some objects are missing or can not be accurately identified, people can still describe images accurately according to commonsense knowledge learned from daily practice. This inspires us to explore how to leverage such prior knowledge to help generate better descriptions. In this paper, we propose a novel method of jointly embedding knowledge graphs and text corpus. We joint knowledge embedding (TransE) [1] and word embedding (Skip-gram) [2] as knowledge representation. The embedding process attempts to add the external knowledge graph information (triplets) into Skip-gram language model to guide the learning of word vectors as the regularization term. The output knowledge embedding vectors learn the inherent relations between objects and commonsense knowledge from the knowledge graph, thus can help improve the caption's qualities.

We experiment on the MSCOCO dataset and evaluate the captions with different automatic metrics. All of the results show the effectiveness of our prior knowledge-based approach.

2   Related Work

Early template-based caption models [3–5] firstly extract visual features, such as the objects and attributes, then use a fixed sentence template to form a complete and semantic sentence.

Recently, neural network-based methods [6–8] primarily based on the encoder-decoder framework which firstly extracts visual features from a CNN and then feeds them into an RNN to generate a sequence of meaningful words. Mao et al. [9] propose a multimodal RNN (m-RNN) for image captioning. The neural image caption (NIC) model [6], an end-to-end framework, has been proposed with a CNN encoding the image to feature vector and an LSTM [10] decoding it to caption. Xu et al. [11] introduce a method that applies stochastic hard attention and deterministic soft attention to generating captions. Anderson et al. [12] propose a combined bottom-up and top-down attention mechanism that enables attention to be calculated at the level of objects and other salient image regions. These deep learning algorithms often do away with feature engineering and learn latent representations directly from raw data that are given as input to the encoder-decoder framework.

Recent remarkable achievements of linguistic knowledge in NLP tasks, such as natural language inference [13], language modeling [14], named entity recognition [15] and relation extraction [16], have validated the crucial role of prior knowledge. This inspires us to explore how to leverage such prior knowledge to help generate better descriptions. Yang et al. [17] propose to employ the scene graph to corporate learned language bias as a language prior for more human-like captions. However, the scene graph in SGAE model [17] ignores the low-frequency category, which is prone to bias.

We notice that word representations can learn the continuously distributed representation of the context. Skip-gram [2] learns word embeddings capturing many syntactic and semantic relations between words. A knowledge graph is embedded into a low-dimensional continuous vector space while certain properties of it are preserved. TransE [1] interprets a relation as a translation from the head entity to the tail entity. Besides, the latest AoA model [18] extends the conventional attention mechanisms to determine the relevance between attention results and queries. Therefore, we combine text method (Skip-gram) and knowledge base method (transE) as knowledge representation to acquire commonsense knowledge, extend the AoA module for better modeling relations among different objects of the image and filtering out irrelevant attention results, thus to help improve the caption's qualities.

3   Our Model

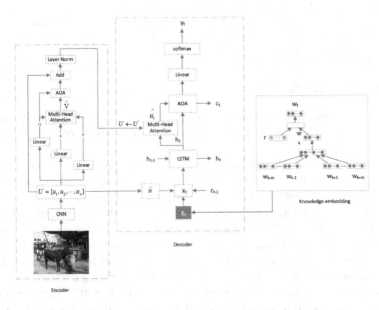

Fig. 1. An illustration of our image captioning model KE-AoA that is comprised of an encoder with the AoA module and multi-head attention, a knowledge powered captioning decoder.

Our KE-AoA model applies the AoA module to the image encoder and combines knowledge embedding with the decoder, as shown in Fig. 1. By combining semantic features from refining knowledge representations with visual features of the image, our model can better model relations among different objects and generate more accurate captions. We first introduce how we jointly embedding knowledge and text, then show details of the image encoder and the caption decoder respectively.

3.1   Jointly Embedding Knowledge and Text

A knowledge graph contains a large number of entities and relations. They are represented in the form of triplets (w_h, r, w_t), each triplet consists of two entities w_h, w_t and a relation r. We refine the conventional knowledge representation model, TransE [1], combining it with Skip-gram [2]. The jointly embedding process attempts to preserve the relations between entities in the knowledge graph and the concurrences of words in the text corpus. The refining knowledge representation adds an external knowledge graph information (triplets) into Skip-gram language model to guide the learning of word vectors as the regularization term.

We use the following combinatorial objective function L to combine text method (Skip-gram) and knowledge base method (a variant of TransE) as knowledge representation:

$$L = \xi + \lambda R \tag{1}$$

where λ is the combination coefficient. Our goal is to maximize the combined objective function L. Traditional neural networks often define the conditional probability $p(y|x)$ in *softmax* function, which is impractical in this task due to the high cost of computing $\nabla \log p(y|x)$ in the case of hundreds of words in the vocabulary. Therefore, we adopt negative sampling (NEG) [19] to solve this problem.

The objective of Skip-gram is to maximize the log probability:

$$\xi = \sum_{w \in C} \log p\left(Context(w)|w\right) \tag{2}$$

where C is the set of words in the vocabulary, $Context(w)$ is the training context $\{w_k - m, ..., w_k - 1, w_k + 1, ..., w_k + m\}$, and we use $2m$ to indicates the context window size.

The basic idea of TransE is that $h + r \approx t$: if (h, r, t) holds, the embedding of the head entity h plus the embedding of the relation r should be close to the tail entity t, otherwise, $h + r$ should be far away from t. Different from TransE that minimizes a margin-based ranking criterion over the training set which resulted in a complex combinatorial optimization problem [20], we adopt an objective function to maximize the probability as below:

$$R = \sum_{r \in R_{w_h}} \log \frac{exp(e_{w_h + r}^T e_{\theta_{w_t}})}{\sum_{j=1}^{|C|} exp(e_{w_h + r}^T e_{\theta_{w_j}})} \tag{3}$$

where R_{w_h} contains all the relations related to w_h, $e_{w_h + r} = e_{w_h} + e_r$, $|C|$ is the number of words in the vocabulary C, and θ is a parameter to be learned.

3.2   Encoder with the AoA Module and Multi-head Attention

Let $U = \{u_1, u_2..., u_n\}$ denote the visual feature vectors, which extracted from an image by Faster R-CNN [21], where $u_i \in \mathbb{R}^D$, n is the number of vectors in

U, and D is the dimension of each vector. Thanks to the superior performance of AoA model [18], we utilize it to refine the representations of U (As shown in the Encoder part of Fig. 1). The basic AoA model [18] can be calculated as:

$$
\begin{aligned}
\text{AoA}\left(f_{att}, \boldsymbol{Q}, \boldsymbol{K}, \boldsymbol{V}\right) = \left(W_q^i \boldsymbol{Q} + W_v^i f_{att}(\boldsymbol{Q}, \boldsymbol{K}, \boldsymbol{V}) + b^i\right) \odot \\
\sigma\left(W_q^g \boldsymbol{Q} + W_v^g f_{att}(\boldsymbol{Q}, \boldsymbol{K}, \boldsymbol{V}) + b^g\right)
\end{aligned}
\tag{4}
$$

where $W_q^i, W_v^i, W_q^g, W_v^g \in \mathbb{R}^{D \times D}, b^i, b^g \in \mathbb{R}^D$ and D is the dimension of $\boldsymbol{Q}$; $f_{att}(\boldsymbol{Q}, \boldsymbol{K}, \boldsymbol{V})$ is the attention result, f_{att} is an attention module and σ denotes the sigmoid activation function.

The AoA model adopts the multi-head attention function [22], where $\boldsymbol{Q}, \boldsymbol{K}$, and $\boldsymbol{V}$ are three individual linear projections of the feature vectors U. Then it's followed by a residual connection [23] and layer normalization [24]:

$$
U' = \text{LayerNorm}\left(U + \text{AoA}\left(f_{mh-att}, W^{Q_e} M, W^{K_e} M, W^{V_e} M\right)\right)
\tag{5}
$$

where $W^{Q_e}, W^{K_e}, W^{V_e} \in \mathbb{R}^{D \times D}$ are three linear transformation matrixes. f_{mh-att} is the multi-head attention function which divides each $\boldsymbol{Q}, \boldsymbol{K}, \boldsymbol{V}$ into $H = 8$ slices along the channel dimension, and employs a scaled dot-product attention function to each slice $\boldsymbol{Q}_i, \boldsymbol{K}_i, \boldsymbol{V}_i$, then concatenates the results of each slice to form the final attended vector.

$$
f_{mh-att}(\boldsymbol{Q}, \boldsymbol{K}, \boldsymbol{V}) = \text{Concat}\left(head_1, \ldots, head_H\right)
\tag{6}
$$

$$
head_i = \text{softmax}\left(\frac{\boldsymbol{Q}_i \boldsymbol{K}_i^T}{\sqrt{d}}\right) \boldsymbol{V}_i
\tag{7}
$$

The self-attentive multi-head attention model seeks the interactions among objects in the image, and the AoA model measures how well they are related. The feature vectors U are updated by U' after refining.

3.3 Knowledge Powered Caption Decoder

The decoder is composed of a LSTM, which models the caption decoding process, and an AoA module, that saves the decoding state and the newly acquired information. The input x_t of LSTM consists of the knowledge vectors $\boldsymbol{E}_t$ of the input word, which make the prior knowledge integrated into LSTM at each time step t, the mean-pooled visual feature vector $\bar{\boldsymbol{u}}$, where $\bar{\boldsymbol{u}} = \frac{1}{n} \sum_i \boldsymbol{u}_i$ and $\boldsymbol{u}_i \in U$, and the previous output $\boldsymbol{c}_{t-1}$ of the AoA module.

$$
\boldsymbol{x}_t = [\boldsymbol{E}_t, \bar{\boldsymbol{u}}, \boldsymbol{c}_{t-1}]
\tag{8}
$$

The input to the AoA module consists of the attended visual feature vector $\hat{\boldsymbol{u}}_t$ and the hidden state $\boldsymbol{h}_t$ of the LSTM, where $\hat{\boldsymbol{u}}_t$ is the attended result from the multi-head attention. The context vector $\boldsymbol{c}_t$ is obtained from the AoA module:

$$
\boldsymbol{c}_t = \text{AoA}\left(f_{mh-att}, W^{Q_d}[\boldsymbol{h}_t], W^{K_d} U, W^{V_d} U\right)
\tag{9}
$$

where $W^{Q_d}, W^{K_d}, W^{V_d} \in \mathbb{R}^{D \times D}; h_t \in \mathbb{R}^D$ is the hidden state of the LSTM, serving as the attention query.

Given the context vector c_t, the probability over a vocabulary of possible words at time step t can be calculated as:

$$p(y_t|y_1, ..., y_{t-1}) = softmax(W_p c_t + b_p) \tag{10}$$

where $W_p \in \mathbb{R}^{D \times |\Sigma|}$ and $b_p \in \mathbb{R}^{|\Sigma|}$, W_p and b_p are the weight parameters to be learned and $|\Sigma|$ is the size of the vocabulary.

4   Experimental Results and Discussion

4.1   Datasets and Data Processing

To evaluate our KE-AoA model, we experiment with MSCOCO [25]. It contains 123,287 images annotated with 5 sentences for each. We use the publicly available Karpathy splits [26] that have been extensively used to report our results. We get 113,287 images for training, 5,000 images for validation, and 5,000 images for testing, respectively.

We convert all sentences to lower case, discarding non-alphanumeric characters. Follow [26], we filter words whose frequency less than 5 in the training set, resulting in 9,487 words for training. We report our results using the standard automatic evaluation metrics, BLEU [27], METEOR [28], ROUGE-L [29], and CIDEr [30] and SPICE [31].

4.2   Implementation Details

The word embedding of the AoANet baseline is randomly initialized. Because of the variance from random initialization, We use "Baseline: AoANet" to represent the experimental results ourselves and evaluate metrics BLEU-2 and BLEU-3 for AoANet baseline as well.

We employ a pre-trained Faster-R CNN [21] model on ImageNet [32] and Visual Genome [33] to extract visual feature vectors of images [12]. As for the training process, Faster R-CNN [21] uses RPN to detect the IoU threshold of positive samples with 0.7 and negative samples 0.3. It allows different images to have different numbers of features, and the number of features in a single image is between 10 and 100. The initial learning rate of the model is $2 \times e^{-4}$, and it decreases by 0.8 times every 3 epochs. Batch size is set to be 10, and the cross-entropy loss function is used to iterate 30 epochs. Then Adam [34] optimization algorithm is used to optimize the CIDEr, beam size is set to 5, and the learning rate is $2 \times e^{-5}$.

4.3   Evaluation

Table 1 shows the experimental results of our model as well as the compared models. All metrics in the row of Cross-Entropy Loss are the results of training

Table 1. Performance of our model and other state-of-the-art methods on the MSCOCO test set of Karpathy's split [26].

	Cross-entropy loss								CIDEr optimization							
	B-1	B-2	B-3	B-4	MT	RG	CD	S	B-1	B-2	B-3	B-4	MT	RG	CD	S
DeepVS [26]	62.5	45.0	32.1	23.0	19.5	–	66.0	–	–	–	–	–	–	–	–	–
gLSTM [7]	67.0	49.1	35.8	26.4	22.7	–	81.3	–	–	–	–	–	–	–	–	–
Soft-attention [11]	70.7	49.2	34.4	24.3	23.9	–	–	–	–	–	–	–	–	–	–	–
Hard-attention [11]	71.8	50.4	35.7	25.0	23.0	–	–	–	–	–	–	–	–	–	–	–
Adaptive [35]	74.2	58.0	43.9	33.2	26.6	54.9	108.5	–	–	–	–	–	–	–	–	–
LSTM [6]	–	–	–	29.6	25.2	52.6	94.0	–	–	–	–	31.9	25.5	54.3	106.3	–
SCST [36]	–	–	–	30.0	25.9	53.4	99.4	–	–	–	–	34.2	26.7	55.7	114.0	–
Up-down [12]	77.2	–	–	36.2	27.0	56.4	113.5	20.3	79.8	–	–	36.3	27.7	56.9	120.1	21.4
SGAE [17]	77.6	–	–	36.9	27.7	57.2	116.7	20.9	80.8	–	–	38.4	28.4	58.6	127.8	22.1
Baseline: AoANet	76.8	61.2	47.3	36.4	28.1	57.1	117.3	21.1	80.1	64.7	50.6	38.4	28.5	58.4	126.7	22.3
Ours	**77.9**	**61.8**	**48.6**	**37.7**	**28.6**	**58.0**	**119.9**	**21.6**	**80.9**	**65.5**	**51.2**	**39.2**	**29.4**	**58.9**	**128.9**	**22.6**

B-n is the abbreviation of BLEU-n, M for METEOR, R for ROUGE-l, and C for CIDEr, and S for SPICE.

Ours: a brown dog laying on the ground next to a pool
Baseline: a brown dog laying next to a pool
GT1: a dog laying on the ground next to a pool
GT2: a brown dog laying next to a pool in the water

Ours: a man and a woman sitting on a bench with a dog
Baseline: a man and a dog sitting on a bench in the street
GT1: a man and a woman sitting on a bench with a dog
GT2: a man and a woman is sitting on a bench with a dog

Ours: a large airplane is parked on the tarmac at an airport
Baseline: a large airplane sitting on the tarmac at an airport
GT1: a large airplane is parked on the tarmac at an airport
GT2: a large airplane is parked on the runway at an airport

Ours: a bedroom with a white bed and curtains on the wall
Baseline: a bedroom with a bed and a television in the wall
GT1: a bedroom with a bed and curtains in the room
GT2: a bedroom with a bed and a television on the wall

Fig. 2. Qualitative captioning results generated by our KE-AoA and a baseline model as well as the corresponding ground truths.

with the standard cross-entropy loss function, and all metrics in the row of CIDEr Optimization are the results of optimization based on the CIDEr. The AoANet baseline and our KE-AoA model are significantly better than the first eight models no matter under cross-entropy loss or optimized for CIDEr score. Relative to the SGAE [17], our AoANet baseline only achieves a considerable performance. After incorporating prior knowledge, our full KE-AoA model shows significant improvements across all metrics regardless of whether cross-entropy loss or CIDEr optimization is used.

Figure 2 shows some examples with images and captions generated by our model and a strong baseline as well as the human-annotated ground truths. From these examples, we find that the baseline model generates captions that are grammatically correct but inaccurate for the image content or the relevance between objects, while our model generates accurate captions in high quality. More specifically, our model is superior in the following three aspects: (1) KE-AoA can seek more interactions among objects and more properties of the object. In the first example, our KE-AoA finds the connections between dog and ground while the baseline can't; in the fourth example, the caption generated by KE-AoA uses "white" to describes the properties of the bed due to the prior knowledge. (2) KE-AoA figures out the missing object. There are a man and a woman with a dog in the image of the second example. However, the baseline model ignores the woman while our KE-AoA discovers correctly. (3) KE-AoA can describe the relations between objects more accurately. In the third example, our KE-AoA describes the connections between airplane and tarmac as "is parked", while the baseline uses "sitting". We can conclude that prior knowledge plays a guiding role in the improvement of the description for image captioning.

5   Conclusion

We have presented the KE-AoA, an approach to image captioning by exploiting prior knowledge available in the form of knowledge graph. It can augment semantic relations between objects and improve the quality of generated captions. Experiments on the MSCOCO data set demonstrate the superiority of our proposed KE-AoA method and the crucial role of prior knowledge.

Acknowledgment. This work was supported by National Key Research and Development Program of China (Grant No. 2020AAA0106600) and National Natural Science Foundation of China (Grant Nos. 61976021, 61672100 and U1811262).

References

1. Bordes, A., Usunier, N., Garcia-Duran, A., Weston, J., Yakhnenko, O.: Translating embeddings for modeling multi-relational data. In: Advances in Neural Information Processing Systems, pp. 2787–2795 (2013)
2. Mikolov, T., Chen, K., Corrado, G., Dean, J.: Efficient estimation of word representations in vector space. In: ICLR (2013)
3. Farhadi, A., et al.: Every picture tells a story: generating sentences from images. In: Daniilidis, K., Maragos, P., Paragios, N. (eds.) ECCV 2010. LNCS, vol. 6314, pp. 15–29. Springer, Heidelberg (2010). https://doi.org/10.1007/978-3-642-15561-1_2
4. Li, S., Kulkarni, G., Berg, T.L., et al.: Composing simple image descriptions using web-scale n-grams. In: Proceedings of the Fifteenth Conference on Computational Natural Language Learning, pp. 220–228 (2011)

5. Kulkarni, G., Premraj, V., Ordonez, V., et al.: BabyTalk: understanding and generating simple image descriptions. IEEE Trans. Pattern Anal. Mach. Intell. **35**(12), 2891–2903 (2013)
6. Vinyals, O., Toshev, A., Bengio, S., Erhan, D.: Show and tell: a neural image caption generator. In: 2015 IEEE Conference on Computer Vision and Pattern Recognition (CVPR), pp. 3156–3164. IEEE (2015)
7. Xu, J., Gavves, E., Fernando, B., Tuytelaars, T.: Guiding the long-short term memory model for image caption generation. In: IEEE International Conference on Computer Vision (2016)
8. Cheng, W., Yang, H., Bartz, C., Meinel, C.: Image captioning with deep bidirectional LSTMs. In: ACM on Multimedia Conference (2016)
9. Mao, J., Wei, X., Yi, Y., Jiang, W., Huang, Z., Yuille, A.: Deep captioning with multimodal recurrent neural networks (m-RNN). In: ICLR (2015)
10. Hochreiter, S., Schmidhuber, J.: Long short-term memory. Neural Comput. **9**(8), 1735–1780 (1997)
11. Xu, K., et al.: Show, attend and tell: Neural image caption generation with visual attention. In: ICML (2015)
12. Anderson, P., He, X., Buehler, C., et al.: Bottom-up and top-down attention for image captioning and visual question answering. In: Proceedings of the IEEE Conference on Computer Vision and Pattern Recognition, pp. 6077–6086 (2018)
13. Mrkšić, N., Vulić, I., Séaghdha, D.Ó., et al.: Semantic specialization of distributional word vector spaces using monolingual and cross-lingual constraints. Trans. Assoc. Comput. Linguist. **5**, 309–324 (2017)
14. Ahn, S., Choi, H., Pärnamaa, T., et al.: A neural knowledge language model. arXiv preprint arXiv:1608.00318 (2016)
15. Ghaddar, A., Langlais, P.: Robust lexical features for improved neural network named-entity recognition. arXiv preprint arXiv:1806.03489 (2018)
16. Vashishth, S., Joshi, R., Prayaga, S.S., et al.: RESIDE: improving distantly-supervised neural relation extraction using side information. arXiv preprint arXiv:1812.04361 (2018)
17. Yang, X., Tang, K., Zhang, H., et al.: Auto-encoding scene graphs for image captioning. In: Proceedings of the IEEE Conference on Computer Vision and Pattern Recognition, pp. 10685–10694 (2019)
18. Huang, L., Wang, W., Chen, J., et al.: Attention on attention for image captioning. In: Proceedings of the IEEE International Conference on Computer Vision, pp. 4634–4643 (2019)
19. Mikolov, T., Sutskever, I., Chen, K., Corrado, G., Dean, J.: Distributed representations of words and phrases and their compositionality. In: 26th International Conference on Advanced Neural Information Processing Systems, pp. 3111–3119 (2013)
20. Xu, C., et al.: RC-NET: a general framework for incorporating knowledge into word representations. In: Proceedings of the 23rd ACM International Conference on Conference on Information and Knowledge Management, pp. 1219–1228. ACM (2014)
21. Ren, S., He, K., Girshick, R., et al.: Faster R-CNN: towards real-time object detection with region proposal networks. In: Advances in Neural Information Processing Systems, pp. 91–99 (2015)
22. Vaswani, A., Shazeer, N., Parmar, N., et al.: Attention is all you need. In: Advances in Neural Information Processing Systems, pp. 5998–6008 (2017)

23. He, K., Zhang, X., Ren, S., et al.: Deep residual learning for image recognition. In: Proceedings of the IEEE Conference on Computer Vision and Pattern Recognition, pp. 770–778 (2016)
24. Ba, J.L., Kiros, J.R., Hinton, G.E.: Layer normalization. arXiv preprint arXiv:1607.06450 (2016)
25. Lin, T.-Y., et al.: Microsoft COCO: common objects in context. In: Fleet, D., Pajdla, T., Schiele, B., Tuytelaars, T. (eds.) ECCV 2014. LNCS, vol. 8693, pp. 740–755. Springer, Cham (2014). https://doi.org/10.1007/978-3-319-10602-1_48
26. Karpathy, A., Fei-Fei, L.: Deep visual-semantic alignments for generating image descriptions. In: The IEEE Conference on Computer Vision and Pattern Recognition (CVPR) (2015)
27. Papineni, K., Roukos, S., Ward, T., Zhu, W.-J.: BLEU: a method for automatic evaluation of machine translation. In: Proceedings of the 40th Annual Meeting on Association for Computational Linguistics, pp. 311–318. Association for Computational Linguistics (2002)
28. Denkowski, M., Lavie, A.: Meteor universal: language specific translation evaluation for any target language. In: Proceedings of the Ninth Workshop on Statistical Machine Translation, pp. 376–380 (2014)
29. Lin, C.-Y.: ROUGE: a package for automatic evaluation of summaries. In: Text Summarization Branches Out (2004)
30. Vedantam, R., Lawrence Zitnick, C., Parikh, D.: CIDEr: consensus-based image description evaluation. In: The IEEE Conference on Computer Vision and Pattern Recognition (CVPR), pp. 4566–4575 (2015)
31. Anderson, P., Fernando, B., Johnson, M., Gould, S.: SPICE: semantic propositional image caption evaluation. In: Leibe, B., Matas, J., Sebe, N., Welling, M. (eds.) ECCV 2016. LNCS, vol. 9909, pp. 382–398. Springer, Cham (2016). https://doi.org/10.1007/978-3-319-46454-1_24
32. Deng, J., Dong, W., Socher, R., et al.: ImageNet: a large-scale hierarchical image database. In: 2009 IEEE Conference on Computer Vision and Pattern Recognition, pp. 248–255 (2009)
33. Vendrov, I., Kiros, R., Fidler, S., et al.: Order-embeddings of images and language. arXiv preprint arXiv:1511.06361 (2015)
34. Kingma, D.P., Ba, J.: Adam: a method for stochastic optimization. arXiv preprint arXiv:1412.6980
35. Lu, J., Xiong, C., Parikh, D., Socher, R.: Knowing when to look: adaptive attention via a visual sentinel for image captioning (2017)
36. Rennie, S.J., Marcheret, E., Mroueh, Y., Ross, J., Goel, V.: Self-critical sequence training for image captioning. In: The IEEE Conference on Computer Vision and Pattern Recognition (CVPR) (2017)

Knowledge-Enhanced Collaborative Meta Learner for Long-Tail Recommendation

Bo Wen[1,2(✉)], Shumin Deng[1,2(✉)], and Huajun Chen[1,2(✉)]

[1] Zhejiang University, Hangzhou, China
{129967,231sm,huajunsir}@zju.edu.cn
[2] AZFT Joint Lab for Knowledge Engine, Hangzhou, China

Abstract. Long-tail effect is common in recommender systems, meaning that a tiny number of users have lots of interaction with items, while the majority of users have extremely little interaction. Most of existing approaches to recommender systems, especially methods based on collaborative filtering, suffer severely from long-tail problems due to the low resource issue. In order to handle the problem of long-tail recommendation, we utilize knowledge graph to enrich the representation of users and items. As auxiliary prior information source, knowledge graph has been popular technology for recommender systems nowadays, while it is rarely exploited with respect to the low resource issue. In this paper, we propose a knowledge-enhanced collaborative meta learner, which combines the priors in knowledge graph and collaborative information between head users to promote long-tail recommendation performance. We conduct several experiments on two real world datasets for long-tail recommendation. The results show that our approach outperforms several commonly used recommendation methods in the long tail scenario.

Keywords: Long-tail recommendation · Knowledge graph embedding · Collaborative filtering

1 Introduction

The personalized recommendation has become an important issue for information retrieval and content discovery in the information explosion age. However, not all users have enriched interaction with items. In fact, the user-item interaction data are sparse on the whole. Moreover, due to privacy issues, collecting personal information is challenging, which might result in the low-resource recommendation problem.

Long-Tail Issues in Recommendation. Traditional recommender systems based on collaborative filtering [8,20,27] are proposed to utilize historical interactions of target users, e.g., clicking and buying records. The recommendations are built upon the existing ratings of other users who have similar ratings as the target user. Such methods are conducive to capture the correlation between

H. Chen et al. (Eds.): CCKS 2020, CCIS 1356, pp. 322–333, 2021.
https://doi.org/10.1007/978-981-16-1964-9_26

users and items, while have difficulty in handling cold-start or long-tail recommendation problems. The *long tail effect* inherently exists in recommender systems for a long time, which reveals the fact that most users have extremely few buying and rating records, called tail users, and really small part of users have much more buying and rating records than tail users, called head users. Due to much fewer historical interactions, it's harder to make a precise recommendation for tail users compared to head users. Furthermore, the tail users make up the majority of users, making it greatly significant to address long-tail recommendation problems, especially for tail users. Some studies have focused on low-resource scenarios in recent years, such as [12], a MAML-based recommender system is proposed to estimate user preferences based on only a small number of items. Nevertheless, Most of these approaches ignored collaborative filtering in recommendation.

Knowledge-Enhanced Recommendation. In order to solve long-tail recommendation problems, knowledge-enhanced recommender systems are introduced to utilize content knowledge [4,16,17]. Such systems use user profile information (e.g., gender, nationality, religion, and political stance) and the contents of the items to make recommendations. The knowledge-enhanced systems are able to alleviate long-tail issues to a certain extent, as tail users may establish associations with others via content information. Topological relationships among items, as well as historical interaction information between items and users can all be stored in knowledge graphs, and these relationships and information is utilized as prior knowledge to promote the procedure of decision and reasoning in recommender systems. However, the systems might have a limitation of recommending the same items to the users who have similar contents regardless of items that the user already rated.

In order to address the aforementioned problems in recommender systems, we propose to resolve the long-tail recommendation problems by decomposing and reassembling. The content information of users and items as well as the interactions among them are all stored in a knowledge graph, which is the most suitable representing methods to deploy the approach of decomposing and reassembling. For example, items are decomposed to attributes they are interacting with users in the knowledge graph, and for long-tail items which have few user interaction information, we can reassemble items back from the attributes representation learned from items with high resource. In this way, we propose a novel **knowledge-enhanced collaborative meta learner** model, termed as *KCML*, to alleviate long-tail effects in recommender systems. *KCML* is able to integrate the advantages of collaborative filtering methods and knowledge-enhanced methods, especially for long-tail recommendation. The main contributions of our work are summarized as follows:

- We fully exploit potential of knowledge graph with decomposing and reassembling approaches for long tail recommendation, and take collaborative filtering into account at the same time.

- We propose a general recommendation framework combing knowledge graph representation learning as well as prototype meta learner to promote long tail recommendation performance.
- By conducting experiments on two real world datasets, we prove that our framework achieve comparative performance in general recommendation scenarios and make a great improvement in especially long tail recommendation.

The rest of this paper is organized as follows. Section 2 introduce several related work in traditional recommendation and graph based recommendation systems. Section 3 shows the details of our proposed method. The experiments related information are introduced in Sect. 4. In the end, conclusions are being displayed and future work are discussed in Sect. 5.

2 Related Work

2.1 Traditional Recommendation

Collaborative filtering(CF) is one of the most well-known techniques for recommendation. CF predict users' interest by collecting preferences information from other users. There are two major categories for CF based methods: memory-based CF and model-based CF.

Memory-based CF is predicting by similar users' or items' ratings, user similarity based methods [18] make use of the ratings of similar users on the same target item, while item similarity based methods [13] leverage ratings of user on similar items of the target item. Model-based CF, on the other hand, is predicting by inferring from an underlying model based on basic hypothesis. Matrix Factorization [11] is the most popular and effective model-based CF method. It supposes that the score of ratings between target user and target item can be estimated by inner product of their latent vector.

Feature-based method is also popular for recommendation and is another exploration direction. Factorization machines is the most popular classical feature-based method, it was first proposed in [14]. It mainly focus on sparse feature combination problem and takes second order crossing interaction into consideration. Field Factorization Machine [9] is also a popular method for feature-based method.

In the latest years, those traditional collaborative filtering method and feature-based method are usually deployed with deep learning designation. Methods like [6,7,21] is beginning to show its popularity and effectiveness for recommendation.

2.2 Graph Based Recommendation

There are two major approaches for graph based recommendation: embedding based method and meta-path based method.

Collaborative knowledge base embedding (CKE) [25] introduces structural embedding representation of knowledge base for recommendation. In the paper, bayesian TransR is used to learn entity and relation embedding and it is one of the classic embedding based method. Thomas described Graph Convolutional Network (GCN) in [10]. GCN is a method for learning an embedding for nodes in graph by convolutional operation over graph in their neighborhood and utilized it in recommendation field in [2].

Meta-path base method mainly leverages path-based semantic relationship as direct feature for recommendation [5,15] or performs some transformation on path-based similarity and utilize the transformed similarity to support item and user representation [24].

3 Methodology

3.1 Problem Formulation

We define our long-tail recommendation problem in this paper as follows: $U = \{U_i | i = 1, 2, \ldots, n\}$ is the user set where $|U|$ represents for the number of users. And $V = \{V_j | j = 1, 2, \ldots, n\}$ where $|V|$ represents for the number of items. U_i and V_j are representation for single user and item respectively. In a common recommendation problem setting, given a combining graph $G = \langle H, P_U, P_V \rangle$ that consists of observable historical interactions H, user property P_U, and item property P_V, we are supposed to predict future interactions between U and V.

$$H = \{R(U_i, V_j) \mid e_{ij} = 1, U_i \in U, V_j \in V\} \tag{1}$$

$$e_{ij} = \begin{cases} 1 & \text{if no interaction between } U_i \text{ and } V_j \\ 0 & \text{if exists interaction between } U_i \text{ and } V_j \end{cases} \tag{2}$$

Historical interactions H are regarded as generalized rating score between given user U_i and V_j, where $R(U_i, V_j) \in \{0, 1\}$ in an implicit feedback condition, such as buying or clicking action user has taken, and $R(U_i, V_j) \in \{1, 2, 3, 4, 5\}$ in an explicit feedback condition, such as scores in five-star rating system.

$$U_{tail} = \{U_i \mid |H_{U_i}| \leq \alpha max(\{|H_{U_i}| \mid U_i \in U\})\} \tag{3}$$

A long-tail recommendation is a problem where recommender system's target user lies in tail users U_{tail} where $|H|$ is the number of observable historical interactions, α is defined as the tail percentage, as describe in Eq. (3).

On the other hand, head users U_{head} is defined as users not in U_{tail}.

3.2 Model Overview

As mentioned above, the knowledge-enhanced method is naturally suitable for the solution to long tail recommendation problem by decomposing and reassembling. Users and items along with their properties and interactions are entities and relations in a knowledge graph. Therefore, we propose a framework of long tail recommendation with knowledge-enhanced representation learning and head user prototype meta learner.

As shown in Fig. 1, our proposed knowledge-enhanced collaborative meta learner framework for long tail recommendation consists of three procedures: (1) *Knowledge Representation Learning of Items*, (2) *Prototype Learning of Users*, and (3) *Long-tail Recommendation*.

3.3 Item Representation Learning

Given a triple in knowledge graph: $\langle h, r, t \rangle$, in which h for head entity, r for relationship, and t for tail entity. The purpose of knowledge representation learning is to learn an embedding for each entity and relation in the knowledge graph. The knowledge representation learning methods, like TransE [3], DistMult [22], and ComplEx [19], are commonly used and achieve good performance. Note that our framework is pluggable, and the knowledge representation method can be substitutable. Here we introduce TransE used in our framework as an example.

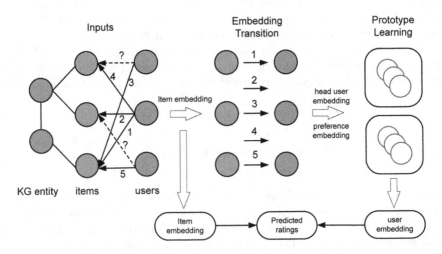

Fig. 1. Model architecture of our proposed framework.

Given a training set $S = \{s_1, s_2, \ldots, s_n\}$, each of the training sample is in the format of triple $\langle h, r, t \rangle$ where $h, t \in E$, E is the entity set and $r \in R$, R is the relationship set. The aim of TransE model is to learn an embedding set $V = \{V_h, V_r, V_t | V_h \in \mathbb{R}^k, V_r \in \mathbb{R}^k, V_t \in \mathbb{R}^k\}$ of entities and relations, where k is the dimension of embedding. TransE contains a hypothesis that:

$$V_h + V_r \approx V_t \qquad (4)$$

If triple $\langle h, r, t \rangle$ exists in knowledge graph, their representation is consistent with Equation (4). According to the energy framework, energy of a triple can be represented as $d(V_h + V_r, V_t)$, where d is some kind of distance indicator. To learn a representation for each entity and relationship, we minimize loss $\mathcal{L}$ on training set S:

$$\mathcal{L} = \sum_{(h,r,t)\in S} \sum_{(h',r,t')\in S'_{(h,r,t)}} [\gamma + d(V_h + V_r, V_t) - d(V_{h'} + V_r, V_{t'})]_+ \qquad (5)$$

$$S'_{(h,r,t)} = \{(h',r,t) \mid h' \in E\} \cup \{(h,r,t') \mid t' \in E\} \qquad (6)$$

In the Eq. (5), $[\cdot]_+$ is a ReLU operation, and $\gamma > 0$ is a hyper parameter that constrains distance between positive samples and negative samples.

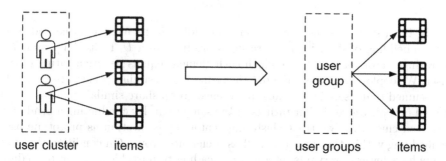

Fig. 2. Demonstration of low resource issue alleviated when treat users in similar preference group as a whole. Left half part shows the situation when treat each user as a single person, some user would have very little interaction with item set. Right half part shows the change of condition when treat them as a user group, the resource of interactions for the group would be rich.

3.4 User Prototype Learning

To alleviate low resource issues of long tail users, which means they have fewer interactions with other items than head users, we make a hypothesis that there exists similar preference groups among users. Similar users gather in similar preference groups where they give similar ratings to same or similar items. So when we regard similar user group as a whole, as showed in Fig. 2, low resource issue of long tail users in these groups can be alleviated. The reason lies in fact that high-resource users can improve the representation of low-resource ones in the same user group. In this way, we can make recommendation to long tail users with higher data utilization efficiency.

User Prototype. In item representation learning module, we deploy TransE model to achieve the embedding $\mathcal{V}$ of item entities. Head user embedding $\mathcal{U}_{head}$ are initialized with standard normal distribution and $\mathcal{L}_{head}$ are used as loss function to update head user embedding. $\mathcal{L}_{head}$ is mean square error loss on dot product predictions against ground truth values on head users.

$$\mathcal{L}_{head} = \left(\mathcal{U}_{head}\mathcal{V}^T - \mathcal{R}_{head}\right)^2 + \lambda \|\mathcal{U}_{head}\| \qquad (7)$$

In Eq. (7), $\mathcal{U}_{head}$ represents head user embedding matrix, $\mathcal{V}$ represents for item embedding matrix, and R_{head} is observable rating matrix for head users. λ is a hyper parameter for head user weight penalty. $\|\cdot\|$ is Frobenius norm of the given matrix.

Representation transition is a procedure to learn hidden embedding in preference space for all of the users. To a certain extent, this hidden representation reflects user similarity in item preference.

$$h_{U_i} = \frac{\sum_{j \in R_u} r_{ij} v_j}{\sum R_{U_i}} \tag{8}$$

In Eq. (8), h_{U_i} is the user hidden embedding in preference space and v_j is the embedding of item V_j, r_{ij} is rating score from user U_i to item V_j, and R_{U_i} is rating vector for user U_i in which each element equals per-item rating of the user. We deploy head user oriented clustering to get similar preference group. It is assumed that users in the same preference group share similar item taste and would probably give similar ratings to the same item. By conducting a head user oriented clustering, we treat clustering center representation as prototype user embedding within the cluster as well as preference group. On condition that tail users have fewer observable interactions, leading to unstable learning procedure of embedding, especially in long tail scenarios, so that tail users are not taken into consideration for clustering operation. By doing so, we obtain preference similar group of head users, as well as prototype users for each group. Next, we will show how items are recommended to tail users with the help of user prototypes.

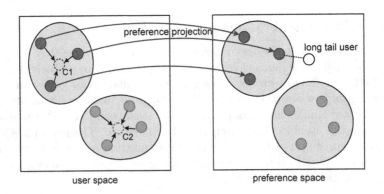

Fig. 3. Illustration of how user embedding is projected from user space to preference space. When long tail user is provided, we are able to find the prototype user embedding for it by: 1) Find the nearest projected preference embedding; 2) With reversed projection, find the corresponding closest user in user space.

3.5 Long-Tail Recommendation

With respect to tail users, we find the most similar preference group for them and treat the prototype user embedding of the group as tail users' embedding, as showed in Fig. 3. In this way, recommendations can be provided to tail users with higher confidence and so as to promote long tail recommendation performance. To find the most similar preference group for a specific item, we use $d = D(u_i, u_j)$ as a metric of user similarity, and $D(\cdot)$ is L2 norm of euclidean dissimilarity. We calculate the user similarity with user rating vectors, denoted by $u_i = R_{U_i}$ and $u_j = R_{U_j}$, as this metric is simple and clear enough to be an assessment for user similarity. However, as the number of items increases in the recommender system, the usage of rating vectors is not good enough.

In this work, we use preference space hidden embedding h_{U_i} as is described in Eq. (8). We regard h_{U_i} as the assessing factor for user similarity for two reasons: First, the more similar two users' tastes are, the closer their preference hidden embedding are. Second, dimension of preference hidden embedding is usually far less than number of items, which making the model simple and efficient. In the end we find proper prototype user for each tail user and choose user prototype embedding as tail user's representation. Finally, we get item-user relativity with dot product operation on item and user embedding, shown in Eq. (9).

$$R_{pred} = \mathcal{U}\mathcal{V}^T, \tag{9}$$

where items are recommended in descending order of predicted item-user relativity.

4 Experiments

In this section, we evaluate our proposed knowledge-enhanced collaborative meta learner framework on two datasets. (1) One is movie recommendation with an explicit feedback where ratings scores from 1 to 5. (2) The other is twitter time-line recommendation with an implicit feedback of interacted or not. Results of experiments on these datasets show a great improvement on long tail recommendation settings compared to several competitive baselines.

4.1 Datasets and Evaluation Metrics

We use two datasets to illustrate the effectiveness of our proposed KCML framework. **MovieLens-1M**[1] is a popular dataset for movie recommendation scenario which consists of 6,040 users and 3,811 movies with 997,580 interactions. Similar to [26], we make an alignment for movies in the dataset and entities in Freebase. To demonstrate long tail effect clearly, we create a limitation on tail users' iterations that the number of tail users' observable interactions is less than or equal to 2, and other interactions of tail users are used for recommending evaluation and

---

[1] https://grouplens.org/datasets/movielens/1m/.

test. **Recsys2020 Twitter dataset**[2][1] are collected from real world twitter timeline data for users. It is a public dataset that consists of 160 million public tweets sampled within two weeks. It contains engagement features, user features as well as tweet features and it shows a strong long tail distribution. Due to huge amount of data in the twitter dataset, we subsample from the dataset according to the original distribution to create a subset that contains 10000 users and 40846 tweets. Detailed statistics are displayed in Table 1.

Table 1. Detailed statistics of two datasets.

Statistics	MovieLens-1M	Twitter Long Tail
#user	6,040	10,000
#item	3,811	40,846
#interactions	997,580	41,162
#KG entities	255,557	100,011
#KG relations	307	14

We evaluate KCML model on several metrics, including (1) *Mean Squared Error (MSE)*, as it is a common used metric in recommendation scenarios. Accordingly, *MSE-tail* is proposed to evaluate the performance merely on tail users; (2) *Area Under the Precision-Recall Curve (PR-AUC)*; and (3) *Relative Cross Entropy (RCE)*, which assesses the improvement of a prediction compared to naive prediction in Cross Entropy.

4.2   Experiment Settings

The model are tuned with Adam optimizer and the learning rate is set as 10^{-4}, weight penalty factor $\lambda = 10^{-4}$. For tail users, we randomly choose 2 interactions to be kept as train data, and the other interactions are split into half as validation data and test data. The datasets are separated into training set, validation set, and test set with the ratio of 0.8, 0.1, and 0.1 respectively, and we set the number of head user prototypes as 6. We set long tail percentage α value as 0.3, margin size as $\gamma = 1.0$, and dimension for user and item representation $k = 50$. For *MovieLens-1M* dataset, we also align Freebase to MovieLens-1M mentioned in [26] as the source of item side knowledge graph. We then sample a sub-graph in one step for aligned entities. Finally, we obtain a knowledge graph with 255,557 entities and 307 relations. For *Recsys2020 Twitter* dataset, setting and scenario are slightly different for the task. The prediction targets are implicit feedbacks for four actions (Reply, Retweet, Retweet with comment, Like). In this condition, we treat the task as four independent binary classification problem. We also create a knowledge graph for tweets from the original tweet features.

---

[2] https://recsys-twitter.com/.

4.3   Experiment Results

The results of the experiment on MovieLens-1M are shown in Table 2. As seen, our proposed KCML method outperforms all the other methods both in general comparison and long-tail part comparison. In Table 2, KCML means the basic framework with TransE as KG embedding method, KCML(Random) is to substitute item embedding learned from KG with gaussian initialized ($\sim N(0,1)$) random embedding, and KCML(GCN) embed items in KG with GCN (Graph Convolutional Network).

Table 2. Experiment result on MovieLens-1M dataset.

Method	MSE	MSE-tail
MF [11]	1.731	2.345
NFM [7]	1.712	2.324
Hitting time [23]	1.753	2.315
MeLU [12]	1.685	2.297
KCML (Random)	1.718	2.240
KCML (GCN)	1.684	2.212
KCML	**1.544**	**2.046**

As for MSE, KCML (Random) achieves competitive performance with traditional collaborative filtering methods (MF and NFM) and long tail recommendation method (Hitting Time and MeLU). As for MSE-tail, KCML (Random) also outperforms traditional collaborative filtering and long tail recommendation methods. These results clearly show the effectiveness of preference user grouping and user prototype learning.

Furthermore, KCML (GCN) achieves better results than KCML (Random) while gets worse performance in comparison with KCML. It shows that compatibility of TransE with our proposed framework is better than graph convolutional network based methods. In general, all of the methods achieve better results on the whole dataset than merely on long tail part, generally showing the effectiveness our proposed framework.

The results of the experiment on Recsys2020 Twitter dataset are shown in Table 3. It can be seen that our proposed KCML outperforms other baseline methods under almost all conditions. Especially the like action of users on RCE, where KCML with the score of 20.542 makes an improvement of 53% than the second placed method MeLU with the score of 9.564. Considering both PR-AUC and RCE, like actions are relatively easier task to predict and retweet with comment actions are actually harder task. Inspired by this, we also find that MF method achieve relatively worse performance in the twitter experiments than results on MovieLens-1M. This could be due to the extreme sparsity of data in Recsys2020 Twitter dataset. Although methods based on collaborative filtering

Table 3. Experiment result on Recsys2020 Twitter dataset.

Method	PR-AUC				RCE			
	Reply	Retweet	Retweet-wc	Like	Reply	Retweet	Retweet-wc	Like
MF [11]	0.011	0.201	0.003	0.794	−940.354	−43.285	−2950.675	−34.538
NFM [7]	0.031	0.284	0.004	0.908	1.617	1.741	0.872	9.220
Hitting time [23]	0.033	0.293	0.006	0.900	1.626	1.728	0.871	9.240
MeLU [12]	0.034	0.221	0.006	0.907	2.235	**2.621**	0.870	9.564
KCML	**0.046**	**0.310**	**0.008**	**0.912**	**4.625**	1.736	**0.881**	**20.542**

could get fairly good performance, they may not work when data is extremely sparse in long tail recommendation, which further proves the effectiveness of our KCML method.

5   Conclusion and Future Work

This paper created a new view of solving long tail recommendation by decomposing and reassembling and correspondingly proposed a knowledge-enhanced collaborative meta learner to alleviate long tail problems in recommender systems. There are two primary modules in the proposed framework. The first one is a knowledge enhanced representation learning module which provides sufficient side information for item representation learning. And the second one is a user prototype module that deploys collaborative clustering. The experiments we have conducted show the effectiveness of our proposed framework. Due to feature sparsity for users, we mainly focus on user-oriented long tail problem and utilize item-based knowledge graph for side information in this paper. However user associated social network can be regarded as a knowledge graph as well, so the item-oriented long tail problem is left for future research.

Acknowledgements. This work is funded by NSFC U19B2027/91846204/61473260, national key research program 2018YFB1402800, and supported by AlibabaZJU Frontier Technology Research Center.

References

1. Belli, L., et al.: Privacy-preserving recommender systems challenge on Twitter's home timeline. CoRR abs/2004.13715 (2020)
2. van den Berg, R., Kipf, T.N., Welling, M.: Graph convolutional matrix completion. CoRR abs/1706.02263 (2017)
3. Bordes, A., Usunier, N., García-Durán, A., Weston, J., Yakhnenko, O.: Translating embeddings for modeling multi-relational data. In: NIPS, pp. 2787–2795 (2013)
4. Domingues, M.A., Rezende, S.O.: The impact of context-aware recommender systems on music in the long tail. In: BRACIS, pp. 119–124. IEEE Computer Society (2013)
5. Feng, W., Wang, J.: Incorporating heterogeneous information for personalized tag recommendation in social tagging systems. In: KDD, pp. 1276–1284. ACM (2012)

6. Guo, H., Tang, R., Ye, Y., Li, Z., He, X.: DeepFM: a factorization-machine based neural network for CTR prediction. In: IJCAI, pp. 1725–1731 (2017). ijcai.org
7. He, X., Chua, T.: Neural factorization machines for sparse predictive analytics. In: SIGIR, pp. 355–364. ACM (2017)
8. He, X., Liao, L., Zhang, H., Nie, L., Hu, X., Chua, T.: Neural collaborative filtering. In: WWW, pp. 173–182. ACM (2017)
9. Juan, Y., Zhuang, Y., Chin, W., Lin, C.: Field-aware factorization machines for CTR prediction. In: RecSys, pp. 43–50. ACM (2016)
10. Kipf, T.N., Welling, M.: Semi-supervised classification with graph convolutional networks. In: ICLR (Poster) (2017). OpenReview.net
11. Koren, Y., Bell, R.M., Volinsky, C.: Matrix factorization techniques for recommender systems. IEEE Comput. **42**(8), 30–37 (2009)
12. Lee, H., Im, J., Jang, S., Cho, H., Chung, S.: MeLU: meta-learned user preference estimator for cold-start recommendation. In: KDD, pp. 1073–1082. ACM (2019)
13. Linden, G., Smith, B., York, J.: Amazon.com recommendations: Item-to-item collaborative filtering. IEEE Internet Comput. **7**(1), 76–80 (2003)
14. Rendle, S.: Factorization machines. In: ICDM, pp. 995–1000. IEEE Computer Society (2010)
15. Shi, C., Zhang, Z., Luo, P., Yu, P.S., Yue, Y., Wu, B.: Semantic path based personalized recommendation on weighted heterogeneous information networks. In: CIKM, pp. 453–462. ACM (2015)
16. Sreepada, R.S., Patra, B.K.: Mitigating long tail effect in recommendations using few shot learning technique. Expert Syst. Appl. **140** (2020)
17. Takama, Y., Chen, Y., Misawa, R., Ishikawa, H.: Analyzing potential of personal values-based user modeling for long tail item recommendation. JACIII **22**(4), 506–513 (2018)
18. Tkalcic, M., Kunaver, M., Tasic, J., Košir, A.: Personality based user similarity measure for a collaborative recommender system. In: Proceedings of the 5th Workshop on Emotion in Human-Computer Interaction-Real world challenges, pp. 30–37 (2009)
19. Trouillon, T., Welbl, J., Riedel, S., Gaussier, É., Bouchard, G.: Complex embeddings for simple link prediction. In: ICML. JMLR Workshop and Conference Proceedings, vol. 48, pp. 2071–2080 (2016). JMLR.org
20. Wang, X., He, X., Wang, M., Feng, F., Chua, T.: Neural graph collaborative filtering. In: SIGIR, pp. 165–174. ACM (2019)
21. Xue, H., Dai, X., Zhang, J., Huang, S., Chen, J.: Deep matrix factorization models for recommender systems. In: IJCAI, pp. 3203–3209 (2017). ijcai.org
22. Yang, B., Yih, W., He, X., Gao, J., Deng, L.: Embedding entities and relations for learning and inference in knowledge bases. In: ICLR (Poster) (2015)
23. Yin, H., Cui, B., Li, J., Yao, J., Chen, C.: Challenging the long tail recommendation. Proc. VLDB Endow. **5**(9), 896–907 (2012)
24. Yu, X., Ren, X., Sun, Y., Gu, Q., Sturt, B., Khandelwal, U., Norick, B., Han, J.: Personalized entity recommendation: a heterogeneous information network approach. In: WSDM, pp. 283–292. ACM (2014)
25. Zhang, F., Yuan, N.J., Lian, D., Xie, X., Ma, W.: Collaborative knowledge base embedding for recommender systems. In: KDD, pp. 353–362. ACM (2016)
26. Zhao, W.X., et al.: KB4Rec: a data set for linking knowledge bases with recommender systems. Data Intell. **1**(2), 121–136 (2019)
27. Zou, L., et al.: Neural interactive collaborative filtering. In: SIGIR, pp. 749–758. ACM (2020)

Author Index

Printed in the United States
by Baker & Taylor Publisher Services